ANAEROBIC DIGESTION
MODERN THEORY AND PRACTICE

ANAEROBIC DIGESTION
MODERN THEORY AND PRACTICE

PETER N. HOBSON

*Department of Molecular and Cell Biology,
University of Aberdeen, Scotland*

and

ANDREW D. WHEATLEY

*Civil Engineering, Loughborough University of Technology,
Leicestershire*

ELSEVIER APPLIED SCIENCE
LONDON and NEW YORK

ELSEVIER SCIENCE PUBLISHERS LTD
Crown House, Linton Road, Barking, Essex IG11 8JU, England

WITH 6 TABLES AND 10 ILLUSTRATIONS

British Library Cataloguing in Publication Data

Anaerobic Digestion: Modern Theory and Practice
I. Hobson, P. N. II. Wheatley, Andrew
665.776
ISBN 1-85166-958-2

Library of Congress Cataloging-in-Publication Data

Anaerobic digestion: modern theory and practice / Peter N. Hobson and Andrew D. Wheatley.
p. cm.
Includes bibliographical references and index.
ISBN 1–85166–958–2
1. Refuse and refuse disposal—Biodegradation. 2. Anaerobic bacteria. I. Hobson, P. N. II. Wheatley, Andrew.
TD796.5.A52 1992
628.4'45—dc20 92-37922
CIP

Printed in Great Britain by Galliard (Printers) Ltd, Great Yarmouth

PREFACE

In 1981 one of the present authors and his colleagues wrote the book 'Methane production from agricultural and domestic wastes'. This described research and development in the field of anaerobic digestion: the conversion by microorganisms of organic wastes into methane. The purpose of the process was originally to reduce pollution from municipal sewage sludges. The gas produced was regarded as a usable by-product, and during wartime some attempts had been made in Germany to use digestion of vegetable and animal wastes primarily as a source of fuel gas for use in tractors, and, later, in India the Gobar Gas scheme envisaged small digesters as a source of pollution-free fuel and fertilizer for peasant farms. The realisation in the early 1970's that reserves of fossil fuels are finite led to increased research in the developed countries on production of fuels from biomass (vegetation) as well as from wind, wave and geothermal power. In the later 1970's and 1980, although laboratory work had shown that wastes from many animals and crops, and most terrestrial and aquatic vegetation, could be a source of biogas, practical digestion was largely confined to the biggest sewage works, although a number of digesters had been built on farms in the West and in India and China. However, work on digestion was still expanding and the process was seen to have a future both as a source of energy and a pollution control.

The period of very high prices and the shortage of oil was followed by low prices and a glut of oil in the 1980's and this changed the outlook on alternative fuels. The purpose of the sewage-works digester is pollution control, its viability is not measured in terms of profit on fuel production. So sewage digestion proceeded in spite of low oil prices. On the other hand, there was in most countries no general pressure to reduce pollution from farms, and once the incentive of a fuel cheaper than oil had been removed digester building on farms more or less stopped and many existing digesters were abandoned. For the same reasons factories were reluctant to put in digesters for treatment of sludge wastes. Although it had been shown that low-solids wastewaters could be treated in anaerobic filters, digestion was still mainly regarded as a process for treating sludges and there was little large-scale treatment of wastewaters.

However, research and development have continued to try to overcome the difficulties found in the earlier digesters. On the one hand, there has been work to develop cheaper and more efficient digesters for use in rural areas in developing countries. This includes the single-farm digester for the waste from a few animals, but more attention has been paid to communal digesters treating wastes from a village or community of farms. In the more developed countries, while there have been improvements in the design and running of digesters which have made digestion of sludges a possibility for the small sewage works and factory, and digesters for solid municipal garbage are in production, the greatest developments have probably been in the anaerobic treatment of wastewaters. This has shown digestion to be a viable, and in some aspects a better alternative to aerobic treatment of low-solids wastes. Work on digestion as a source of alternative energy from energy-crops and wastes has also continued. While some of these processes may be for the future in many countries, the recovery of gaseous fuel from the natural digestion of wastes in municipal land-fills is becoming an important business in industrial areas.

Combined with the R & D on digesters themselves there has been an explosion in research on the methanogenic bacteria and to a lesser extent on the other bacteria in the communities responsible for digestion. Following on work on the rumen system has been work on treatment of feedstocks to improve their digestion and the testing of many new feedstocks. Many aspects of the use of digester effluents as fertilizer or animal feed have been investigated.

There is now an increasing concern in many countries about pollution of land, water and the atmosphere by effluents from factories, farms and domestic sources. It would seem that not only will dumping of raw wastes to watercourses, even seas, be forbidden but that treated effluents will have to be of a higher standard than previously. Digestion offers a complete treatment for some wastes in some situations and in other cases can be a part of a multi-stage treatment. Pollution control by anaerobic digestion will probably be of more importance than energy production in the near future.

The number of papers on digester feedstocks and effluents, on the design and running of digesters of all forms, and on the biochemistry, microbiology and theoretical aspects of digestion has increased exponentially since the last book was written and there are now some thousands of papers on the subjects. It seemed to us that it was time for a new book on digestion in which, although it would be impossible to refer to and summarise all the papers available, it would be possible to review the developments in the theory and practice of digestion during the 1980's to give the reader the basic facts about the process and to guide him to relevant papers from which he could proceed further if needed.

It is hoped that the book will be useful to the microbiology and biochemistry student and lecturer in describing the application of microorganisms to what is still the largest biotechnology industry; waste treatment. The scientist working on one aspect of digestion should find information on other applications of the process, and, while engineering aspects are not dealt with in detail, the engineer should find information of use in design and running of plants.

CONTENTS

Chapter 1

INTRODUCTION

The Introduction to the previous book on anaerobic digestion by one of the present authors and his colleagues discussed the forms of energy used by man through the ages and the dependence of the world on oil for much of its energy needs. This was written at a time when the price of oil had increased drastically and it seemed as if the world supplies of oil would be limited in the not far distant future. The production of fuels from biomass then seemed a viable and necessary alternative to dependence on oil, and anaerobic digestion was a means of producing fuel from biomass of all kinds. Thus, although the pollution-control role of digestion was not overlooked, most people regarded digesters as a source of fuel for farms and factories and only secondarily as a means of reducing pollution. The European Commission, for example, predicted that 10% of Europe's energy needs could be met by renewable energy and invested £100M in research and development. The earlier book to some extent reflected this outlook.

Now the wheel has turned full circle. There is a surplus of oil and the price has plummeted and forecasters have increased the life span of oil reserves. Research and development on alternative sources of energy has slowed down as the stocks of oil have grown larger and people have forgotten that oil will run out, even if not as early as had been suggested. The research on other renewble sources has also slowed because experience showed that production of 'alternative' energy, whether from biomass, waves or wind or other sources, was not as simple and reliable as had been at first thought by many. There was no immediate economic reason for alternative energies and the provision of systems for use in the future was not considered by many authorities.

Another problem which militated against the production of electricity from alternative sources was the difficulty, in Britain at least, of selling surplus power to the electricity supply Boards. Biogas is normally produced continuously, while the demand for electricity by the factory or farm attached to the biogas plant may have considerable diurnal or longer variations. If the higher demands are met then there can be a surplus of power at some times which can only be avoided by shutting down the generator and flaring the biogas if there is no alternative use for it. This can adversely affect the economics of the biogas plant. The 1983 Energy Act required Electricity Boards to publish tariffs for buying of electricity from private generators and facilitated the sale of electricity to the Boards. Although the 'buying' price of electricity has remained low compared to the selling price by the Boards, and long-term contracts with the Boards have not been possible, the Act has led to an increase in development of Combined Heat and Power systems and in their use in may different areas. In other countries the electricty services have always bought in power from outside generators.

There was also the argument, particularly by proponents of nuclear power, that 'alternative' sources could never supply significant amounts of the energy needed and could not compete with the bigger power stations which were replacing the smaller plants. Its proponents argued for atomic energy as the only source of power for the future. However, the bigger power stations have not been as reliable and successful as expected, particularly the nuclear stations, many of which have had very poor records for reliabilty. Also, the two recent large disasters at nuclear power stations, the smaller incidents and the problems of nuclear-waste disposal have combined to turn a large section of the public against nuclear power generation. Increasing costs of nuclear power stations as planned and the neccessity to add even more stringent safety equipment in the wake of the two disasters has led to some nuclear stations being declared uneconomic and to a halt in their construction. Privatisation of the power industry in the UK will mean that there will be no additional nuclear stations and the problems of nuclear waste disposal and security remain largely unsolved. But while there are arguments against the very large power station whatever the power source, it is true that demands for energy in industrialised countries are very high and that, in general, the present alternative energy sources are suited only to relatively small production units. However, large overall demands for power are made up of many small demands, and alternative sources, such as biogas, can supply the needs of a factory or other consumer unit. In this way demands on the central power production systems can be decreased, with long-term savings in the use of the fossil fuels for static power generation.

Although there is, perhaps, not the massive interest of earlier years, alternative generation of power is not forgotten and, indeed, there is a resurgence of development in many countries. Landfill gas is an example of biogas use, and it was recently calculated that production of power from biofuels such as alcohol exceeds the atomic power production in the USA. In many areas of the world the terrain makes large-scale power transmission impossible. Small power units are required and, as discussed later, biogas can be used to produce this power. So, research on, and development of, anaerobic digestion as a source of power has continued. But there have been other reasons for the continuing research on anaerobic digestion. Over the last ten years there has been a growing awareness of environmental pollution by both scientists and the public; the latter calling for more stringent controls on disposal of wastes. This has led to increased emphasis on treatment of wastes and to higher standards for discharge of treated wastes. Developments in the last ten years or so have led to improvements in the anaerobic digestion of sludges and to the use of anaerobic digestion for low-solids wastewaters. In all these cases digestion is regarded as a pollution-control system, not primarily as a source of fuel.

A number of countries, particularly those in tropical or semitropical areas, have no sources of fossil fuels and little currency to buy-in oil. Such oil as is imported is too expensive for use for cooking and lighting by most of the population who keep to the traditional fuels, wood and cowdung. But reliance on wood for fuel is rapidly denuding the countryside of trees, from bushes to forests, and not only is the source of fuel decreasing but the deforestation has caused soil erosion, floods and other problems. Cowdung is an inefficient fuel, and in burning most of its value as a fertilizer, and all of its value as a soil conditioner, is lost. There is thus a need in these countries to develop alternative sources of energy and anaerobic digestion provides such a source from animal, human, municipal and factory wastes, as well as vegetation. The digestion also reduces the possibilities of spread of disease by the wastes and other pollution problems, while leaving a residue that has value as a fertilizer and soil conditioner. The development of reliable digesters at a price which can be afforded by the peasant populations of these countries is an important part of the worldwide work on digestion. Thus, it seems likely that anaerobic digestion of wastes for pollution control, and energy and fertilizer production, has a very good future. Its use has increased rapidly in many countries.

There has been a very large amount of work done on the development of practical digesters and the exploitation of new feedstocks in the last ten years, and this has recently been paralleled by an increased interest by the microbiologists in the microbes of digesters. There has been a particular emphasis on the methanogenic bacteria, although these may not always be the controlling bacteria of the digester population. While most of the work to improve digestion has been done without knowledge of the bacteria concerned and, indeed, a knowledge of the bacteria is not needed, in the longer term information on the bacteria could lead to advances in digester technology. Genetic engineering is the way to improvements in all branches of biotechnology, we are told: but apart from the intrinsic difficulties of genetic manipulations with the anaerobic bacteria and the enumeration of the populations, it is not certain that improved strains of bacteria could be successfully introduced into the natural, mixed, populations of digesters. Better fitting of the operational parameters of digester plants to the needs of the microbes to aid in developing and retaining an optimal natural flora seems to be a more immediate way forward.

The earlier digesters being built ten or more years ago, particularly on farms, were being added to existing buildings and one of the major problems that occurred was that this was not always conducive to optimising digester feed and biogas use. Along with the research on the digestion itself has been more planning and modelling to integrate digesters into various situations. The increase in knowledge of the digestion process and the ancillary equipment has helped to make such planning and forecasting more accurate.

Any book about anaerobic digestion must, then, endeavour to take these considerations into account. The book begins with a discussion of the microbiology and biochemistry of digestion. However, the bacterial populations of digesters are developed from natural populations in, for instance, the soil and putrifying wastes, and digesters are not inoculated with defined bacteria as are industrial fermentations. The necessary bacteria are developed, by experiment with the particular waste and different digester configurations and running parameters, from an undefined inoculum in a non-sterile system. It is probably true to say that in no case is the bacterial population of a digester defined in the sense that all the species and their numbers and their roles in the reactions are known, indeed, these may change with time. Thus, detailed knowledge of microbiology is not needed in designing a digester. On the other hand, knowledge of the overall reactions occurring in digestion and the possible rate-limiting reactions has helped in the design and running of the more modern sludge digesters and in the design of the multi-phase and retained-biomass digesters for wastewaters. So the book starts with a general account of the microbiology of digestion.

Increasing knowledge of the reactions in digesters and increasing data from experimental digestion of different feedstocks has enabled more detailed mathematical models of digestion and digesters to be made. Although experiment is the final test of a digestion process, the mathematical model can predict to some extent the running parameters which will optimise the digestion of at least certain feedstocks and so reduce the experiments needed in design of a digester plant. Models can also point to limitations in our knowledge of the biochemistry of the process, of the composition of feedstocks and the effects of certain parameters on the process kinetics, and so can point the way to relevant experiments. The model of digestion depends on experimental data for its actual running, and such data are dealt with in later chapters, but it is based on the microbiology of the process and so it seemed reasonable to consider such models after discussing the microbiology. Modelling is also helpful in construction and engineering work. The economic model sets out to quantify the costs and benefits of the proposed plant. This type of model also includes the 'engineering model' in which the design and construction of the components of the plant are defined and costed. The benefits to be set against the costs of a digester may vary from situation to situation, although the design of the actual plant may be the same. On the whole, the 'engineering' and 'economic' models are extensions of standard engineering practice in design and construction of any factory: so, while some computer modelling for construction of digester plants has been done, its use in practice has been limited.

Anaerobic digestion began with the sewage-sludge digester, which was a development of the septic tank used in domestic sewage treatment. For many years digestion was confined to feedstocks of high solids content, and although there were mechanical improvements in digesters the basic design remained the same. As in most areas of science or engineering technology the latest inovations have usually been foreseen in earlier work, but possibly forgotten about, the last decade has seen the development not only of different designs for sludge digesters but also the design of digesters for different feedstocks, particularly low-solids wastewaters. These different designs are considered in the next chapters.

The design of digesters is fundamentally based on the feedstock to be digested, and so it seemed reasonable to describe the detailed working of digesters on the basis of the physical characteristics of the feedstocks now being used or considered for digestion. The essential physical characteristic is whether the feedstock has a high or low suspended solids content, and this dictates the basic type of digester to be used.

However, treatment of the feedstock may result in fractions of different solids contents. These may then be treated in different types of digester, or one fraction may be treated and the other discarded. Although most organic matter could be said to be digestible, it may be possible to improve the extent of digestion by some chemical or physical treatment. In this way the extent of pollutant reduction may be increased, or the biogas yield from the substrate improved.

In general, digesters are used to reduce wastes of high polluting power, and even if the digestion is nearly 100% efficient in removing the pollutants the resulting effluent may be still too polluting for discharge. Thus, very often a secondary treatment has to be applied to the digester effluent. Such treatment may also allow the recovery or production of a commercially valuable material and so not only is the polluting power of the final effluent reduced but the overall economics of the process are improved. But whether or not useful products are made out of the liquid effluent, the biogas is a constant product that can be used to the advantage of the process running, or economics, or both. The uses of the products of digestion are, thus, considered.

An immense amount of work has been done on anaerobic digestion in the last ten years. Some of this may not lead to practical results for some years, but some has been immediately exploited in practice. As in any science, some experiments have more or less repeated earlier work, but much information has been obtained on new feedstocks and new digesters. We have tried to summarise this information in the book.

Chapter 2

THE MICROBIOLOGY AND BIOCHEMISTRY OF ANAEROBIC DIGESTION

INTRODUCTION

Pollution of the land and waters is generally regarded as change brought about in the systems by man's activities. Pollution may be caused by industrial discharges that change the colour of waters and cause foul smells. These discharges may or may not be toxic to animal and plant life, but other discharges that do not make a visual impact on waters or land can be quite toxic to the natural flora and fauna. A major source of pollution, though, is material which actually encourages growth of microbial and plant and animal life, but because the material is discharged in excessive amounts the natural ecosystems are replaced by excessive growth of undesirable organisms with visual, olefactory and other effects on the systems. Since growth of organisms is encouraged the harmful effects of many pollutants can be removed by allowing growth of micro- and/or macro-organisms to a point where the pollutant no longer contains growth substrates and it can be discharged without further effects on the natural ecosystems. This biological treatment of potentially-polluting wastes is mainly carried out by growth of microorganisms under controlled conditions. The microorganisms can grow and metabolise the waste using oxygen or in the absence of oxygen, and it is this latter microbial metabolism and waste treatment – anaerobic digestion – which is the subject of this book.

Anaerobic digestion is a microbiological process but it is not a unique, man-made process as is the industrial fermentation: the same reactions are carried on in nature, in soils and streams and in the oceans. The objective of the sewage engineer and, later, the microbiologist has been to confine the natural organisms in a man-made system and to optimise the rates and extents of the natural reactions so that polluting substances will be destroyed.

The first application of the system was in the natural decay of human excreta in earth closets, but having observed the overall process men provided conditions in which the process could be more nearly optimised in the septic tank and variations of this. From this came the more efficient anaerobic digester, at first running at ambient temperature and later at elevated temperature. The history of digestion is described in a number of papers, for instance the one by McCarty (1982), and will not be covered here, but it might be relevant to note that although the presence of microbial reactions was determined quite early on, the 'high-rate' sludge digester was developed without any detailed knowledge of the microbial flora and the digester reactions.

The microorganisms carrying out the reactions in anaerobic digestion are bacteria, and that class of bacteria known as 'anaerobes'; bacteria that live without oxygen and may, indeed, be killed by oxygen. There is a gradation in tolerance to oxygen among anaerobic bacteria, but many of the digester bacteria are amongst the anaerobes which are the least tolerant of oxygen. The investigation of these bacteria has needed the development of special techniques and media for culture, and these were largely initiated by work on the bacteria of the rumen. The specialised stomach systems of herbivorous animals such as cattle and sheep are akin to the anaerobic digester in the reactions carried out. The modern phase of rumen microbiology, where the reactions and the bacteria responsible for them have been elucidated and the bacteria isolated and grown in pure culture, started about forty years ago. The advances in digester microbiology and biochemistry have come about since then. The rumen reactions are described in many papers, but a review of the earlier work on the rumen is contained in the book by Hungate (1966) and later work is described in the review by Hobson & Wallace (1982), and in the recent book edited by one of the present authors (Hobson, 1988*a*) which also contains chapters outlining the cultural methods for anaerobes with references for further reading.

Since the metabolic pathways in digesters are similar to those in the rumen, the methods of biochemical investigation of digesters have followed in many ways those of the rumen biochemists. The main biochemical pathways in anaerobic digestion have now been elucidated, but, perhaps with the exception of the methanogenic bacteria, the linkage of these pathways to specific bacteria is nowhere near as complete as it is for the rumen system. Work on the methanogenic bacteria in digesters and other habitats has increased remarkably over the last few years and many new species and genera have been recently isolated. However, although many isolates of methanogenic and other bacteria have been obtained from digesters, there has been comparatively little work on determining the numbers in which the bacteria exist, the variations in these numbers with time and operational factors of the digesters, and the relationships of the different types of bacteria with each other. There is a vast amount of knowledge of the microbial ecology of the rumen system, there is nothing like this information available about digesters.

In the next sections the reactions and the bacteria concerned in anaerobic digestion will be discussed in more detail. However, this is not meant to be an exhaustive treatise on microbiology and digester microbiology. The microbiologist will be familiar with the concepts of microbial growth and culture introduced here and can find further details of the application of these to the particular environment of the digester in the text or references. For the non-microbiologist it is hoped that sufficient basic microbiology has been given for him to understand the principles upon which digesters are designed and work.

General considerations

The feedstocks of anaerobic digesters vary in physical form as well as in chemical composition. They can be sludges formed mainly from the particulate matter settled from municipal sewage plus the excess microbial suspensions from the aerobic treatment of the sewage water: or they can be sludges of farm animal excreta: or suspensions of macerated terrestrial or aquatic plants, either residues from food crops or crops grown for digestion and biogas production. All these contain insoluble, cellulosic, fibrous plant material or the residues of plants that have been partially digested in human or animal digestive tracts, and the main primary substrate of the digester bacteria is these plant fibres. Feedstocks with particulate substrates may also come from industrial processing and these may contain material other than plant fibres; for instance, the residues from animal slaughter houses will contain particulate proteinaceous and fatty tissues, while the residues from potato processing will contain particulate starch as well as cellulosic material, and distillery wastes will contain yeast as well as plant residues. On the other hand, the wastes from many vegetable- and fruit-processing factories will be largely wastewaters with little particulate matter, but with sugars in solution.

The main bacterial population and the reactions of the digester will be determined by the nature of the feedstock. Figure 1 shows the overall reactions that can take place in digesters with a feedstock containing particulate plant residues. As can be seen, there is a series of linked reactions providing energy and substrates for bacterial growth and leading to the final formation of methane, carbon dioxide and microbial cells. Here, the primary energy-producing reactions are breakdown of carbohydrates. If the feed is particulate animal tissues the reactions will be similar but the balance will be moved to more fat and protein degradation. On the other hand, if the feed contains mainly dissolved sugars the primary hydrolytic reactions will not be needed and the reaction chain will start at the fermentation of simple sugars. However, in each case there is a degradation of substances that could be metabolised by aerobic or anaerobic bacteria and other microorganisms in the open environment. This latter metabolism is, as previously mentioned, a basic cause of pollution, but if the microbial substrates are degraded in the controlled environment of the digester then the digester effluent will not contain substrates for the terrestrial organisms and so will be non-polluting. This, of course, is the ideal case: for various reasons connected with microbial growth and the practical running of digesters one hundred percent removal of pollutants is not attained, but this will be discussed later.

At (b) in Figure 1 there is shown 'phenolics'. This is an example of types of compounds which can be wastes from chemical factories, wood pulp production and other processes. Such compounds are normally classed as bacterial poisons, but anaerobic flora can be developed which will degrade these compounds and so they can form the feedstocks for digesters.

In the cases mentioned above all the reactions are taking place in one tank, which contains the complete bacterial population. But, as will be seen later, it is possible to separate the reactions so that the primary hydrolysis of polymers and/or the fermentation of monomers occurs in one tank and the acidic fermentation products are pumped to another tank where the last step in the process, the conversion of acids to methane, takes place. The first tank should then, in theory, contain no methanogenic bacteria as these form a separate population in the second tank (Fig. 2).

Fig. 1. The main microbial reactions in anaerobic digesters.

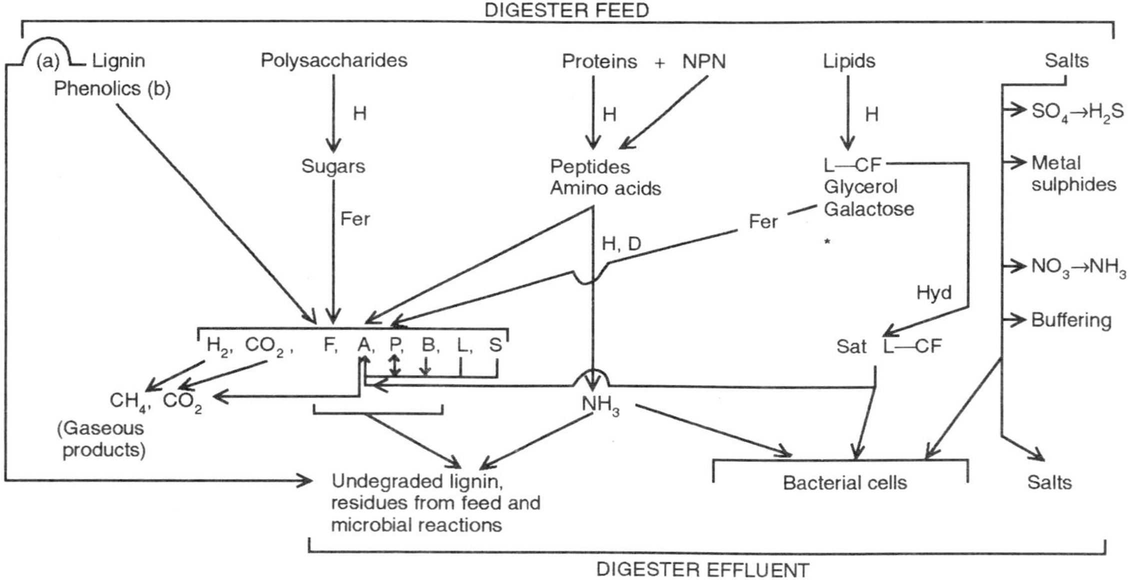

(a) Primary substrates for bacteria in particulate plant fibres. Lignin is associated with polysaccharides.
(b) See text.
*, plus other lipid residues.
NPN, Non-Protein-Nitrogenous compounds; amino acids, peptides, nucleic acid bases, amino sugars, ammonia, etc. Peptides are degraded to amino acids, other NPN compounds to ammonia, plus VFA etc.
F, Formic; A, Acetic; P, Propionic; B, Butyric (n, iso); L, Lactic; S, Succinic acids. (F, A, P, B; Volatile Fatty Acids, VFA).
L–CF, Long-Chain Fatty Acids.
Sat L–CF, Saturated Long-Chain Fatty acids.
Fer, Fermentation of mono- or di-saccharides from hydrolysis of polysaccharides or lipids.
H, Hydrolysis of polysaccharides, proteins, peptides, lipids.
Hyd, Hydrogenation of unsaturated long-chain fatty acids.
D, Deamination of amino acids.
Part of the carbohydrate carbon and amino acids is used in bacterial cell synthesis. Carbon dioxide and VFA can also contribute to synthesis of some bacteria. Digester feeds often contain particulate inorganic material which is generally unattacked and passes out with the effluent.

In connection with Figure 1, various metabolic types of bacteria have been mentioned. These are only a portion of the bacterial types carrying out the reactions that convert the feedstock to gas and cells, while some of the bacteria in digesters may not be involved in the reactions. Anaerobic digestion is not a fermentation carried out with a sterilised fermentor and feedstock and a defined bacterial inoculum. The nearest to a sterile feedstock which is fed to digesters is probably waters from fruit- and vegetable-processing which are used at boiling point. These may be almost sterile as they leave the point of use, but after that they are run through ordinary drainage channels to the digester and can pick up bacteria from the surroundings. Wastes from sewage treatment, farms, slaughter houses, etc., all contain very large numbers of the bacteria in human and animal faeces and gut contents, and, again, they are collected and fed to the digester under non-sterile conditions.

Digester feedstocks thus contain varying numbers of many species of bacteria which may or may not multiply in the digester. A further point is that even if the feed is more or less sterile there is the inoculum to be considered. Start-up of digesters will be dealt with in more detail later, but faecal wastes contain at least small numbers of the bacteria needed to carry out the digestion and so the active bacterial population of a sewage or farm digester can be developed from the feed, but along with the 'active' bacteria other species may develop small populations although they have no particular role in the digester reactions. Another way of inoculating a digester is to use some of the waste which has been allowed to stand in a tank or lagoon and to begin to putrify, but factory-waste digesters are also inoculated from sewage digesters. As the inoculum can be ten percent or more of the digester volume there is thus a good possibility that the many types

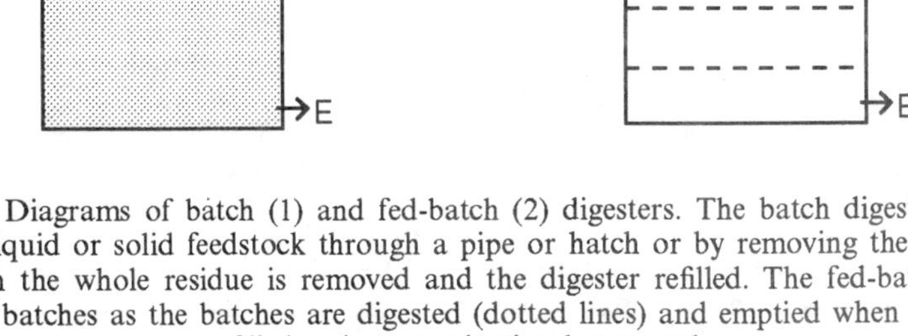

Fig. 2a. Diagrams of batch (1) and fed-batch (2) digesters. The batch digester is filled with a liquid or solid feedstock through a pipe or hatch or by removing the top. After digestion the whole residue is removed and the digester refilled. The fed-batch tank is filled in batches as the batches are digested (dotted lines) and emptied when the tank is filled and gas production has ceased.

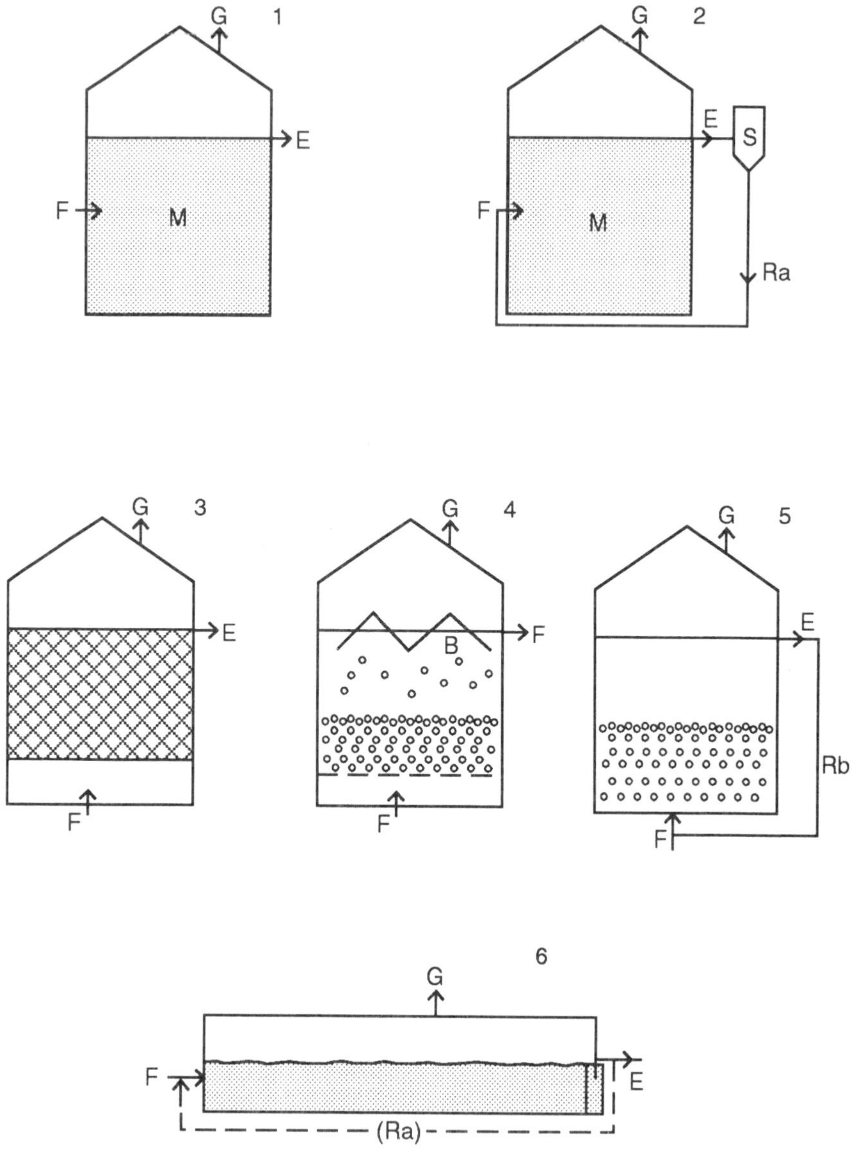

Fig. 2b. Diagrams of the basic types of continuous-culture digesters. 1, stirred-tank: 2, Contact: 3, upflow anaerobic filter: 4, sludge-blanket: 5, fluidised- or expanded-bed: 6, plug-flow or horizontal-tank. This may have bacterial recylce. F, feed flow. E, effluent flow. G, gas flow. Ra, effluent bacteria recycle flow. (Ra), optional return flow. Rb, effluent liquid recycle flow. B, baffles to retain bacteria (a similar system can also be in 5). M, mixers; mechanical or gas recirculation. All digesters can be heated by internal or external systems. For more details of digesters see Fig. 9.10. Two-phase digesters can be two stirred tanks with the effluent from one being the feed of the other, or a stirred-tank coupled to 3, 4 or 5.

of bacteria in the sewage digester or lagoon will be introduced into the factory digester. Overall, then, a sample of digester contents will be likely to contain anaerobic and facultatively anaerobic bacteria which do not have any specific role in the digester reactions. These may have been introduced in the inoculum and have managed to continue to grow as a minor component of the population by utilising a minor component of the feed or a metabolic product of other bacteria. On the other hand, they may not be actually multiplying but as they are being continuously introduced in the feed there will always a certain number of viable cells to be picked up in a bacterial count. Aerobic bacteria may be among those continuously introduced. Mah & Susman (1968), for instance, obtained counts of about 10^8 anaerobic bacteria per ml in samples from a sewage sludge digester. But they also cultured about 10^6 aerobic bacteria per ml. on nutrient agar, EMB or a sludge-supernatant medium with glucose. These aerobic bacteria may have a limited metabolic activity metabolising sugars and other simple substrates produced by the main bacterial population while scavenging small amounts of oxygen introduced into the digester dissolved in the feed-water or in other ways. In this way the aerobes and facultatives can help to provide the reduced conditions necessary for the growth of the main bacterial population. But continuous introduction in the feed may give a false impression of the numbers of active aerobically-metabolising bacteria or, indeed, of other groups or species, in a digester. The fact that bacteria may remain viable, although not growing, in a hostile environment and may then begin to divide when placed in a suitable medium makes analysis of the functional populations of systems like anaerobic digesters difficult. As an example of the presence in digesters of bacteria from the feed the work of Hobson & Shaw (1973) may be quoted. A large part of faecal material, especially that from farm animals where there is a fermentation of the animal's feed in the caecum and large intestine, is made up of dead and living bacterial cells. Amongst these are hydrolytic and methanogenic anaerobes from which the active flora of a digester can develop. But there are also facultative anaerobic bacteria which proliferate in the intestines and these bacteria can form a large proportion of the bacteria in a digester. The contents of a domestic sewage digester used to start a piggery-waste digestion contained *Escherichia coli* as the predominant facultatively anaerobic bacterial species. *Escherichia coli* are found in large numbers in the human intestine. When the digester had been adapted to a feed of pig excreta over many weeks the coli had been replaced by Streptococci, the predominant facultatively anaerobic bacteria of pig and other animal intestines. The Streptococci formed 50% or more of the culturable bacteria in the digester, but they were present in such large numbers in the pig excreta feedstock that this, rather than growth in the digester, could have been mainly responsible for the digester population. Indeed, their properties were such that it was difficult to assign to them a role in the digester reactions. Ueki *et al* (1978) also found that Streptococci made up a large proportion, up to 45%, of the bacteria in the liquid of pig waste and that Streptococci were the predominant group of bacteria isolated from digesters having this pig waste as feedstock.

Mention has been made above of the continuous introduction of bacteria into anaerobic digesters with the feed. Industrial bacterial fermentations can be carried out as 'batch' or 'fed-batch' or 'continuous' cultures and these types of culture systems are used in anaerobic digestion (Fig. 2a). In the batch culture a tank of substrate is inoculated with bacteria and kept at a suitable temperature. The bacteria begin to grow, and, conditions being suitable, they continue to grow at a regular rate until the substrate is used up or they alter the conditions in the tank (for example by acid formation) to ones inimicable to growth. The bacteria then enter a phase of slower and, finally, no growth and the culture has then reached its end (Fig. 3). The residual substrate and bacterial cells have then to be cleared out and the tank refilled with fresh substrate and inoculum bacteria; or the inoculum may be provided by leaving a little of the previous culture residues in the tank. The fed-batch culture is similar except that new substrate is added after a certain time of growth of the bacteria when the first lot has been almost used up, and more substrate is added at further intervals until the

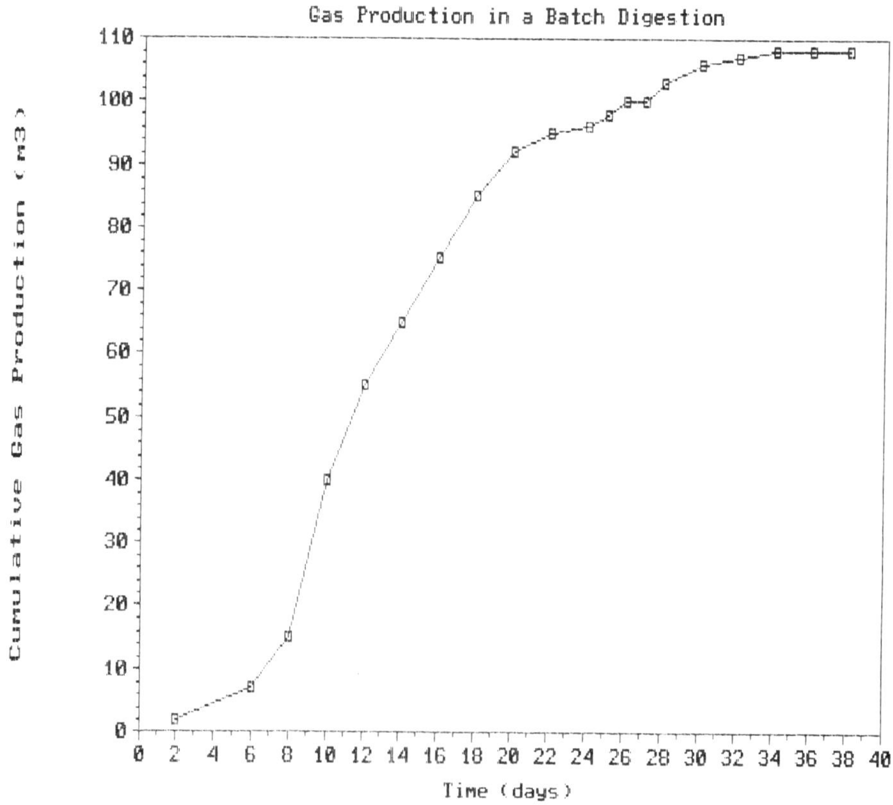

Fig. 3. Time-course of gas production in a batch digester. After 36-38 days the digester would be almost emptied leaving an inoculum to start the digestion of the next filling of the tank.

fermentor tank is filled. A large proportion of the fermentor volume is then pumped out leaving a small volume containing bacteria to inoculate the next initial batch of substrate and the process is repeated. Thus, the batch fermentor produces products at an increasing and then steady and then decreasing rate before production stops altogether. The fed-batch produces continuously, although at varying rates for a time, until the tank is filled. The batch anaerobic digester has only the one inoculation of bacteria with the feed, but the fed-batch has a semicontinuous inoculation which could be sufficient to keep up numbers of viable but non-growing bacteria in the digester. However, most wastes are produced continuously and have to be treated continuously and the purpose of energy-crop digestion is to provide continuous supplies of biogas energy. With a batch or fed-batch digester there is a period when the digester is being refilled and starting up when gas production is nil or small, and during the active phase of the digester gas production is not at constant rate. A feedstock being supplied continuously would have to be stored until the loading of the next batch. Only by using a number of batch digesters running out of phase, so that one is producing gas while the others are being loaded or just starting up production, can a constant feed be dealt with and constant energy produced.

If towards the end of the growth phase in a batch culture (Fig. 3) fresh medium is added and an equal volume of the spent medium and bacterial cells is taken out, then the remaining bacteria can continue to grow on the new medium (Fig. 4). If this addition and removal of medium and removal of bacteria is repeated every few minutes or continuously then the bacteria can settle down to grow at a rate equal to the rate of medium addition; that is, they double in numbers in the time taken for one fermentor volume of medium to pass through the fermentor. The number of bacteria in the fermentor then remains constant with time and so the rates of dissimilation of substrates and of production of products remain constant. One fermentor can then deal with a regular flow of feedstock and produce a regular quantity of product. This is a 'continuous-culture' (Fig. 2b). The laboratory and industrial continuous-cultures usually are supplied with a sterile feedstock of dissolved substrates and the continuous- culture theory to be described later deals with dissolved substrates with no bacterial inoculum in the feed. The degradation of solid substrates in a stirred-tank digester is not exactly the same, and some digesters are continuously inoculated from the feed, but the stirred-tank digester can be approximated in dynamics to the theoretical continuous-culture and is the same in physical form. The other types of digester to be described later are also continuous- cultures but differ in theory from the stirred-tank. Digesters running as continuous-cultures can then deal with continuous flows of wastes or other feedstocks and can produce gas continuously. Most digesters are run on the continuous-culture principle.

Bacteria require supplies of carbon- and nitrogen-containing substrates as well as sulphur and metal salts and possibly carbon dioxide, for growth. In an ideal medium these compounds would be in related concentrations so that all the compounds would be used up during bacterial growth in a batch culture. In practice this is seldom, if ever, the case. Apart from other problems, with natural substrates although it may be known that the substrate is a nitrogen source, say a protein, it may not be known exactly what part of the protein is being used by the bacteria. It may be that one or two amino acids only are used and so the other amino acids will be left unused when the bacterial growth stops. However, even if all of a particular substrate can be used, some other substrate may be used up before that one and so some of the first substrate will be left at the end of the bacterial growth. The second substrate, which is used up, is termed the 'growth-limiting' substrate and bacterial growth in continuous cultures is regulated by the concentration of such substrates, which can be sources of carbon or nitrogen, or sulphide or other ions. So, the microbial growth is controlled at a constant rate by the metabolism of one substrate which is in constant concentration in the feed and which is used up almost completely by the bacteria. This type of continuous-culture is known as a 'chemostat' continuous-culture to distinguish it from the 'turbidostat' culture where growth is regulated 'mechanically' by external systems, not internally by bacterial metabolism. The 'chemostat' is really a continuous culture in a stirred tank like the sludge digester. The 'retained-biomass' types of digesters to be described later are forms of continuous-culture in that they metabolise a continuous flow of feedstock with bacteria that are continuously growing and do not need reinoculation. But here the bacteria do not all flow out with the effluent, they are kept in the digester tank and so increase in numbers until the washing out of a proportion of the bacteria with the effluent, natural death of a proportion of the cells, and the limitation of growth of cells by the concentration of a substrate, produces an equilibrium of almost constant bacterial numbers which can degrade a constant flow of feedstock (Fig. 2). There are many papers on continuous culture of bacteria and the subject is dealt with in chapters in books on microbiology, but one of the best descriptions of microbial growth and culture, batch and continuous, is the book by Pirt (1975).

The theoretical continuous-culture is one bacterial species growing in a medium with defined sources of carbon, nitrogen and other elements, such as simple sugars, ammonia or urea. The medium is also added and removed continuously. The 'medium' (feedstock) for the anaerobic digester is seldom, if ever, defined in composition and the individual sources of carbon, nitrogen and other elements are generally complex compounds. Even if some constituents are given a name, this name may refer to a substance which is defined more by method of analysis (eg., a substance that is soluble in some acid under certain conditions) than by composition and such methods of analysis may give results that are not comparable from feedstock to feedstock. The same name may cover a variety of materials, eg., 'cellulose', which can vary from chemically- and physically-treated soft and partially-soluble toilet papers to the original insoluble, lignified wood. The limiting substrate in digestions may not be accurately known, but this does not matter as the bacteria will settle into a 'steady-state' as the flora develops during digester start-up. However, since the digester contains not one bacterial species but many, all interacting with many depending on substrates produced by other bacteria, while the initial substrates may be complex, digesters do not settle down to the theoretical steady-state: there is always some oscillation in gas production and substrate degradations, but observed over periods of weeks metabolism is more or less constant (Fig. 4). A further factor which leads to oscillations in bacterial numbers and metabolic activity is the fact that because the feed rate to a digester is often low and the feedstocks may be difficult to pump, the feeding has to be intermittent (say, once or twice a day) rather than continuous.

However, the reactions in the digester are determined by the substrates present in the feed, not the physical conformation of the digester, and the degradation of substrates will be discussed in the next sections.

Substrate degradations

Physical factors
It was said before that bacteria need compounds containing carbon, nitrogen and various other elements, including hydrogen and oxygen, to grow. They also

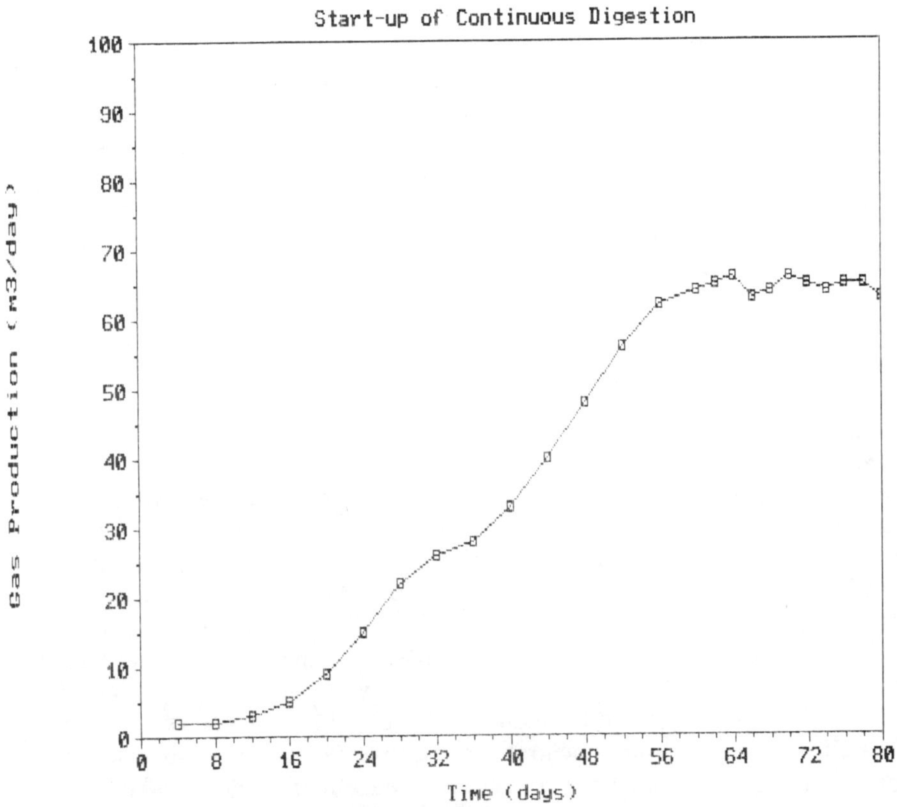

Fig. 4. Generalised diagram of the gas production during start-up of a continuous-culture, stirred-tank digestion. The system might be run as a batch digestion until day 32 and then as a continuous digester, the loading being gradually increased until the steady-state at design loading is attained after 65 days.

need water, as a large part of any organism is water which preserves the cell's spacial integrity and acts as a medium for transfer of nutrients into the cell and metabolic products out of the cell. With bacterial cultures the water may not be obvious to the eye, as in the case of the 'solid substrate' fermentations described later. But even these 'solid' municipal and vegetable wastes contain 40 to 80 % water, which is sufficient for the microbial needs.

Apart from substrates and water, bacteria require suitable conditions of pH and temperature to grow optimally. Different bacteria can grow over a wide range of pH values from quite acid (pH about 3) to alkaline (pH 10 or even above). The bacteria concerned in the reactions in anaerobic digesters vary in optimum pH for growth but are generally in the middle of this pH range. However, as digestion is carried out by mixed, interdependent, cultures the pH of the digester must be held at some value which allows of activity of all the bacteria although it may not be optimum for all the reactions. In general, the fermentation of sugars can take place at relatively low pH values of 5 to 6. However, with the exception of starch, the hydrolysis and fermentation of polysaccharides takes place optimally at higher pH values of 6·5 to 7 or over, and the optimum for degradation of fatty acids and production of methane is about 6·5 to 7·5. This has been shown, for instance, by Buchholz *et al.* (1986) who found the optimum pH range for hydrolysis and fermentation of sugar-beet pulp polysaccharides by a natural consortium of digester bacteria to be 6·2 to 6·8. The cellulolytic Clostridia isolated by Sharma & Hobson (1985) grew over a pH range from 6 to 8 but had optima about 7 to 7·5. Some other presumptive Clostridia grew from 6·5 to 8 with, again, an optimum about 7 to 7·5, while the cellulolytic Sporolactobacilli grew from 5 to 8 with an optimum for different strains of 7 to 7·5 (Sharma & Hobson, 1986*a*). The cellulolytic Clostridia isolated by Maki (1954) grew from pH 6·2 to 7·8, with an optimum at 6·8. The pH optima for methanogenic bacteria in pure culture lie generally in the range from 6·5 to 7·8, with a few having higher optima (Jain *et al.*, 1988). Attal *et al.* (1988) found that methanogenesis from acetate or propionate by mixed bacteria in a sludge adapted to glucose digestion was more inhibited by media pH below 7 than was methanogenesis from hydrogen plus carbon dioxide. They suggested that inhibition of the acid degradations was caused by the increasing concentrations of unionised substrate acids as pH was lowered; the unionised acids penetrated the cells and disturbed the internal pH. The inhibition mechanism for methanogenesis from hydrogen must have been different, but was obscure.

Single-stage digesters, in which all the reactions take place in one tank, are, therefore, run at a pH of 7 to 7·5. Unlike many industrial fermentations where the pH has to be controlled by addition of acid and alkali, most digester systems are self-buffering at around neutral pH (7). In some cases it is possible that the pH of the feedstock may need correcting, if it is initially highly acid or alkaline, but most feedstocks are near neutral. The buffering of the digestion is carried out by a combination of the removal of acid fermentation products by conversion to methane, and chemical buffering by ammonia, volatile fatty acids, bicarbonate-carbon dioxide, phosphate and other salts. The buffering activity of the first three is described by Georgacakis *et al.* (1982). The maintenance of a suitable pH then depends on correct functioning of the bacterial reactions, and sources of instability in pH and possible digester failure are mainly found during start-up of digesters and rapid increases in loading rate, when reactions may get out of balance and the digester contents become too acidic for digestion to continue. Such problems will be discussed later.

Bacteria have three ranges of temperature at which they can grow and, generally speaking, different strains or species of bacteria grow in the different temperature ranges. Two ranges for growth are from 0°C to about 15°C and from 15° to 45°C. In the temperature range below 45°C (the 'mesophilic' and 'psychrophilic' ranges) digestion becomes slower as temperature decreases and so far as the normal 'mesophilic' digestion is concerned digestion virtually ceases about 15°C. However, the bacteria are not completely inactivated and digestion will take place if the bacteria are allowed sufficient time. It is not certain whether the bacteria concerned in this 'psychrophilic' digestion at temperatures below about 18°C are different from the bacteria active in the mesophilic temperature range or are the same bacteria growing at slow rates, but, in any case, 'psychrophilic' digestions are always run with very long retention times, particularly if the feedstock contains mainly solid substrates. Above 45°C there is about a 5°C gap in digestion then above 50°C digestion takes place at an increasing rate up to about 65° to 70°C and this is the range of 'thermophilic' bacteria and 'thermophilic digestion'. The main bacteriological work has been done on mesophilic digesters, but it is probable that the bacteria concerned in thermophilic digestion are different varieties, if not different species, from those active in mesophilic digesters, although the reactions in the thermophilic digester are basically the same as in mesophilic and psychrophilic digestions. The reasons why thermophilic digestion has been little used and investigated will be discussed later. Since the metabolic heat generated in anaerobic digestions is small, digesters have to be heated to run at temperatures above ambient. Feilden (1981) showed that taking into consideration gas production and gas utilisation for heating a stirred-tank digester there was a critical feed-solids concentration (3·5% for pig waste; solids are related to gas production) below which it was optimal to operate at 25–30°C and above which it was optimal to operate at 40–44°C. There was also a second, higher, solids concentration (7·5% for cattle waste) above which it was optimal to operate in the thermophilic range. These calculations were concerned only with the maximum use of the gas produced and with the retention times needed and not with other considerations.

The optimum temperature for mesophilic digestion is generally accepted to be about 35°C and the thermophilic optimum at about 55 to 60°C. However, increase in temperature increases the rate of polysaccharide hydroysis and fermentation but does not increase the extent of hydrolysis of fibres, so, as found for instance by Chen (1983) and Hashimoto *et al.* (1981), the ultimate degradation of manure solids (and so gas production) at long digester retention times is the same whatever the temperature. Only at short retention times does the difference in rates show up (Fig. 5). The previously mentioned cellulolytic Clostridia from

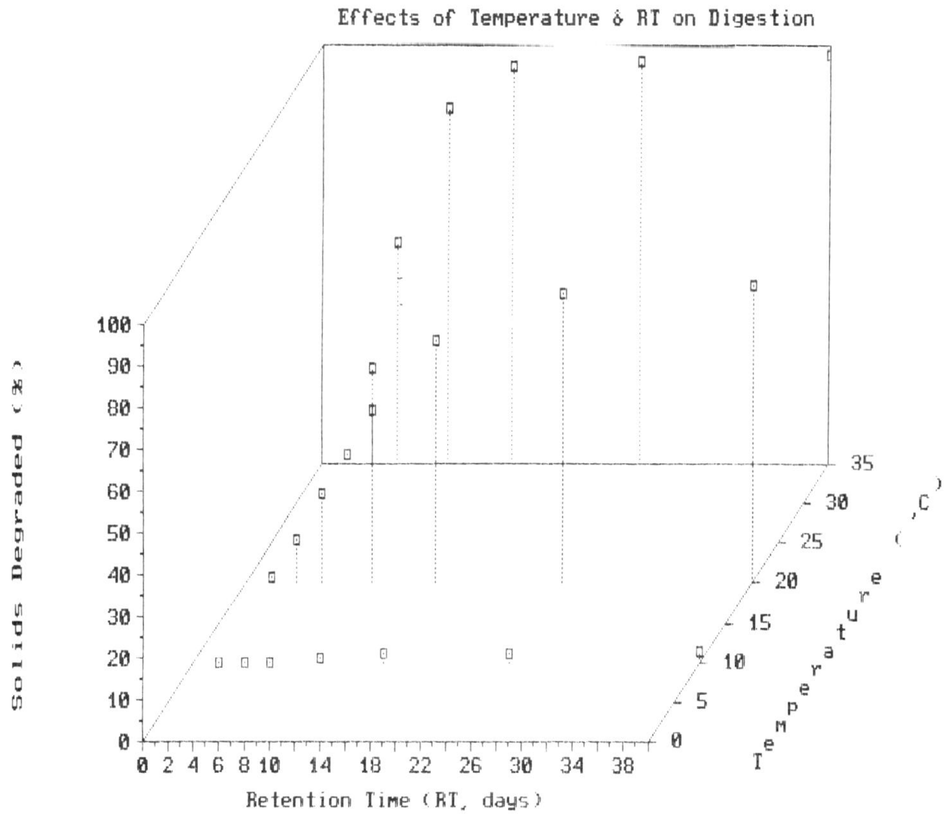

Fig. 5. Generalised diagram of the effects of temperature and retention time on digestion. 'Solids' could be dissolved or suspended substrate.

a mesophilic digester grew optimally at 35°C with no growth detectable at 20° and 50°C. The other Clostridia (Sharma & Hobson 1986*a*) had optimum growth at 35° to 39°C, with no growth at 45°C but a slow growth at 20°C. The Sporolactobacilli showed no growth at 50°C, but depending on strain had optima at 39°C or between 27 and 35°C. Maki's (1954) cellulolytic bacteria had a temperature for growth of from 23° to 40°C with optimum at 30° to 37°C. All these bacteria were isolated from mesophilic digesters run at about 35°C. *Clostridium thermocellum*, which has been isolated from thermophilic digesters, has an optimum temperature of 60° to 64°C (Ng *et al.*, 1977). The Methanobacteria, even those isolated from soils and muds, generally have temperature optima about 35°C, but some from habitats of higher temperature have much higher optima (Jain *et al.*, 1988). The rate of digestion in the mesophilic range increases rapidly from about 15°C to 25°C and then there is a smaller rate of increase from 25 to 40°C (Summers & Bousfield, 1980). Rajasekeran *et al.*, (1986) found numbers of cellulolytic, proteolytic, lipolytic and methanogenic bacteria were the same in batch digesters run at 28°C or 37°C with various feedstocks. Gas productions were higher at 37°C, but this would seem to be determined by the rates of enzymic activities not absolute numbers of bacteria.

Enumerating the bacteria

Wastes composed mainly of human or animal excreta contain all the necessary bacterial nutrients in the form of intestinal bacteria, intestinal secretions and food residues undigested in the gut. In addition, domestic sewage contains kitchen wastes and material from drainage water. Feedstocks composed of vegetable matter such as crop wastes or actual crops also contain all the nutrients for the bacteria.

However, although these feedstocks contain compounds of carbon, nitrogen and other elements which supply the nutrients needed by the bacteria, these may not be in balanced proportions. For instance, straws contain little nitrogenous compounds, and are composed mainly of carbohydrates such as cellulose. Bacterial growth could be limited by the nitrogen content and so all the carbohydrate would not be fermented and would remain in the digester effluent. This would lead to low gas production in a digester run for energy production and to a residue which could be relatively easily degraded by soil organisms and so cause pollution if the residue was used as a fertilizer. To ensure that there was sufficient microbial growth in the digester to degrade all the available carbohydrate it would be necessary to add more nitrogen. This could be added as a chemical like urea or it could be added in the form of a waste containing excess nitrogen compounds, such as animal excreta or sewage sludge. The balancing of nutrients in feedstocks will be described in a later chapter, but it is sufficient to note here that a better and more complete digestion may be obtained by mixing wastes. If a feedstock has none of a particular microbial nutrient or is very unbalanced in composition it will be undigestible or nearly so. Some factory wastes could, for instance, be almost pure sugar solutions or suspensions of polymeric carbohydrates. However, although when a potential feedstock will not digest because of lack of nutrients addition of the required nutrients is needed if digestion is to take place at all, partial digestion of a nutrient-unbalanced feed may have to be accepted because the economics or mechanical aspects of obtaining additional nutrients make the manipulation of the feedstock impossible. The manipulation of feedstocks will be dealt with in more detail later.

The analysis of the reactions and the functional microbial populations of anaerobic digesters involves the cooperation of the methods of bacteriology and biochemistry. Chemical analysis can determine the compositions of the in-puts to, and the end-products of, digestion, although such analyses may vary in the detail they provide. A paper by Colin *et al.* (1983) suggested the analyses needed to describe an anaerobic digester feedstock and the reactions going on in a digester, and gave a guide to the methods that could best be used in such analyses. The paper also suggested the information that should be given to describe the working of a digester in terms of pollution control and energy production. The units in which the data were to be presented were also given. However, it was realised that in practice the analyses that could be made could be limited by the facilities available; not all full-scale digesters would be near laboratories or be equipped with sophisticated analytical apparatus. Other techniques, such as the use of isotopically-labelled compounds or of defined chemical compounds as feedstock can help to identify some of the intermediates between feed constituents and end-products, but these have usually to be used with small-scale bench digesters. By analysis and by analogy with other similar systems possible reaction pathways between these initial substrates and end-products may be worked out and then tested by more detailed analysis. The bacteriologist isolates bacteria, using medium constituents suggested by the chemical analyses, and determines their properties, and these may show the role the isolates might have in the reaction schemes. The numbers in which the bacteria occur can indicate whether they could play a leading part in the postulated reactions. Another criterion of whether or not a bacterial species is of importance in digester reactions is whether it can be isolated at different times from the same digester and whether it can be isolated from other digesters treating the same or similar feedstock. A mumber of tests is needed, though, as the limited numbers of multiple cultures done from digesters suggest that, similar to the rumen system, the numbers of particular species fluctuate from day to day or week to week (Sharma & Hobson, 1985). But serological analysis of rumen bacterial populations has shown that even when a species of bacterium is a major, long-term constituent of the flora different varieties of the species may be dominant at different times (Hobson, Mann & Oxford, 1958). Cultures from both the rumen and anaerobic digesters show that isolates classified as the same species may differ in fermentative and other properties (e.g. Sharma & Hobson, 1986*a*; Maki, 1954). Although total numbers of bacteria may remain relatively constant with time, the composition of the bacterial population of an anaerobic digester is, thus, not constant with time, it is a dynamic system in which relative numbers of species and varieties change even when the feedstock and its rate of feeding are unvarying. Serological and other analyses have shown that the bacterial populations in granules from UASB (see later) digesters treating solutions of acetic or propionic acids were, as might be expected, simpler than in granules from a digester treating sugar-processing wastewaters (Grotenhuis *et al.*, 1991). Methanogens in granules from UASB on potato-processing wastewaters changed in species and relative numbers with time after a shift from mesophilic to thermophilic running (Visser *et al.*, 1991), and they varied in different digesters and with substrate (Macario *et al.*, 1991). Such experiments show not only changes in populations but also that species can be 'dormant' in undetectable numbers until conditions become suitable for them to become major components of the population. Also, different mixtures of species and varieties can carry out the same reaction(s).

If a digester is fed continuously, or at intervals which are short compared with the time taken for degradation of the feed components, then almost constant total bacterial numbers should be obtained in cultures made from samples taken at different times of day. If, however, the feeding is at intervals of some hours or a day or more, then numbers of culturable bacteria will vary with the time after the last feed at which the sample for culture was taken. In comparing bacterial numbers over periods of days or weeks it is better in all cases to take samples at the same time each day.

There can be antagonisms due to production by one bacterial strain of substances inhibitory to other strains of the same species or to bacteria of different species, although these seem unlikely to result in total dominance by one strain but to fluctuations in numbers of different strains (Sharma & Hobson, 1986*b*). Mutations can take place which may lead to diminished numbers or elimination of a strain or species from the system. Bacteriophages are common in the rumen and so probably in digesters. Prensier *et al.* (1988) found virus particles associated with empty cells of methanogenic bacteria in sludge granules, and their work also suggested that a Spirochaete preyed on the methanogens. Work on the rumen has suggested that bacterial strains may have a limited life in a mixed-population system and natural reinoculation from the feedstock may be needed to carry on the species (Hobson, 1988*a*). Since there is constant variation in bacteria, and the rumen and intestines contain a large number of strains of particular species varying slightly in properties, the constant reinoculation from the feedstock of a digester fed partly or entirely on faecal wastes, as well as mutation in the existing bacteria, may be a cause of variations in the populations of such digesters. Although a factory-waste or an alternative-energy digester fed on treated vegetation is open to inoculation from the surroundings, the numbers of 'foreign' bacteria entering the system will be less than those entering the faecal-waste digester and mutation of the bacteria developed from the initial inoculum should be of greatest importance in causing strain variations in the digester flora. Generalisations are difficult to make, though, as most of the long-term bacteriological work has been done on faecal-waste digestions. A further problem is that the initial inoculum of plant-biomass and factory-waste digesters has been varied and the populations in the digesters may have been affected by the initial inoculum and for how long the digesters had been running.

The methods and media used for isolating digester bacteria are based on those of rumen microbiology, and these are described in, for instance, the book edited by Hobson (1988*a*) and in other reviews mentioned therein. These methods involve the use of highly-reduced media, prepared under an oxygen-free atmosphere, with constituents based on analyses of digester liquids and, for initial isolations or total counts where the nutrient requirements of the bacteria are not known, digester fluid as a source of growth factors. Ammonia is the usual nitrogen source, but the energy source can be a mixture of carbohydrates for a total-count medium, or a particular carbohydrate where counts of, say, cellulolytic, bacteria are being attempted. The carbohydrate can be replaced by, say, an organic acid or a gaseous substrate of hydrogen-carbon dioxide when methanogenic, lipolytic, or other bacteria are to be cultured. Details of media are given in the papers describing the bacteria concerned in the different reactions to be discussed later.

There is, however, one problem common to bacteriological investigations of the rumen and digesters. That is, the difficulty of properly enumerating the bacteria. Counts of the total numbers of bacteria are made by the standard methods of counting the cells visible in a stained film of digester contents or in a layer of the liquid in a counting chamber. Counts of viable bacteria are made by standard dilution and culture methods using liquid or solid media. The methods for both total and viable counts were developed for enumerating bacteria growing freely suspended in liquid media. In such cases the cells can be seen as discrete entities and so counted under a microscope. Suspensions of free bacteria will be evenly distributed in the dilutions of the sample made for viable counts. Thus, some reasonable measure of the concentrations of bacteria may be made. In micro-ecosystems such as digesters, though, the bacteria are not evenly suspended in a liquid.

The bacteria hydrolysing solid substrates such as plant fibres are strongly attached to the surface of the fibre. Surrounding these, but not attached, are the bacteria utilising the excess sugars liberated by the hydrolytic bacteria and those, such as the methanogenic bacteria, using gaseous and volatile fatty acid fermentation products of the former organisms. Bacteria in retained-biomass digesters form colonies attached to the particulate support medium, or, in the case of 'sludge-blanket' types of digester form dense granules of mineral matter and bacterial cells (e.g. Dubourguier *et al.*, 1988*a*). The attached hydrolytic bacteria and the granule bacteria cannot be easily dislodged and broken up into discrete cells in a medium used for microscopic or cultural counts. Thus, it is impossible to count the actual number of cells in a granule or in and on a fibre. When dilutions are made of such material for viable counts the fibres and granules tend to remain in the initial dilutions and the bacteria are not evenly distributed into the succeeding dilutions. Various methods are used to try to overcome the problems, but none are entirely successful as the operations needed to liberate all the cells are so violent that they result in the death of the cells and so culture is no longer possible, and in disintegration such that microscopic identification of cells and debris is difficult or impossible. For cultural counts of bacteria in digestions of plant material the methods used by Sharma & Hobson (1985) for culture of cellulolytic bacteria are representative. The digester sample was initially diluted by a factor of ten in an anaerobic dilution fluid and mixed for 40s in a Waring Blendor under an inert atmosphere. The homogenised suspension was then diluted again and mixed for a few seconds in a test-tube vibrator. Further dilutions into culture media were then made from this suspension. The diluting fluid also contained a surface-active agent; Tween 80. Triton X-100 gave similar results, as did methyl cellulose, an agent recommended by Minato & Suto (1981). Treatment with dilute HCl, NaOH, EDTA or by boiling or autoclaving did not disintegrate granular sludge, and ultrasonics removed only 9% of the weight of the granules into discrete suspension (Dubourgier, 1988).

The numbers of bacteria on and in plant fibres and in sludge granules will, thus, tend to be underestimated and it will not be certain if the relative proportions of those counted are the same as the proportions in the actual digester.

The bacteria in suspension should, though, be accurately counted. These will consist of hydrolytic and associated bacteria which have been dislodged from fibres by shear stresses from mixing or by the digestion of the fibre leaving no further substrate. Some bacteria are also being continuously removed from sludge granules. In addition, there will be the previously mentioned bacteria which are not active in the digestion, including aerobic bacteria from the feed. The 'active' bacteria in the liquid will contribute to the counts of cellulolytic and other bacteria, but, again, it may be that those in suspension are not, for various possible reasons, in the same proportions as they are on the substrate-solids or granules. The non- active bacteria may also be overestimated in comparison with the active bacteria. However, it is usually assumed that the counts are representative of the types and numbers of bacteria in the digester and that counts done in the same way, on digesters with the same or similar substrates, but at different times reflect diurnal or longer-term changes in the digester populations; for instance, in the commissioning of a digester. Measuring total mass of bacteria will not, of course, identify species, but Fynn & Ibba (1988) found twice the total mass of bacteria on an anaerobic-filter support as free in the liquid.

Although microscopic observation may not allow of the accurate counting of bacteria on solids or in granules, it has been extensively used to examine the make-up of the micro-colonies of bacteria on solids or in granules. Electron microscopy, particularly, is much used nowadays; both transmission and scanning EM are used. The identification of the bacteria depends on morphology, and while the morphology of some bacteria may be distinctive enough to allow of identification, in other cases some doubt will remain. However, even if species identification is not possible the relative numbers of different morphological types may give a guide to changes in digester populations with changes in substrate or operating conditions. Electron microscopy has been the main means of showing the bacterial structure of micro-colonies of bacteria attached to and near the surface of solids and the physical microstructure of sludge granules (e.g. Hobson, 1985; Lequeriea & Toldra, 1986; Dubourguier *et al.*, 1988). The methanogenic bacteria contain a biochemical component of methane formation, factor F_{420}. This factor fluoresces in UV light and blue-fluorescing cells can be seen in a UV microscope. This has been used for visual assessment of methanogenic bacteria, but use of measurements of the fluorescence for quantitative determination of numbers of methanogens and methanogenic activity seems to be doubtful (Dolfing & Mulder, 1985). The yellow-fluorescing factor F_{350} has also been suggested for enumerating methanogenic bacteria (Doddema & Vogels, 1978). Hobson *et al.* (1955) and Hobson & Mann (1957) introduced the use of fluorescent antibodies to identify bacteria in samples of rumen contents. They showed that specific antisera could be obtained by absorption to remove cross-reactions, but, in these and other papers, they showed that there were many variations with time in serological types of a species in the habitat. The apparent absence of a species in a particular sample might then mean only that the particular serological type present had not been isolated and so no antiserum to it existed. Others confirmed the wide variation in serological types of species present in the rumen. With the advent of monoclonal antibodies, labelled antisera are again being tested, particularly with methanogens, in digester samples (Macario & Conway de Macario, 1983; Kemp *et al.*, 1986; Prensier *et al.*, 1988; Macario *et al.*, 1989). As in the rumen, there seems to be considerable immunological diversity (Macario & Conway de Macario, 1988). Determinations of the concentrations in digester or other fermentation liquids of various constituents of bacterial cells by chemical or spectroscopic analysis have been tested for calculating the quantities of bacteria present or the variations in particular species. Henson *et al.* (1985) tested the analysis of bacterial lipids as a method of following changes in thermophilic digesters. However, many constituents of bacteria vary in amount, or presence, with growth rate and other metabolic changes in bacteria, so interpretation of analyses may be difficult.

The breakdown of carbohydrates

Carbohydrates in feedstocks

Carbohydrates constitute a large fraction of the feedstocks of most waste digesters and are the main feedstocks of nearly all digesters whose primary functon is energy production. Most of this carbohydrate is in the form of insoluble polysaccharides from residues of vegetable matter or whole plants, with sometimes soluble polysaccharides and sugars from the same sources. Polysaccharides, generally more complex in structure and less biodegradable than the plant carbohydrates, are also found in the tissue residues from slaughter houses and in the intestinal secretions and bacteria in faeces.

Sewage-sludge feedstocks of digesters contain the primary sludge from the sewage works and sludge from the aerobic treatment systems. The primary sludge contains faecal particles and pieces of toilet papers and other paper and cotton-wool materials as well as rags and papers from street debris. Most of these papers have been chemically treated to make them ‘soft’ and ‘soluble’ and the cellulose in them will be easily degraded by the digester bacteria. The aerobic sludge contains the finer faecal and other particles from the wastewater and the microbial cells from the activated-sludge- or filter-plants. Microorganisms from faeces or aerobic treatment plants can contain starch-like storage polysaccharides as well as more complex polysaccharides in different cell structures. Gut secretions contain complex mucopolysaccharides. Some of these complex polysaccharides may be virtually undegraded in the digestion process even if they are liberated by death and lysis of faecal bacteria and intestinal cells. Overall, they probably do not contribute to a large proportion of the gas generation in digesters. The most important source of gas in digestion of faecal material from farm animals is the residues of vegetable matter from the feedstuffs passing through the gut. In sewage sludge digestion lipids also contribute to a large proportion of the gas.

Human beings have a limited ability to digest polysaccharides and only starch can be degraded by gut enzymes. Starch, particularly if it is cooked, is almost completely utilised in passage through the digestive tract, while the structural polysaccharides of plants, the 'fibre' fraction, are almost unattacked and so voided in the faeces. Farm animals have digestive tracts which allow of microbial degradation of starch and the plant structural polysaccharides. However, degradation of the latter is not complete and so residues remain in the faeces. Depending on the animal and feed, the plant residues in animal faeces will tend to form a poorer substrate for, and require a longer time for degradation by, the digester bacteria than the residues in human faeces. Farm wastes may also contain animal beddings. These, being straws, wood shavings and similar materials, will be relatively difficult to degrade in a digester as they are highly-lignified plant structural polysaccharides (Fig. 6).

Plants contain some soluble sugars, but the amount varies with the stage of growth of the plant. Ordinary temperate-climate grasses may contain some 6–7%

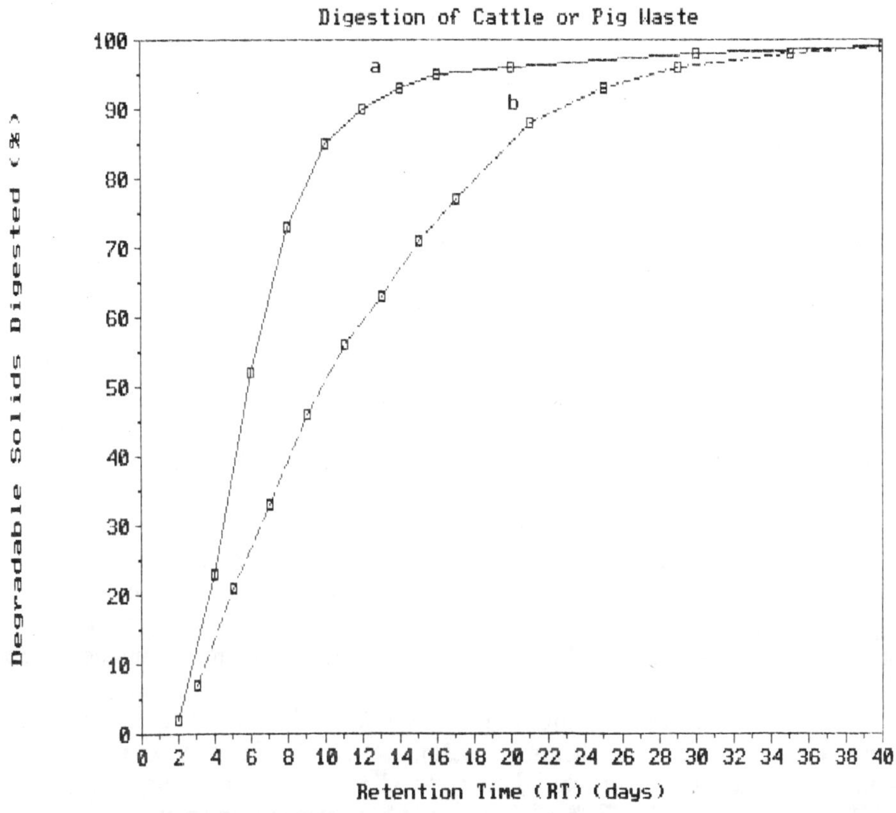

Fig. 6. Generalised diagram of the continuous-flow digestion of cattle or pig wastes at the same mesophilic temperature. Full line (a) pig waste; dotted line (b) cattle waste containing less-easily degradable solids than pig waste. Note that the graph shows digestion of metabolisable solids not of total solids. The solids degraded in steady-state running at each RT asymptotically approaches 100% of those metabolisable as RT is increased: there will be a residue of biochemically-inert solids.

of the dry matter as soluble sugars (Waite & Gorrod, 1959), but other stems and tubers may contain more. Sugar beet and turnips contain 60% and 35%, while the stems of kale contain about 25% sugars, and the stems of sugar-cane contain some 50% sugars. Some of these plants are used in manufacture of sugar as a human feed and some are used as animal feeds because sucrose and other soluble sugars are rapidly metabolised by gut and tissue enzymes and by bacteria. The plants containing high amounts of sugar are not likely to be used directly as feedstocks for anaerobic digestion as they are more important in sugar manufacture or as animal feeds, but the residues from sugar manufacture may be digested, and various low-sugar plants may be digester feedstocks. The pathways of metabolism of the sugars start where shown in Fig. 1.

Overall, the different bacterial species contain enzymes that can metabolise all the common sugars and this metabolism is rapid because the sugars are in solution and in a form in which they can be absorbed by the cells. Plant polysaccharides are made up from these sugars but the sugars are not directly available to bacteria as the sugar polymers are either insoluble in water or, if dissolved, are too large to enter the bacterial cell. Before the sugars can be absorbed and metabolised they have to be generated by hydrolysis of the polysaccharides, so sugar metabolism in this case starts at 'polysaccharide' in Fig. 1.

The principal plant storage polysaccharide, found in grains and tubers, is starch. Here it is a food for the germinating plant and so must be readily degradable by the plant enzymes to utilisable sugar. It follows that it will be degradable by bacterial enzymes. Starch is a polymer of glucose residues linked mainly in the α-1–4 position. It occurs in the form of granules which vary in size and shape with the plant species, from about 2 to 150 μm in diameter. Because of their surface structure the isolated, intact, starch granules are, in general, largely insoluble in cold water and not easily attacked by enzymes. However, processing of starchy materials for human or animal foods, and chewing of the ingested food, damages the surface of the granules and makes the interior starch available to degradative enzymes which separate the glucose units by a hydrolytic mechanism. Thus, little starch will escape the action of intestinal enzymes and bacteria and be found in the faeces. Large amounts of starch will generally be present only in digester feedstocks from food manufacture.

Polymers of fructose, the 'fructosans,' also occur as storage polysaccharides in some plants and in leaves and grains. These are soluble and are readily degraded by bacteria.

The structure of plants is maintained by polysaccharides. Thus these structural polysaccharides, together with lignin, waxes and other materials, are designed to protect the plant from damage by climate or by predatory organisms. This means that they are not easily degraded by bacteria and not at all by gut enzymes. Thus even with herbivorous animals, which have a microbe-containing stomach system for the digestion of plants, large amounts of plant structural material can escape digestion and be found in the faeces.

The principal structural polysaccharide is cellulose. This is a polymer of glucose units linked β-1-4. The α-linked glucose units in starch allow the formation of rather amorphous molecules which are accessible to enzymes which can hydrolyse the inter-glucose bonds. The β-linked glucose units of cellulose form linear chains which can lie side-by-side to form microfibres which cannot easily be penetrated by hydrolytic enzymes. Bacterial degradation of cellulose is, therefore, slow and is influenced by the type of plant fibre containing the cellulose. For instance, Stanogias *et al.* (1985) found that the rate and extent of fibre destruction in digesting pig faeces varied with the plant fibre in the pig feed. Fischer *et al.* (1981) found less gas production from faeces of pigs fed on corn and milo than on corn and wheat.

Cellulose is associated in the plant cell wall with hemicelluloses and pectic substances. While starch and cellulose are polymers of one sugar, the other carbohydrates are polymers made up from a number of different sugars. Hemicelluloses are not soluble in water and are a mixture of polymers made up from xylose, arabinose, glucuronic acid and glucose with mainly β-1-4 and β-1-3 linkages. The type of polymer and the amount varries with plant and age of plant. The pectic substances are chains of α-1-4-linked galacturonic acid units which are partly esterified with methanol and linked at the 2 position to acetyl groups. The chain also contains some rhamnose. These polysaccharides with varying sugar residues and branched chains are not easily degraded by bacterial enzymes.

Lignin is a three-dimensional polymer of three complex aromatic alcohols joined by some twenty different kinds of linkage. It is uncertain whether lignins are random structures or whether lignins have some similarity within the same plant even though they vary between plants. The various linkages are resistant to both chemical and bacterial degradation and it is usually assumed that lignin is not degraded in anaerobic digesters, although it can be degraded by some aerobic microorganisms. There have, however, been some reports of lignin depolymerisation by anaerobes. For instance, Ziomek & Williams (1989) reported that *Desulfovibrio desulfuricans* partially depolymerised Kraft lignin and lignosulfonates and could utilise the sulphate of the latter. Lignin is associated with the cell-wall polysaccharides and the amount and complexity of the lignin molecules vary with the part of the plant and the age of the plant, older plants being more lignified. Since it is not itself degraded, lignin is generally regarded as the main constituent which acts as a barrier to microbial degradation of the plant cell walls, although degradation may be impeded by other fibre components Thus, in the herbivore digestive tract only the less-lignified parts of the food vegetation are degraded and the faecal residue is the resistant lignified-fibre. This residue can be further degraded in anaerobic digesters only by allowing the bacteria a long time of action. Similarly, straws, being the dry, lignified residues left after plant growth has ceased, will take a long time for only partial degradation to occur. On the other hand, fresh, young plant material will be easily, and comparatively extensively, digested.

Seaweeds, which are being tested as energy crops for gas production by anaerobic digestion, are giant algae. Like the microscopic algae, which are also possible digester feedstocks, the cell walls are composed of cellulose as in higher plants. But they also contain large amounts of gummy, gelatinous polysaccharides such as alginic acid, a polymer of mannuronic acid, and agar which is a sulphuric acid ester of a polymer of the sugar galactose. King *et al.* (1985) found that the sulphated polysaccharide carrageenan, from seaweed, was degraded by anaerobic digestion. The brown seaweeds *Laminaria* spp. and *Ascophyllum nodosum* were also anaerobically digested (Hanssen *et al.*, 1987). The polysaccharides from *Ulva* sp. were digested after chemical hydrolysis (Briand *et al.*, 1987).

Examples of the effects of lignification on the digestion of plant materials are found in earlier work on digestion. A retention time of about 1·7 days was suggested for degradation of a purified, virtually 100% degradable, cellulose by Ghosh & Klass (1978). The cellulose in pig excreta was only 41% degraded at a retention time of 10 days (Hobson, 1976), while barley straw added to pig waste was only 35% digested at a retention time of 20 days (Summers, 1978; Hobson, 1979). These digestions were all run at 35°C. Pfeffer (1978) found that wheat straw was only 38% digested at a retention time of 13·7 days at a temperature of 59°C, and calculated that at infinite retention time the (maximum possible) degradation would be only about 50%, a figure similar to that calculated for the maximum digestion of barley straw at 35°C (Hobson, 1979). Robbins *et al.* (1979) added different cellulosic substrates to a cattle-waste digestion. Straw gave about the same gas production as the cattle waste, but straw delignified with acid chlorite, and purified cellulose (filter paper) gave much higher gas productions. It was calculated that the native straw was only about 32% digestible while the delignified straw and paper were almost 100% digestible. Chandler *et al.* (1980) found an inverse relationship between the lignin content of wood samples and the degradation of the wood in digesters.

Bacteria concerned in polysaccharide digestion

It was remarked previously that the number of particular bacteria (per unit weight or volume) in a digester population can give an idea of the importance of the bacteria in degradation of a substrate. The number of bacteria may also reflect the concentration of a particular substrate in a digester feedstock. The cellulose content of sewage sludges used as digester feedstocks is about 1 to 4% (Buswell & Neave, 1930; Pohland, 1962). Hungate (1950) found $0{\cdot}8$ to 2×10^3 cellulolytic bacteria per ml and Maki (1954) found $1{\cdot}6 \times 10^4$ to $9{\cdot}7 \times 10^5$ per ml in sewage digesters. Rajasekaran *et al.* (1986) found between 1 and 4×10^5 cellulolytic bacteria in batch digesters using different mixtures of cowdung, poultry droppings and silk-worm litter. The variation probably reflected the cellulose content of the sludges. Hobson & Shaw (1974) found 4×10^3 cellulolytic bacteria in digesting sewage sludge, but when the bacterial population of this sludge had been adapted to digestion of pig waste with about 15% cellulose the cellulolytic bacteria increased to 4×10^4 to 4×10^5 per ml. (The chemically-determined cellulose in the pig waste would not be all digestible, as previously mentioned, so the numbers of bacteria would not be in proportion to the analysed cellulose content). Sharma & Hobson (1985) isolated nearly 400 strains of cellulolytic bacteria from a mesophilic (35°C) digester which had been running on cattle waste for some 5 years. The digester was fed every 20 min with 5·8% TS cattle slurry, at a retention time of 21 days. The TS contained 76% VS and 32·7% cellulose, 15·8% hemicellulose, 11·8% lignin, 1·6% nitrogen and 8·2% ash. The digester effluent contained only acetic as the residual acid and was producing gas of about 62% methane (i.e., it was running correctly). Samples were obtained mid-morning over a period of 15 days. The bacteria were cultured on anaerobic media containing either barley straw, NaOH-delignified straw or powdered, HCl-treated, filter paper. These gave a range of celluloses of increasing degradabilities. The isolates were quantitatively screened for cellulolytic activity on filter paper, cotton and straw. The bacteria were cultured in numbers of 10^5 to 10^6 per ml of digester contents. The numbers of cellulolytic bacteria were rather higher than in the pig-waste digesters of Hobson & Shaw (1974), but the feedstock had a higher cellulose content than the pig waste.

In the work of Sharma & Hobson (above), there was, overall, little correlation between the substrate in the isolation medium and the ability of the isolates to degrade different cellulose preparations. There was no pattern in types of bacteria isolated on the different media. The number of isolates with high or low cellulolytic activities varied from day to day. The specific gas production (gas per kilogram of solids fed to the digester) varied a little from day to day, but there was no more than a suggestion that this might be linked to the cellulolytic activities of the bacteria present in the digester on those days. The isolates tested, although able to hydrolyse filter paper, varied in their ability to hydrolyse straw (0 to 50% degradation), where both cellulolytic and hemicellulolytic activities could be involved. Most isolates had xylanolytic (hemicellulolytic) activity. Buchholz *et al.* (1986) said that in the digestion of sugar beet-pulp, pectins and pentosans (xylans) were hydrolysed before the cellulose in the beet solids.

Hungate (1950) isolated three, non-sporing, Gram-negative cellulolytic rods from a domestic sewage digester. Maki (1954) isolated ten different rods, divided into two groups of high and low cellulolytic activities from a sewage digester. One isolate formed spores. Hobson & Shaw (1971, 1974) isolated eleven types of cellulolytic bacteria from a pig-waste digester. Ten were Gram-negative rods and cocco-bacilli and one was a Gram-positive rod. All the digesters were mesophilic and the isolates were made on powdered filter paper. The bacteria isolated by Sharma and Hobson from the cattle-waste digester were divided into fourteen different, named, genera and species plus some unnamed isolates (Sharma, 1983; Hobson, Reid & Sharma, 1984). Most were rods, of which the predominant isolates could be classified as known or new species of *Clostridium* or *Sporolactobacillus* (Sharma & Hobson, 1985; Sharma, 1983; Sharma & Hobson, 1986*a*). A small proportion of the isolates were cocci, which were assigned to the genus *Sarcina* or were sporing streptococci (Sharma & Hobson, 1990, 1991). Khan *et al.* (1981) obtained cellulolytic sporing rods from samples of a digester started from sewage sludge and fed with powdered cellulose. The samples were heat-treated at 80°C for 10 min before culture and this would have killed all except sporing bacteria. Patel *et al.* (1980) indirectly isolated a non-sporing cellulolytic rod (named *Acetivibrio cellulolyticus*) from a sewage digester. Eighty seven percent of the bacteria isolated by Jain *et al.* (1986) from cattle-waste digestions were Gram-positive sporing rods and 13% were Gram-negative rods. Cellulolytic activities in each group varied considerably. *Clostridium populati, Cl. cellulovorans*, and *Cl. aldrichii* were isolated from digestions of poplar wood (Sleat & Mah, 1985; Sleat *et al.*, 1984; Yang *et al.*, 1990).

The results of all these experiments suggest that digesters contain a much more varied population of cellulollytic bacteria than does the rumen, where three or four genera and species seem to comprise the main cellulolytic population and clostridia are uncommon. In mixed culture some of the digester bacteria showed both more and less cellulolytic activity than the individual bacteria, but most showed no interactions. Some *Clostridium* isolates produced factors which inhibited other isolates. However, it was suggested that these interactions were probably confined to localised areas of plant fibres colonised by bacteria as the whole population seemed to remain in a 'dynamic' steady-state with no species, or group of species, becoming dominant (Sharma & Hobson, 1986*b*). In the same way, the bacteria need not, overall, be competing for substrates. The bacteria can have different requirements for nitrogen sources and growth factors which could be the factors controlling their growth rather than carbohydrates. If there is competition for sugars, then this could be localised by the growth on solid substrate and in two nearby colonies different bacteria could be growing (Hobson, 1979).

Unlike the facultative *E. coli* and *Streptococcus* spp. previously mentioned, these cellulolytic bacteria were not predominant members of the human or animal gut and were not entering the digesters in large numbers. The cellulolytic bacteria seemed to have been selected by growth in the digester from bacteria making up a small proportion of the bacteria in the digester feedstocks. No rumen cellulolytic bacteria were found in the pig- or cattle-waste digesters and previous experiments had shown that cellulolytic rumen bacteria introduced into a digester treating pig-waste and straw survived only transitorily (Summers, 1978). However, the later tests with mixed cultures of bacteria from the cattle-waste digester and rumen cellulolytic bacteria showed no definite antagonism between the digester and rumen bacteria (Sharma & Hobson, 1986*b*). Many of the cellulolytic (and other, see later) digester bacteria are sporing forms and it seems likely that these would survive better in a digester with a long retention time and substrates of poorly-degradable fibre than would the non-sporing rumen bacteria adapted to a system of short retention and substrates of much higher degradability. Novaes *et al.* (1988) cultured cellulolytic and cellobiose-fermenting *Butyrivibrio* and *Clostridium* species from a UASB digester treating domestic sewage waters at 30°C. The Clostridia were non-cellulolytic. *Butyrivibrio* is generally the least active of the rumen cellulolytic bacteria, and whether it was actually cellulolytic in the digester is not known: a wide range of sugars is fermented by the genus.

The cellulolytic bacteria mentioned above grew and hydrolysed cellulose optimally at about 35°C and pH 6·5 to 7; i.e., under the normal mesophilic digester conditions. *Clostridium thermocellum* found in thermophilic digesters degrades cellulose and hemicellulose (Ng *et al.*, 1977) at higher temperatures but similar pH. The bacteria fermented cellulose and hemicellulose to mixtures of lower volatile fatty acids, lactic and succinic acids, carbon dioxide, hydrogen and ethanol. All of these are intermediates in the digester metabolic pathways, or are products of pure cultures which are modified in the digester mixed culture. The isolates also fermented a variety of mono- and di-saccharides, as would be expected. These bacteria can, then, carry out the reactions of the first two stages of carbohydrate digestion shown in Fig. 1. In these cultures and cultures mentioned later, starch-fermentation was found to be a common property even though there was little starch in any of the feedstocks. Starchy wastes are usually easily digested, but the wastes may contain some bacterial inhibitors or be deficient in nitrogen. Segretain & Bories (1987) had to add nitrogen-containing material to starch from cassava processing to allow digestion, and had to experiment to reduce the toxicity of cyanogenic glucosides in the feedstock. Pectic substances are not found in the faecal feedstocks of the digesters used in the bacterial isolations experiments. The pectins are digested in the animal digestive tract and any in vegetable wastes in sewage would tend to be leached out into the sewage water and lost to the sludge. In most of the tests no pectin-degrading bacteria were looked for but the *Cl. populati* mentioned before secreted a pectate lyase. More bacterial cultures should be done on digesters fed on raw plant material.

As some of the cellulolytic bacteria isolated in the experiments mentioned above had hemicellulolytic activity, they could, potentially, degrade the two main polysaccharides in the plant structure. Hobson & Shaw (1971, 1974) isolated non-cellulolytic xylan-degrading bacteria and found coccobacilli the majority of which were similar to the rumen *Bacteroides ruminicola*, although strains differed. Sewell *et al.* (1988) isolated xylan degrading *Clostridium* spp. and *Butyrivibrio fibrisolvens*, the latter similar to rumen isolates, from a laboratory digester fed on Napier grass. Although, as previously mentioned, seaweed polysaccharides have been digested, little is known about the bacteria responsible for the degradation in digesters. Forro & Lesinski (1985) isolated *Cytophaga* spp. which degraded algin, laminarin and cellulose, while Sowers & Ferry (1985) obtained a consortiun of bacteria from the sea which would degrade alginate and mannitol to methane in the presence of high concentrations of salt.

Cox (1978) made 50 isolations of *Clostridium* from digesting sewage sludge. These 50 isolates contained representatives of 15–20 strains, or species, of the genus; which agrees with the work previously quoted showing the large variety of bacteria present in digesters. These isolates fermented sugars and some fermented starch, but none were cellulolytic. This latter would not be unexpected, as both Hobson & Shaw (1974) and Iannotti *et al* (1982) showed that cellulolytic bacteria could not be isolated from primary dilution cultures grown in media containing mixtures of sugars. This was probably due to the fact that the sugar-fermenting bacteria grew in the highest dilutions of the digester samples, and the cellulolytic bacteria, initially in smaller numbers, were lost in the mixed growth in the lower dilutions. Mah & Susman (1968) and Kirsch (1969) found from 1 to

7×10^8 sugar-fermenting bacteria in sewage-sludge digesters and Hobson & Shaw (1971, 1974) found between $6{\cdot}4 \times 10^5$ and $8{\cdot}4 \times 10^6$ such bacteria in a pig-waste digester. Iannotti *et al* (1982) cultured 7×10^9 sugar-fermenting bacteria from another pig-waste digester. The identified isolates belonged to the genera *Peptotreptococcus, Eubacterium, Lactobacillus, Peptococcus, Clostridium,* and *Streptococcus.* None were cellulolytic or hemicellulolytic. Dubourquier *et al* (1988) cultured 10^9–10^{10} glucose-fermenting bacteria per ml of samples from a UASB digester treating starch- and sugar-producing wastewaters.

A number of defined co-cultures of mesophilic and thermophilic polysaccharide- or sugar-fermenting bacteria with methanogens utilising VFA or hydrogen have shown that the pathways of carbohydrate digestion are those suggested from other bacteriological work and analysis of digester contents (Jack-Jones *et al.*, 1984; Khan, 1980; Laube & Martin, 1981; Patel & Roth, 1978; Weimer & Zeikus, 1977; Winter & Wolfe, 1979, 1980; Yang *et al*, 1988).

These results show that there is a large population of bacteria in digesters which must be living on the sugars formed by the hydrolytic bacteria. However, there is the one polysaccharide hydrolysing ability in many of the isolates of digester bacteria which does not appear to be related to the composition of the feedstock or the intermediary products of the fermentation of feedstocks, and that is starch hydrolysis. The only carbohydrase activity noted by Cox (1978) in the isolates mentioned above was amylolysis. Amylolytic bacteria were found in numbers of about 10^4 to 10^5 per ml in pig-waste digesters (Hobson & Shaw, 1971, 1974) Torien (1967) also found amylolytic bacteria in sewage digesters and a number of the isolates of Iannotti *et al.* (1982) from a pig-waste digestion were amylolytic. A number of the the cellulolytic isolates of Sharma & Hobson (see previous references) from a cattle-waste digester were also amylolytic. There was no starch in any of these digester feedstocks: amylolysis seems to be a common property of bacteria.

Recently, two cellulolytic fungi (*Curvularia* sp. and *Penicillium* sp.) have been shown to convert grass to methane and carbon dioxide in pure culture at 33°C (Ekwenchi *et al.*, 1991), but, so far as the authors are aware, no such fungi have ever been isolated from digesters.

The metabolism of nitrogen compounds

The fermentation of carbohydrates just described provides energy for the growth of bacterial cells. Degradation of organic acids and the reduction of carbon dioxide by hydrogen can also result in generation of energy for cell growth. In the case of some bacteria, growth energy can also be obtained by fermentation of amino acids. It is uncertain how much this last mechanism contributes to the growth of bacteria in anaerobic digesters, but with most digester feedstocks the principal source of growth energy is in the fermentation of carbohydrates and further metabolism of some of the fermentation products. However, besides energy the bacteria need sources of nitrogen for synthesis of cell constituents, and the metabolism of nitrogenous compounds will be considered next.

Metabolic pathways and the bacteria involved in protein metabolism

Proteins are found in the leaves and other parts of plants. They also form a large proportion of the food of humans either in the form of plant proteins or as meat and fish products. Plant, animal and fish proteins also form parts of the diets of farm animals. Domestic and farm wastes will, therefore, contain proteins from feeds not digested in the alimentary tract and from kitchen wastes or spilt animal feeds. In addition, faeces will contain proteins from intestinal secretions, sloughed-off intestinal cells and bacteria of the digestive tract. Many factory-waste feedstocks will contain proteins from vegetable or animal materials used in the factory, and feedstocks for digesters run on 'energy crops' will, naturally, contain plant proteins. Proteins are polymers of amino acid residues and so have to be hydrolysed to make the amino acids, or the derived ammonia, available to digester bacteria. The main pathways of protein digestion are shown in Fig. 1. There will also be recycling of nitrogen in the digester as some of the viable digester bacteria die and lyse and their constituents are reincorporated into growing cells.

As well as proteins there can be non-protein nitrogenous substances in digester feedstocks. Among these are ammonia from human and animal faeces, urea and uric acid from animal and poultry urines, and free amino acids and amino sugars from plants and from lysis of bacteria and epithelial cells and secretions in the distal portions of animal and human large intestines. There may be nitrates from crops used as feedstocks and other nitrogenous compounds in particular factory wastes. These nitrogenous compounds are attacked by digester bacteria to give ammonia as a final product. Rivard *et al.* (1988) demonstrated the quantitative reduction of nitrate to ammonia, nitrogen gas was not formed. Ammonia, however formed, is probably the principal nitrogen source for digester bacteria. Shaw (1971) found that many of the carbohydrate-fermenting bacteria isolated from a pig-waste digester could use ammonia as N source, and the methanogenic bacteria use ammonia (Mah & Smith, 1981, quoted by Jain *et al.*, 1988). However, some bacteria need amino acids and others may be able to use amino acids instead of ammonia (Hobson & Shaw, 1974). The digester feedstocks also contain nucleic acids from plant, animal and bacterial sources. These can be degraded and their constituents utilised in production of bacterial nucleic acids or, after degradation of the purines and pyrimidines, of amino acids and proteins.

Not a lot of work has been done with the express idea of isolating proteolytic bacteria and testing their protease and peptidase activities. Bacteria isolated for other reasons, for instance to test for cellulolytic activity, have been tested for proteolytic activity during the course of normal bacteriological typing, although this is often done with gelatin or casein (proteins not normally found in digester feedstocks) as the substrate and the result is recorded only qualitatively.

Sharma & Hobson (1986*a*) found evidence in analysis of culture gases for possible nitrogen-fixation by cellulolytic Clostridia from a cattle-waste digester. Leschine *et al.* (1988) demonstrated acetylene reduction by seven strains of cellulolytic bacteria from muds and wet woods (possibly Clostridia) and weak activity was previously reported for *Cl. thermocellum* (Bogdahn & Kleiner, 1986). However, it seems likely that in many digesters nitrogenase activity would be repressed by the high ammonia content and that little, if any, nitrogen fixation would take place.

The general results of recorded investigations suggest that the principal proteolytic bacteria in sewage and animal-waste digesters are Clostridia. Rajasekaran *et al.* (1986) reported about 2 to 5×10^4 proteolytic bacteria per ml in batch digesters treating cattle-, poultry- and silkworm-waste mixtures, but the bacteria were not identified. Harkness (1966) found about 7×10^4 proteolytic, Gram-positive bacteria in digesting sewage sludge. Torien (1967) found proteolytic activity to be associated mainly with Gram-positive sporing rods, although there were some other types. Kotze *et al.* (1968) found proteolytic enzyme activity to vary with the type of waste being digested, although this was probably a reflection of the substrate levels and total bacterial activity rather than a specific change in numbers of proteolytic bacteria. Siebert & Torien (1969) isolated proteolytic bacteria in numbers of about 5×10^7/ml from a sewage digester. The principal types were, again, Clostridia, but there were also other Gram-positive rods, and cocci, including *Peptococcus* and *Bifidobacterium*. The principal proteolytic bacteria isolated from a pig-waste digester by Hobson & Shaw (1971, 1974) were also Clostridia, in numbers about 4×10^4/ml. The Clostridia isolated by Iannotti *et al.* (1982) were proteolytic, but several strains of the other isolates (see above for tentative identification), including possible *Bacteroides* and Peptococci, also hydrolysed casein and/or gelatin in the identification tests. The cellulolytic Clostridia isolated by Sharma & Hobson (1985) and the presumptive Clostridia and Sporolactobacilli (Sharma & Hobson, 1986a) hydrolysed gelatin in standard media, although the latter were only slightly active. The proteolytic bacteria isolated by Singh *et al.* (1986) from digesting cattle waste were mainly Gram-positive sporing rods, with some cocci. Jain & Zeikus (1985) isolated a proteolytic *Clostridium* from a sewage digester. The cellulolytic Clostridia isolated by Maki (1954) had no action on gelatin. One group of the Clostridia isolated by Cox (1978) hydrolysed casein or gelatin, but he suggested that the Clostridia could be more important in hydrolysis of peptides than in the primary hydrolysis of proteins. The formation by these isolates of iso fatty acids in a low-carbohydrate Tryptone-yeast extract medium suggests that they could deaminate amino acids. The hemicellulose-hydrolysing *Bacteroides* isolated by Hobson & Shaw (1974) could, by analogy with similar rumen types, be involved in deamination of amino acids. In some digestions ammonia and protein concentrations in influent and effluent are not appreciably different (e.g. Hobson & Shaw, 1971; Summers & Bousfield 1980), suggesting that proteolytic activity is small or, more likely, that protein degradation and protein production as new bacterial cells are approximately balanced. Mackie & Bryant (1990) measured protein production in cattle-waste digesters. The digesters were loaded at 3, 6, 9, 12 g VS/l/d with 4–8% VS slurry at RT of 13 to 5 days. The digesters were fed once per day for 4 days and twice on the 5th day. Cell production from ammonia was 3·97 to 8·72 mg dry cells/l/h 1 to 5h after feeding at 40°C. At 60°C 5·46 to 9·77 mg were produced. The rates decreased as time from feeding increased whether feeding was once or twice a day. Ammonia concentrations remained fairly constant, but slightly higher in the first few hours after feeding than later. However, there is no doubt that protein is degraded by digester bacteria. Breure & van Andel (1988), for instance, showed the similar time courses of degradation of gelatin added to a digester treating a synthetic proteinaceous waste and of the formation of ammonia. Normal- and *iso*-volatile fatty acids (VFA) up to valeric, and then methane, were the other products, with acetic and propionic acids predominating. In this case the anaerobic digestion of proteinaceous wastewaters, such as might be obtained from food processing, was being investigated. With such wastes protein might be expected to be the major primary substrate for methane production. However, Breure *et al.* (1986*a,b*) found that high concentrations of fermentable carbohydrates inhibited protein degradation to fatty acids and so to methane (the VFA were not inhibitory). Thus the contribution of protein to methane production would be expected to vary with the composition of the feed and be reduced in digestions of plant materials or food-processing wastes high in carbohydrates. Iannotti *et al.* (1979) found that ammonia increased and protein decreased when comparing feed and effluent in their pig-waste digestions. Amino acid analysis of the feed and effluent showed that the proportions of some amino acids increased and proportions of others decreased. This was presumably a balance between degradation and synthesis of proteins, as suggested before. Carbohydrates were the largest components of the pig waste (above) and they contributed most to the biogas formation. Lipids, although they were in lower amount than protein (15% compared with 20%), were the next largest contributor to methane production. Sixty percent in total of the Volatile Solids were destroyed, and of the components, 47% of the protein, 65% of the hemicellulose, 64% of the cellulose and 69% of the lipids were destroyed. Ninety four percent of the starch was destroyed, but starch was a very small percentage (1·5) of the organic matter of the waste. In domestic sewage sludge lipids were 28% of the sludge organic matter and were 76% destroyed during digestion, while carbohydrates and protein (34 and 29% of the organics) were 13 and 35% decomposed (Chynoweth & Mah, 1971).

Some work has been done with defined cultures which shows that proteins can be converted to amino acids and the amino acids fermented to VFA,which in turn are converted to methane. Jain & Zeikus (1989) cocultured *Clostridium collagenovorans* with *Methanosarcina barkeri* and the mesophilic coculture converted gelatin to methane and carbon dioxide with traces of hydrogen. The *Clostridium* fermented the amino acids of the gelatin to, mainly, acetate and carbon dioxide, and the methanogen utilised the acetate. Earlier, Ollivier *et al.* (1986) had shown that *Thermobacillus proteolyticus* with a hydrogen-utilising *Methanobacterium* sp. and an acetate-using *Methanosarcina* converted gelatin to methane and carbon dioxide at thermophilic temperatures. All the bacteria concerned in these transformations have been isolated from digesters. However, although the metabolic pathways in the laboratory experiments and in digesters may be the same, the rates and extents of protein degradations may be quite different in the two systems. Proteins differ in structure and proteases differ in ability to degrade them, and in many digester feedstocks the proteins, like the polysaccharides, will be part of a complex structure and not entirely, or easily, available to enzyme attack.

Recently, Bhadra *et al.* (1987) measured the conversion of a synthetic mixture of volatile fatty acids and the acids formed from digestion of a paper-mill sludge in methanisers packed with ceramic Raschig rings, hardwood chips or charcoal. The wood chips were the optimum support and acetic acid was degraded fastest. However, they also deduced from nitrogen balances that urea, added as an N-source to the acids was converted to methane and gaseous nitrogen. Urea was converted to ammonia and the ammonia to methane as below:

$$3HCO_3^- + 8NH_4 \rightarrow 3CH_4 + 4N_2 + 5H^+ + 9H_2O$$

Nitrogen was not lost as volatilised ammonia, so was presumably converted to N_2. So far as is known, these observations have not been repeated.

The metabolism of fatty acids, lipids and hydrogen, and production of methane

Although the fermentation of cellulose and other carbohydrates can be affected by hydrogen metabolism, in that the removal of hydrogen can increase the rate or extent of the fermentation, carbohydrate fermentations can proceed in the absence of hydrogen-utilising bacteria. On the other hand, the complete anaerobic metabolism of lipids can proceed only in the presence of a suitable hydrogen-utilising bacterium.

Hydrogen is mainly utilised in three ways in anaerobic digesters: by combination with CO_2 to form CH_4; by combination with CO_2 to form acetic acid; by reaction with SO_4 to form S^{--} and H_2O. In most digesters the first is the most important, but in some estuarine muds the third reaction can predominate. As the reactions of production and utilisation of hydrogen are carried out by different bacteria, these reactions have come to be known under the collective name of 'inter-species hydrogen transfer'.

Hydrogen can also be used in the hydrogenation of unsaturated long-chain fatty acids, a reaction that occurs in all digesters treating feedstocks containing lipid, and in increasing the production of short-chain acids more reduced than acetic. Henderson (1980) experimented with three species of rumen bacteria that produced acetate, succinate and propionate or acetate, succinate and lactate when fermenting glucose or fructose under a carbon dioxide atmosphere. Under hydrogen plus carbon dioxide the proportion of succinate in the fermentation products was increased, as was propionate. It was shown that extraneous hydrogen was taken up in the reduction of fumarate to succinate. The Km for the reaction appears, however, to be higher than that for hydrogen uptake by the methanogens, so it seems likely that this reaction could not normally compete for hydrogen with methanogenesis. However, if methanogenesis were partially inhibited for some reason this reaction could come into play as there are similar propionate- and succinate-producing bacteria in digesters and it would, either directly, or indirectly through the conversion of succinate to propionate, increase the amount of propionate in the digester fluid. Since propionate inhibits methanogenesis from hydrogen plus carbon dioxide the extra production of propionate would further inhibit methanogenesis and so increase the chances of digester failure.

The conversion of propionate itself to acetate and hydrogen is inhibited by hydrogen plus acetate or by propionate (Fukusaki *et al.*, 1990*a*), and, similarly, degradation of acetate to methane is inhibited by the (undissociated) acid (Fukusaki *et al.*, 1990*b*). So again there is the possibility of a feed-back mechanisms increasing digester inhibition. Smith & McCarty (1989) found that longer-chain fatty acids (up to heptanoate) were formed from acetic and propionic acids in a digester fed on ethanol and propionate when the hydrogen partial pressure was increased, and n-propanol was also formed from propionate.

The reduction of sulphate to sulphide removes hydrogen and also forms sulphide which can aid in the chemical reduction of the digester habitat to an Eh suitable for the growth of the strict anaerobes. Sulphide can also play a role in precipitating and so detoxifying heavy metals which could otherwise inhibit the digestion. On the other hand excess sulphide (above about 100 mg per litre) can be toxic to digester bacteria and methane formation (McCarty, 1964). However, sulphate, unless from seawater or industrial processing, is not usually found in digester feedstocks; probably most of the sulphide in animal wastes, for instance, comes from the degradation of sulphur amino acids in proteins of bacterial or animal origin. In many cases any sulphate initially in the feed would be reduced by bacterial action in storage tanks. Nitrate in feedstocks can be reduced in the digester to nitrite and then to ammonia, but, again, nitrate is not a common component of feedstocks.

Methane production

The main substrates for methane production in anaerobic digesters are acetic acid and hydrogen plus carbon dioxide. Of these, acetic acid has been generally regarded as the main precursor since Smith & Mah (1966) showed that 73% of the methane in their domestic sewage digester came from acetate. Two moles of methane are produced from acetate and one from hydrogen in glucose digestion (see later). However, domestic sewage sludge usually contains a relatively large amount of lipid material and this may have contributed significantly to the acetic acid in Smith & Mah's digester, as acetic acid can be generated by the degradation of the long-chain fatty acids of lipids, as will be discussed later, as well as being a primary product of carbohydrate fermentation. It can also be generated as a secondary product of carbohydrate fermentations from the degradation of the primary fermentation products propionic, butyric or higher volatile fatty acids by the same mechanism as is involved in the degradation of long-chain fatty acids. Acetic and propionic acids are also formed by fermentation of lactic and succinic acids by bacteria such as the *Veillonella*. Hobson *et al.*

(1974) found 3×10^7 lactic-fermenting bacteria per ml of digesting piggery waste. Sulphate-reducing bacteria can also convert lactate to acetate. The lactic and succinic acids are primary products of carbohydrate fermentations. A relatively small amount of acetic acid will be generated by the combination of carbon dioxide and hydrogen, possibly 1 to 4% of the total acetate production (Mackie & Bryant, 1981). Mylroie (1954) observed uptake of hydrogen and carbon dioxide by digesting sludge and isolated a *Clostridium* which formed acetate but not methane from these substrates. This observation was followed up by Ohwaki & Hungate (1977) who isolated similar Clostridia which were similar to, but not identical with, *Clostridium aceticum*, and the combination of hydrogen and carbon dioxide provided energy for growth of the bacteria. They also reported between 0·04 and 700×10^4 acetate-producing bacteria in different sewage digesters. Lorowitz & Bryant (1984) isolated *Peptostreptococcus productus* from a sewage digester: this formed acetate from hydrogen and carbon dioxide and it also formed acetate from carbon monoxide (and water). The bacteria were present in numbers of about $7{\cdot}5 \times 10^6$ per gram of digester sludge, but they were present at about $1{\cdot}1 \times 10^5$ per gram of human faeces as well, so it was supposed that this was the source of the bacteria. The relative importance of acetate production from carbon monoxide in digesters seems likely to be small and the bacteria would probably be growing on carbon dioxide and hydrogen. However, Hickey *et al.* (1987) reported that trace amounts of carbon monoxide were found in sludge digesters, and the CO production was associated with methanogens growing on acetate or methanol in additional tests. Acetate also appears to be an intermediate in carbon dioxide uptake by methanogens and some seems to leak from the cells, as acetate production by methanogens has been reported (Kenealy & Zeikus, 1982; Westerman *et al.*, 1989).

The fermentation of carbohydrates by pure cultures of digester bacteria results in the formation of mixtures of products from amongst the following: acetic, propionic, butyric, valeric, caproic, lactic and succinic acids, and ethanol. Deamination of amino acids can result in some of these acids as well as branched-chain volatile fatty acids such as isovaleric. The formation of fermentation acids is often accompanied by the formation of hydrogen and carbon dioxide. The maximum amount of hydrogen is formed when acetic acid is the only fermentation acid, as follows (formic acid is the equivalent of hydrogen plus carbon dioxide);

$$C_6H_{12}O_6 + 2H_2O \rightarrow 2CH_3COOH + 2CO_2 + 4H_2$$

When fatty acids higher than acetic, or lactic or succinic acids, are the products the production of hydrogen is lower as some of the hydrogen is combined into the acids, which are more reduced than acetic. This has been shown by fermentation balances on pure cultures and also by analysis of the fermentation products of the mixed cultures in the acidogenic phase of two-phase digestions where methane formation in the first phase is suppressed by a low pH and/or a retention time too low for growth of methanogens. One of the first demonstrations of this was by Cohen *et al.*(1979) in the two-phase digestion of glucose. In the first phase degradation of glucose was to hydrogen, carbon dioxide, butyric and acetic acids, with some ethanol and lactic and propionic acids. When hydrogen is removed from such a fermentation system by formation of methane or in the other ways mentioned above, a mass-action effect tends to bias the fermentation towards the formation of acetic and hydrogen rather than the other products; the hydrogen then being removed. Hydrogen can be inhibitory to the growth of bacteria forming the gas and removal of hydrogen in mixed culture can overcome this product inhibition. This was demonstrated in a defined mixed culture of *Clostridium cellobioparum* and *Methanobacterium ruminantium* by Chung (1972). The *Clostridium* growing alone on glucose produced a mixture of acetic, lactic and butyric acids and ethanol with hydrogen, carbon dioxide and formate. Growing with the *Methanobacterium* it produced some lactic and butyric acids and ethanol, but more acetic with methane instead of hydrogen, and growth was increased. The shift of the fermentation towards hydrogen and less-reduced acids can be demonstrated even with bacteria that do not form normally detectable hydrogen in pure culture. Scheifinger *et al.* (1975) showed that hydrogen was detectable in the fermentation products of *Selenomonas ruminantium* only by the most sensitive gas chromatography, but large amounts of hydrogen (immediately converted to methane) were formed in co-culture with a methanogenic bacterium. Chen & Wolin (1977) showed that the main fermentation product of the *Selenomonas* growing alone on glucose was lactic acid, but with the methanogen acetate and hydrogen (converted to methane) were formed. Growing on lactate the *Selenomonas* produced propionate and acetate, but with the methanobacterium only acetate was, again, produced. *Clostridium thermocellum* alone formed ethanol, acetate, hydrogen and carbon dioxide: with *Methanobacterium thermoautotrophicum* the Clostridium produced more hydrogen and less ethanol, and the hydrogen was converted to methane (Weimer & Zeikus, 1977). Hydrogen production and removal (not necessarily by methane formation) has also been shown with other co-cultures. The hydrogen does not necessarily come from sugar fermentations, degradation of aromatic-ring compounds such as form the monomers of lignin occurs with hydrogen transfer, as was demonstrated with benzoate by Mountfort & Bryant (1982) and Mountford *et al.* (1984). Unidentified consortia of bacteria have been shown to degrade many of the monomers of lignin, and similar compounds (Colberg & Young, 1985; Healy & Young, 1979; Benner *et al.*, 1984; Grbic-Galic, 1983).

The hydrogen can also be removed by sulphate-reduction, and in the isolation of the benzoate-decomposing bacterium mentioned above the organism (named *Syntrophus buswellii*) was grown in the presence of sulphate and a *Desulfovibrio*. It also grew when hydrogen was removed by methanogenesis by *Methanospirillum hungatei* or by reduction of fumarate in the presence of formate by *Wolinella succinogenes*, a non-carbohydrate-fermenting bacterium. As mentioned before, some sulphate-reducing bacteria can utilise acids and McInnerney & Bryant (1981) showed degradation of lactate to methane by a co-culture of *Desulfovibrio* sp. and *Methanosarcina barkeri*. Phelps *et al.* (1985) found that *Desulfovibrio vulgaris* could not use methanol or acetate in pure culture but could grow on these substrates in co-culture with *Methanosarcina barkeri*. The methanogen produced carbon dioxide and methane, with some hydrogen, in pure culture. In co-culture the sulphate-reducer reduced sulphate with the hydrogen and less methane was formed along with the carbon dioxide. Ethanol, found in digesters as a minor product of carbohydrate fermentations or in distillery-waste feedstocks, can also be degraded by some sulphate-reducers and Tatton *et al.* (1989) formed methane from ethanol with a *Desulfovibrio* sp. producing hydrogen and acetate which were used by two methanogens.

In anaerobic digesters the *Wolinella* has not been demonstrated and sulphate-reducing bacteria are usually in smaller numbers than methanogens as there is, as said before, little sulphate in most digester feedstocks. Labat & Garcia (1986) found only 10% of the total anaerobic bacteria were methanogens and only 0·1% were sulphate-reducers in a digester treating sugar-beet pulp. Torien *et al.* (1968) demonstrated the effect of sulphate in the feedstock on numbers of sulphate-reducing bacteria. When a mine water containing high concentrations of sulphate was added to a sewage digester feedstock the numbers of sulphate reducing bacteria rose from $6{\cdot}6 \times 10^3$ to $9{\cdot}5 \times 10^7$ per ml of digester contents. Petersen & Ahring (1988) found that a laboratory thermophilic digester treating sewage sludge with added sulphate reduced the sulphate at a rate some 50 times faster than an unacclimated digester. Novaes *et al.* (1988) isolated *Desulfovibrio* and *Desulfotomaculum* from a UASB treating sewage waters.

Some species of sulphate-reducing bacteria can use lower fatty acids as substrates for sulphate reduction and in some circumstances, particularly in marine sediments, sulphate reducers can outcompete methanogens for acetate when this is limiting. Winfrey & Zeikus (1977) found that sulphate-induced inhibition of methanogenesis in muds could be removed by addition of extra hydrogen or acetate.

A further source of hydrogen for methanogenesis, or sulphate reduction, is long-chain fatty acids. Lipids in digester feedstocks can come from animal residues, as in slaughter-house and meat-processing effluents, and from faecal and kitchen residues in domestic sewage. Lipids are also present in plant materials and in microorganisms whether from the intestines or from algae grown as digester feedstocks, and they are present in intestinal secretions in faecal residues. It should be noted, though, that hydrocarbons as well as true lipids are sometimes included in the term 'oils and grease' used in the analysis of sewage digesters and hydrocarbon oils are not degraded by anaerobic microorganisms. Plant waxes are also probably little degraded in anaerobic digesters.

The degradation of triglycerides and other classes of lipids to give long-chain fatty acids before metabolism of the acids has been most extensively studied in the rumen system (see Hobson, 1988), but Hobson *et al.* (1974), using trilaurin as the substrate, isolated about 10^4 to 10^5 lipolytic bacteria per ml of digesting piggery waste where about 64% of the lipid in the waste was being degraded. Torien (1967) had used sunflower oil to show the presence of lipolytic bacteria in sewage digesters. Most of these bacteria were vibrios probably related to the *Anaerovibrio lipolytica* first isolated from the rumen (Hobson & Mann, 1961).

Rajasekaran *et al.* (1986) found about 8 to 12×10^3 lipolytic bacteria per ml in batch digesters treating mixtures of poultry-, cattle- and silkworm-wastes when using the same medium as Hobson *et al.* (1974). In addition to truly lipolytic bacteria, bacteria with esterase activities are usually common in these anaerobic systems, although some of the esterase-containing bacteria can be facultative anaerobes such as bacilli.

The hydrolysis of lipid to long-chain fatty acids and a glycerol or other residue is a prerequisite for complete metabolism of lipids. Heukelekian & Mueller (1958) demonstrated hydrogenation of free unsaturated fatty acids by analysis of digester contents, a typical reaction being conversion of oleic or linoleic acid to stearic. Iannotti *et al.* (1979) also showed a decrease in the proportion of unsaturated long-chain fatty acids during digestion of pig-waste, as did Morelli *et al.* (1990) in digestions of olive-oil mill wastewaters. Hydrogenation was first demonstrated in rumen contents, but even here the isolation of the bacteria carrying out the reactions is not complete, although Vibrio species seem to play a part. Similar bacteria presumably grow in digester contents. The resultant fatty acids can then be incorporated into digester bacteria lipids or be degraded.

The results of experiments by, for example, Chynoweth & Mah (1971) and Heukelekian & Mueller (1958) showed that saturated long-chain fatty acids were broken down in digester contents by a process akin to β-oxidation, two carbon atoms at a time being split from the acid chain to give acetic acid and a shorter-chain fatty acid:

$$CH_3\text{-}CH_2\text{-}//\text{-}CH_2\text{-}CH_2\text{-}CH_2COOH + 2H_2O \longrightarrow$$

$$CH_3\text{-}CH_2\text{-}//\text{-}CH_2\text{-}COOH + CH_3COOH + 4H$$

This reaction generates hydrogen and is thermodynamically impossible unless the partial pressure of hydrogen is kept very low. In the digester the hydrogen-utilising reactions described above keep the hydrogen pressure low and the fatty acid-degrading bacteria can be isolated only in co-culture with a hydrogen-utilising bacterium. The first isolate of a bacterium that degraded fatty acids grew in co-culture with a methanogen utilising hydrogen and carbon dioxide. This fatty-acid degrading bacterium was later called *Syntrophomonas wolfei.* It degraded odd- and even-numbered carbon chains up to C7 to leave propionic acid as a residue from odd-numbered chains, but it did not degrade higher fatty acids (McInerney *et al.*, 1979).

It was shown in the 1960's that long-chain fatty acids could inhibit the rumen fermentation (e.g. Demeyer & Henderickx, 1967) and Henderson (1973) showed that these acids, particularly oleic, could inhibit growth of pure cultures of rumen bacteria, especially the Gram-positive species which includes methanogens, when in certain concentrations: at other concentrations they could have no effect or be stimulatory. McCarty (1964) showed that oleic acid was similarly toxic to anaerobic digestion and Hanaki *et al.* (1981) investigated the inhibition in more detail. It was shown both in the rumen and anaerobic digester that Ca, by forming insoluble soaps with the fatty acids could prevent the bacterial inhibition (e.g. McCarty, 1964; Galbraith *et al.*, 1971). McCarty (1964) showed that while 500mg of oleic acid per litre inhibited an anaerobic digestion, in the presence of sufficient calcium 2000–3000 mg oleic acid per litre had no effect. Koster (1987) found that lauric acid inhibited methane production from acetate by digester sludge and that addition of calcium prevented this inhibition.

Although *Syntrophomonas wolfei* even in the presence of calcium would not degrade fatty acids longer than C7, Roy *et al.* (1985) isolated two undefined mixed cultures from a sewage digester. These cultures contained probably two or three morphological types, including a methanogen and short, curved, Gram-negative rods, and grew on oleic and C10 to C18 (stearic) saturated acids with the formation of acetate and methane. The addition of calcium ions kept concentrations of free fatty acids below inhibitory values.

In all the reactions above, hydrogen transfer plays a leading part. However, it has been suggested that formate (which can be used by many methanogenic bacteria), not hydrogen, is transferred. Thiele & Zeikus (1988) took granules or free bacteria from a UASB digester treating whey: the main bacteria metabolising lactate and ethanol were *Methanobacterium formicicum* and *Desulfovibrio vulgaris*. The whey lactose was initially converted to formate, lactate and ethanol. The authors said that formate, rather than hydrogen, was transferred to the methanogen. Boone *et al.* (1989) later said they had confirmed this latter suggestion. They grew *Syntrophomonas wolfei* with *M. formicicum* on butyrate and said that the rate of diffusion of hydrogen was not rapid enough for hydrogen to be transferred between bacteria, but that formate was transferred rapidly, so formate transfer was the preferred mechanism in digesters. It should be noted, however, that these were specialised systems. Boone *et al.* used liquid cultures in which the bacteria were dispersed: in digesters bacteria grow as mixed colonies on substrate or other surfaces and the bacteria are in close proximity and possibly connected by gelatinous polymers rather than liquid. The system of Thiele & Zeikus was biased towards formate as a primary fermentation product of the sugar. In many of the mixed-culture systems mentioned in the preceding paragraphs the production of hydrogen rather than formate as the primary fermentation product has been demonstrated. Some years ago Hungate *et al.* (1970) examined the role of formate in rumen methane production and concluded that only 18% of the rumen methane was produced from formate, the major portion came from hydrogen. The K_m for hydrogen metabolism was much lower than that for formate conversion to methane and the rate of methane formation could be correlated with dissolved hydrogen. Most of the formate produced as a primary fermentation product was converted to hydrogen and carbon dioxide by non-methanogenic bacteria.

The numbers of many types of bacteria and the variation in numbers with different digester feedstocks and digester operating conditions have not been investigated; however, bacteria capable of performing most of the reactions in digestion deduced from biochemical analyses have been isolated. Thus a start has been made on the investigation of the microbial ecology of digestion. The importance of the removal of hydrogen from fermentative and other reactions has been stressed in the previous paragraphs and the most important digester bacteria involved in hydrogen removal are the methanogenic bacteria.

In the last few years there has been an 'explosion' in research on methanogenesis and the number of species of methanogens now recorded is large. Some of these have been isolated from anaerobic digesters, many have their origins in muds, sediments and natural waters. Mesophilic and thermophilic species and vareties are known. There is not space here for a listing and discussion of the methane bacteria, the reader is referred to reviews, a recent one being that of Jain *et al.* (1988). Zinder (1988) has reviewed thermophilic acetate-utilising methanogens. A previous review, which started the reclassification of the methanogens, was that of Balch *et al.* (1979) in which it was recognised that the methanogens belonged to the kingdom of the Archaebacteria and differed in many aspects of composition from the true Eubacteria (Woese, Margum & Fox, 1978). Anaerobic fermentative Mycoplasmas are known, in the rumen for instance (Robinson & Hungate, 1973) and Rose & Pirt (1981) described two Mycoplasmas from a sewage-sludge digester one of which, *Anaeroplasma* sp. converted glucose to acids and hydrogen while the, other *Methanoplasma elizabethi*, converted hydrogen and carbon dioxide to methane. Iannotti *et al.* (1982) isolated Mycoplasma-like organisms from a pig-waste digester.

Media and methods for isolation of methanogenic bacteria are described in the review previously mentioned (Jain, Bhatnagar & Zeikus, 1988) and in the papers mentioned therein. A further short review is that by Ferguson & Mah (1987). Methods are also included in papers describing the bacteria of digesters of which the title may not mention methanogens particularly. Hydrogen plus carbon dioxide is used by the majority of methanogens and many of these also use the equivalent formate, although Belay *et al.* (1986) said that formate utilisation was an adaptive process. Acetate is used by *Methanosarcina* and *Methanothrix* species. Mono-, di- and tri-methylamine or methanol can be used by some species. These latter substrates, particularly methanol, are probably of little importance in anaerobic digestion. In some cases the presence of hydrogen inhibits the utilisation of acetate (e.g. Ferguson & Mah, 1983). Zehnder *et al.* (1981) reviewed the kinetics, yield factors, etc., of methane formation from hydrogen or acetate. There have been papers since then dealing with the kinetics of hydrogen, acetate and propionate conversions (e.g., Fukuzaki *et al.*, 1990*b*; Yang & Okos, 1990) and some are mentioned in the chapter on mathematical modelling.

The methanogens have varied requirements for growth factors (see e.g. the reviews mentioned previously), amongst which nickel, which is a component of F_{430} (Dickert *et al.*, 1980), has been suggested as a promotor of methane formation in digesters. However, metals can show stimulatory or toxic effects on digestion according to their concentrations and the presence of more than one metal. For instance, Kelly & Switzenbaum (1984) found nickel, iron and cobalt to be limiting the digestion of cheese whey. Conavas-Diaz & Howell (1986) also found that nickel enhanced cheese-whey degradation in a filter and increased butyrate degradation. Dar & Tandon (1987) found nickel both stimulated and inhibited digestion of cow dung and there was an optimum concentration of 4·2 to 6·2 μg nickel per gram of dry cow dung for enhanced biogas yield in batch digestions. Above 8 μg per gram nickel was inhibitory. Ashley *et al.* (1982) found that methanogenesis was inhibited by greater than 1mg nickel/l. Increasing concentrations of nickel inhibited carbohydrate hydrolysis and fermentation and protein-hydrolysing and amino acid-fermenting bacteria were selected. Nickel caused a build-up of fatty acids, some of which were formed only by amino-acid fermentation. Counts of various groups of bacteria were also done, with rather complex results as nickel concentration varied. The toxicity of nickel to bacteria is discussed by Babich & Stotzky (1983). As another example, Wilkie *et al.* (1986) found that addition of Ni,Co, Mo, Se to a Napier-grass digester (which had added N and P) improved digestion: the elements were not present in the grass. Like the rumen bacteria, methanogens (Perski *et al.*, 1982) and probably other digester bacteria (e.g. some involved in production of acetate from hydrogen and carbon dioxide; Yang & Drake, 1990) seem to be moderately halophilic and require sodium. The optimum concentration of sodium for digester operation has been reported as 230 mg/l (Kugelman & Chin, 1970). Sodium can reverse the inhibitory effects of high potassium concentrations. The interactive and other effects of metals are reviewed by Oleszkiewicz & Sharma (1990).

Although methanogens have been isolated from digesters there have been few studies on the enumeration of methanogenic populations of digesters. Hobson & Shaw (1974) reported counts of 2×10^5 to 2×10^6 hydrogen-utilising methanogens per ml liquid in a series of samples from mesophilic piggery-waste, stirred-tank digesters, and about 2×10^4 formate-utilisers and similar numbers of bacteria using butyrate. In a domestic digester sludge lower numbers of formate- and hydrogen-utilisers were found (2×10^2 and 2×10^3) and no butyrate-utilising bacteria. Acetate-utilising bacteria could not be detected in counting media containing acetate. However, the apparent lack of bacteria using acetate was probably an artefact due to deficiencies in the media used at that time as acetate was used during the start-up of the pig-waste digesters as methane production built up (Hobson & Shaw, 1973). The hydrogen-utilising bacteria all appeared to be of the same species, *Methanobacterium formicicum*, and this has been isolated from a number of digesters. Later counts of hydrogen-using methanogens in more highly loaded pig-waste mesophilic digesters were of the order of 10^4 to 10^7 per ml (Hobson & Richardson, 1985), and similar numbers were found in a thermophilic pig-waste digester. A series of counts of hydrogen-using bacteria in mesophilic digesters treating whole and separated dairy-cattle wastes gave numbers of about 10^4 or 10^5 per ml at different retention times, although there was no correlation between numbers and RT (Summers *et al.*, 1987). Rajasekeran *et al.* (1986) reported counts of the order of 10^4 hydrogen-utilising methanogens per ml in mesophilic batch digesters fed on mixtures of cattle, poultry and silkworm wastes.

Diversity of species and varieties of digester bacteria carrying out a specific reaction, including methanogenesis, has been mentioned, and Visser *et al.* (1991)

also showed that there was a greater diversity in species in mesophilic than in thermophilic methanogenic granules from a wastewater digester and that the species differed in the two types of granules.

The fatty-acid degrading co-culture of McInerney *et al.* (1979, 1981) did not degrade branched-chain fatty acids. Shelton & Tiedje (1984) found a consortium of two bacteria which could grow on 2-methyl butyrate. Richardson *et al.* (1987) found a consortium of two bacteria which in the presence of acetate, which it utilised first, degraded the branched-chain 2-methyl butyric acid to propionate and acetate, the latter being further metabolised to methane. The similar branched-chain volatile fatty acids 2-methyl and 3-methyl valeric acids were not used. This breakdown of 2-methyl butyric acid was later confirmed with mixed bacteria from an anaerobic filter by Jimeno *et al.* (1990).

The branched-chain volatile fatty acids can arise from fermentations of carbohydrates where they form a small proportion of the products of some bacteria, or they can arise from deammination of amino acids. In a properly-functioning digester such acids are in very low concentrations or are undetectable. It was recently suggested that if the levels of iso-butyric and iso-valeric acids were below 5 mg per litre a digester was functioning properly; if concentrations were between 5 and 15 mg digester failure was imminent and above 15 mg failure had occurred and the digester flora had been damaged in some way and a long period of recovery would be needed after the fault in the digester feed or operating parameters had been corrected. The increase in acid concentrations to 5 to 15 mg was said to give up to a week's notice before gas production fell seriously or other signs of failure became obvious (Hill & Bolte, 1989; Hill *et al.*, 1987). Ashley & Hurst (1981) had earlier said that increases in activities of phosphatases were observed 10 days before iso-butyric and iso-valeric acids were detected in failing digesters. Peek & Chynoweth (1990) said that the fluorescence of reduced pyridine nucleotide (NADPH) decreased 24 hours before the reduction in methane production caused by overloading of a glucose-fed digester.

Some inhibitions of digester bacterial reactions

There is a number of possible inhibitors of the reactions of anaerobic digestion. In most cases inhibition increases with concentration of inhibitor and the effects on digesters can be either continued running but at suboptimal efficiency, a gradual onset of failure, or a rapid and complete digestion failure. Inhibitors can be grouped into two classes; end-products of microbial reactions normally part of the digestion process and organic or inorganic substances introduced into the digester in the feedstock. The former usually produce an inefficient digestion; among the latter compounds are those that induce catastrophic digester failure.

Some of the first class, the endproducts, such as hydrogen, of intermediate reactions, have already been mentioned but may be referred to again. The final gaseous endproducts are methane and carbon dioxide. Carbon dioxide is a growth factor for many bacteria and is utilised in the formation of succinic acid in carbohydrate fermentations. It is also a source of cell carbon for growth of some fermentative and methanogenic bacteria. Nevertheless, because it is an endproduct of carbohydrate fermentations and the production of methane from acetate it can act as an inhibitor of these reactions. For instance, Hansson & Molin (1981*a*) used a mixed bacterial culture originally from a sewage digester but cultured for many years in a glucose medium. This culture fermented glucose about 30% faster under nitrogen or low carbon dioxide than it did under a 100% carbon dioxide atmosphere. Conversion of propionate to methane was affected by carbon dioxide: the rate under 100% CO_2 being 60 mg acid converted per litre/day while under 20% CO_2 it was 200 mg. At less than 20% CO_2 conversion decreased, showing the need for some of the gas as a growth factor for the bacteria. Carbon dioxide also inhibited methane production from acetate, the rate of acetate utilisation being 1000–1500 mg per litre/day at low levels of CO_2 and about 350 mg under 100% of the gas (Hansson & Molin, 1981*b*). However, butyrate degradation was little affected by CO_2, as the degradation converts butyrate to acetate and hydrogen without carbon dioxide. The partial pressure of methane in the fermentor atmosphere had little or no effect on the rates of methane production from glucose or acetate (Hansson, 1979; Hansson & Molin, 1981*b*).

The atmospheric concentration of the gases is related to the concentration in solution, and it is the gas in solution which is in contact with the bacteria. Methane is only slightly soluble in water and is a chemically inert gas so far as reactions with constituents of digester liquid are concerned and the concentration of the gas in solution will depend largely on the temperature of the liquid and pressure of gas in the digester headspace. Carbon dioxide is quite soluble in pure water in which it forms bicarbonate ions. These ions can then react with alkaline constitutents of the digester liquid to form metal (e.g. sodium) or ammonium bicarbonates and carbonates, which contribute to the buffering of the digester contents at about neutral pH, and carbon dioxide is always being taken up in bacterial metabolism, so the solution of CO_2 is a dynamic process not just a result of temperature and pressure. However, although the gas will be used by some of these reactions and so will be continually being replaced from the atmosphere, the partial pressure of carbon dioxide reqired for equilibriun with bicarbonate and the bacteria is not large, so that the partial pressure of the gas could be reduced to decrease inhibitions without problems being caused. Hansson (1979) reduced the dissolved carbon dioxide by nitrogen sparging and by recirculating the digester gas after absorbing the CO_2. This latter would be possible on a full-scale digester as the carbon dioxide is sometimes removed from digester gas before it is used, and gas sparging is the most common method of mixing digester contents. However, the actual formation of methane may not be the rate-limiting reaction in a digestion and sparging with nitrogen would be impossible on a large scale, while sparging with biogas of reduced CO_2 content would only be possible if the methane were being purified for some other reason. Probably, for most practical-scale digesters removal of CO_2 to increase the rate of methanogenesis would not be economically worth the costs involved.

The digestion process is carried out by bacteria which need strict anaerobiosis (negligible pO_2 and Eh −100 to −400 mv) for growth in pure culture. On the other hand, with the mixed culture of the digester there is no need for the methods of the laboratory culture to maintain highly-reduced conditions in the digesting liquid. Sewage digesters have been run in open tanks, the liquid a few centimetres below the air interface remaining highly reduced. As explained before, aerobic metabolism by facultative bacteria and chemical removal of oxygen by sulphide, etc., ensure that any oxygen getting into the liquid is rapidly removed. The fact that digesters are generally closed systems has as much, or more, to do with the collection and utilisation of biogas and the prevention of explosive mixtures of air and biogas being formed as it has to do with keeping the bacteria anaerobic. But Pirt & Lee (1983) suggested that small amounts of air enhanced anaerobic digestion. They fed air into small batch digesters at the rate at which the oxygen was utilised (in this case 0·1 vol. air per 1 vol. digester liquid per 3 days) and found that the digesters fed with air produced biogas faster than the ones maintained under anaerobic conditions. They suggested that the facultative anaerobes could maintain the low Eh required and that enhanced growth of these bacteria could supply more growth factors for the strict anaerobes. Experiments with defined populations of anaerobic bacteria in gnotobiotic lambs have suggested that the complex mixture of facultative bacteria which have no apparent role in the rumen reactions may be providing some necessary conditions for optimum growth of the cellulolytic and other anaerobes (Hobson, 1988).

The ability of the digester flora to take up oxygen means that intake of air into a digester, particularly into the liquid itself, need not necessarily result in either cessation of digestion or an explosive mixture of air and biogas in the gas-utilising system. Hobson *et al.* (1974) quoted an instance of a pilot-scale pig-waste digester in which the boiler burner went out and it was found that the gas consisted mainly of nitrogen with some carbon dioxide and methane. Air had been sucked into the digester through an undetected hole in a sludge-heater circulation pipe and the oxygen in the air had been metabolised in the sludge leaving nitrogen which had displaced the digester gas. However, although the oxygen had been metabolised it must have damaged the fermentative or methanogenic bacteria as it took some days running after the hole had been repaired before the gas production returned to normal and the nitrogen was displaced.

Ammonia is generally formed by bacterial action from microbial degradation of proteins or non-protein nitrogenous compounds. With human and animal excreta, ammonia is formed in the animal digestive tract and also in the excreta as they are collected in tanks and pipes for delivery to the digester. In the digester the utilisation of ammonia in formation of new bacterial cells (ammonia is the main N-source for all types of digester bacteria) seems often to equal the formation of ammonia by degradation of nitrogenous compounds in the feedstock and there is little change in the ammonia content of the digested feedstock while the total nitrogen content remains virtually unchanged (there may be some loss of ammonia-N by volatilisation). This has been particularly noted in the case of faecal feedstocks where there are usually high concentrations of ammonia and other non-protein N-compounds as well as protein (e.g. Summers & Bousfield, 1980; Iannotti *et al.*, 1979).

Ammonia exists as ammonium ions (NH_4^+) or as ammonia gas (NH_3) in solution. These forms are in equilibrium and the relative concentrations of each depend on the pH of the solution. At low pH (high hydrogen ion concentration) ammonium predominates, but ammonia is more inhibitory than the ionic form. However, in analysis of digester contents total 'ammonia' or 'ammonia-N' is determined and so the behaviour of digesters is usually related to this total ammonia measurement, while the pH may or may not be recorded. Sprott *et al.* (1984, 1985) and Sprott & Patel (1986) working with pure cultures of methane bacteria suggested that ammonia toxicity was linked to high pH and that the ammonia facilitated hydrogen-ion toxicity, the free ammonia deregulating the internal pH of the cell. Koster & Koomen (1988) and Hunik *et al.* (1990), investigating the effects of pH and temperature on ammonia inhibition of the natural consortia of methanogenic bacteria in a digester supported this theory.

It was realised early on that a high concentration of ammonia could be related to digester failure and McCarty (1964) found that at any pH (within the usual range of digester pH values, near to 7) 3000 mg ammonia-N per litre was inhibitory to digestion and that at pH 7·4 and above 1500 to 3000 mg was inhibitory. In conformity with this, Hobson & Shaw (1976) showed that a strain of *Methanobacterium formicicum* growing on hydrogen plus carbon dioxide was inhibited by ammonia-N concentrations of 3000 mg per litre at pH 7·1 and its growth was completely stopped at 4000 mg per litre. Overall, experiments suggested that at the normal pH of just above 7 inhibition of mesophilic digestion began at ammonia concentrations about 1700mg or 1800 mg per litre and increased rapidly as the concentration increased. Hobson (1983) showed that the inhibition by ammonia of methane formation from acetate by sewage sludge (McCarty & McKinney, 1961) and from hydrogen by *M. formicicum* followed essentially the same course with increase in ammonia concentration, with a sharp increase in inhibition above about 1800 mg ammonia-N. Melbinger & Donnellon (1971) reported that in sewage digesters ammonia-N above 1700–1800 mg per litre could be toxic or inhibitory. Van Velsen (1979) found a large increase in lag phase for gas production and VFA removal in batch sewage digestions when ammonia-N was increased from 1210 to 2360 mg per litre and the start-up of digestion of piggery waste was affected by similar ammonia concentrations. Bousfield, Hobson & Summers (1979) and Hobson *et al.* (1980) showed effects of ammonia on digestions of cattle, pig and poultry wastes, poultry waste being of high ammonia concentration. Koster & Lettinga (1984), again, found a change in the effects of ammonia on the methanogenic activity of a pelletised sludge from a UASB digester treating sugar beet-processing wastewater at an ammonia concentration of about 1700 mg per litre. Van Velsen (1979) found that ammonia inhibition was more pronounced in thermophilic than in mesophilic digestions of pig waste, as the proportion of free ammonia in solution increases with temperature as well as with pH. Hobson *et al.* (unpublished) attributed the difficulties of obtaining thermophilic digestion of piggery wastes to the effects of ammonia; Temper *et al.* (1985) found digestion of pig waste at 56°C was impossible because of high ammonia concentrations. Hashimoto (1983) did get thermophilic digestion of pig waste at relatively low ammonia concentrations, although ammonia was thought to be the reason for instability at high loading rates. Cail & Barford (1985) reported successful thermophilic (57°C) digestion of palm oil mill effluent with a total nitrogen content of only 920 mg per litre. Cattle wastes, in general, have lower concentrations of ammonia than do piggery- and, especially, poultry-wastes. Thermophilic digestion of cattle wastes has, therefore, been more successful than that of pig wastes (Varel *et al.*, 1977; Shelef *et al.*, 1980) when ammonia concentrations were normally low. However, Hashimoto (1983) found ammonia inhibition of both meso- and thermo-philic cattle waste digestions to begin about 2500 mg ammonia-N per litre and Zeeman *et al.* (1985) found the thermophilic digestion of cattle manure containing 3000 mg ammonia-N per litre to be much inhibited and difficult to stabilise.

Van Velsen (1979) found that the mixed bacteria in digesters could apparently adapt to high ammonia concentrations and this is probably the explanation for the differing effects of ammonia reported by a number of workers. Adaptation is mentioned in some of the work quoted above and Webb & Hawkes (1985) found that bacteria in their poultry-waste digesters could adapt to high ammonia concentrations although inhibition of unadapted digestions by shock loading of ammonia could occur. Pechan *et al.* (1987) reported adaptation of bacteria in poultry-waste digesters to as much as 7500 mg ammonia-N per litre. Robbins *et al.* (1988) reported adaptation in cattle-waste digesters.

Summers *et al.* (1987) reported on digestions of whole and separated cattle wastes carried out over some years in the same digesters. The wastes were generally of low ammonia content and the digester bacteria were generally adapted to low ammonia. However, for some periods of the work ammonia concentrations in the slurries were higher or lower than normal. The effects of ammonia on the digestions could be studied by removing the effects of retention time, all digestions being the same pH, about 7·2. Plotting methane production against ammonia-N concentrations gave a biphasic line almost identical to that found by Koster & Lettinga (1984) in studying the effects of ammonia on batch digestions of a mixture of acetic, propionic and butyric acids by sludge from a UASB digester treating beet-sugar wastewater. The rates of digestion decreased with increasing ammonia concentrations, but there was a change in slope of the lines at about 1600 to 1800 mg ammonia-N per litre. Ammonia is generally thought to inhibit primarily the methanogenic bacteria, as found for *M. formicium* (above) and the volatile fatty acid substrates in Koster & Lettinga's experiment, with hydrogen utilisation being most affected (Wiegant & Zeeman, 1986). However, the primary substrate in the digestions of Summers *et al.* (above) was the cattle-waste fibre, so either the inhibition of the methanogenic bacteria has a feed-back effect on the fermentative bacteria or the fermentative bacteria are inhibited by ammonia in a similar fashion to the methanogens. However, Koster & Lettinga

(1988) found that the production of acids from carbohydrates and proteins of potato juice was hardly affected by increasing concentrations of ammonia which increasingly inhibited methane production. In Koster & Lettinga's 1984 experiments the ammonia concentration did not go below about 700 mg per litre. Since ammonia is a primary nitrogen source for all digester bacteria, at some concentration available ammonia will begin to limit the growth of the bacteria, assuming carbon substrates are available. To obtain maximium methane production from any carbon substrate the limiting ammonia concentration should be related to the available carbon substrate concentration. Summers *et al.* (1987) related the methane production from the solids in the feed to the concentration of ammonia-N per unit of solids fed. From the data a K_s value for ammonia-N and solids degradation could be obtained. The value, and the value for maximum solids degradation, converted to nitrogen and carbon ratios in the ammonia and solids, were in the range of C:N values in feedstocks often suggested as required for optimum digestion (see Chapter 5). There are, thus, two effects of ammonia. Inhibition is related to the concentration of ammonia in solution while nutrient limitation is related to the relative concentrations of ammonia-N and available carbon substrate. The results could also explain the decrease in specific gas production with increasing concentrations of pig-waste solids in a digester feed at non-inhibitory ammonia concentrations found by Bousfield *et al.* (1980) and similar apparent 'solids inhibitions' found by other workers (e.g. Hashimoto, 1984). The apparent solids inhibition in the pig-waste experiments could have been due to ammonia-N becoming limiting as the solids concentration increased.

The volatile fatty acids (VFA), acetic, propionic and butyric can be inhibitory to methanogenesis as well as being substrates for the reaction (see earlier). Accumulation of these acids could inhibit by a mass-action effect the fermentation reactions producing them, but such inhibition seems unlikely in most cases of digester failure. Digester failure is often accompanied by a build-up of VFA which suggests that the utilisation of the VFA (as previously mentioned), not the production, is first inhibited. The mechanism, explained before, is that if methanogenesis from hydrogen is reduced, perhaps by some toxic compound in the feedstock, then this inhibits breakdown of the VFA and concentrations of acids can increase. Experiments showing inhibition by the acids of acetate and propionate degradation have been mentioned previously, and Bolle *et al.* (1983); de Zeeuw (1984): Fukuzaki *et al.*, (1990) have shown similar effects, but, generally, inhibition of methanogenesis by acetic and butyric acids seems to be not significant. Hobson & Shaw (1976) found that growth of *M. formicicum* was not inhibited by concentrations of up to 10000 mg per litre of acetic or butyric acids and McCarty & McKinney (1961) had earlier observed that acetic acid in high concentration did not inhibit digestion as a whole. However, the experiments of Andrews (1969) suggested that propionic acid was inhibitory to laboratory digestions and Hobson & Shaw (1976) found that 1000 mg per litre (measured as acetic acid, as is usual for digestions) of propionic acid partly inhibited growth of *M. formicicum* on hydrogen plus carbon dioxide. This was the lowest concentration of propionic acid used in these experiments. As mentioned previously, propionic acid has been shown to also inhibit methanogenesis from acetate and propionate (Zinder *et al.*, 1984; Markl *et al.*, 1984; Lin Chou *et al.*, 1978; Fukuzaki *et al.*, 1990*a*). Hydrogen inhibits methanogenesis from acetate and propionate (Zinder & Mah, 1979; McInerney & Bryant, 1981; Heyes & Hall, 1981; Gujer & Zehnder, 1983; Fukuzaki *et al.*, 1990*b*). Thus, if utilisation of hydrogen is reduced by inhibition of bacteria forming methane from hydrogen and carbon dioxide, inhibition of acetate and propionate conversion to methane will in turn occur, and the increased partial pressure of hydrogen will cause more propionate formation in fermentations of sugars. Increase in propionate in particular will increase the overall inhibition and so it can be seen (as previously suggested) that a type of 'auto-catalytic' inhibition will set in, rapidly leading to total digestion failure. Wiegant & Zeeman (1986) postulated a similar series of reactions, beginning with ammonia inhibition of methane formation from hydrogen and carbon dioxide, to explain ammonia inhibition of a thermophilic digestion of cow manure.

As previously mentioned, the digester bacteria, like all other bacteria, have a range of pH over which they can grow, and this is near pH 7. The mechanisms of pH inhibition of bacteria are complex, but here it is only necessary to note that the primary inhibition of hydrogen utilisation and subsequent increase in acids concentration can lead to reduction of the digester pH ('souring' of a digester). The increasingly acid conditions lead to inhibition of growth and activity of bacteria concerned not only in methanogenesis but also in carbohydrate, protein and lipid breakdowns. In addition, the unionised VFA, which increase in proportion as the pH is reduced, have a greater inhibitory action on both fermentative and acid-utilising methanogenic bacteria than the ionised forms (Hobson, 1969; Attal *et al.*, 1988). Increase in pH above about 8 is not so likely to occur as is a decrease in pH, but such an increase in pH could also lead to digester inhibition. It should be noted, however, that although a high concentration of VFA in a digester is inhibitory and a sign of digester failure, a high concentration of VFA in a digester feedstock is not necessarily dangerous. Hobson and co-workers (unpublished; Summers & Bousfield, 1980) ran farm-waste digesters with feed concentrations of VFA of up to 10 000 mg per litre without problems. In these cases the digester was running as a continuous-culture and acid-metabolising bacteria were active. Thus, the feed acids, even when concentration changes were quite rapid, were quickly metabolised and the concentration of VFA in the digester remained very low. On the other hand, although the digester bacteria can adapt to high ammonia concentrations if the concentration is increased slowly, ammonia-N in feedstocks such as animal excreta is usually far in excess of the amount needed for forming bacterial cells using the carbon and energy available in the feed. The ammonia is only slowly and partially metabolised, so that inhibitory concentrations of ammonia in feedstocks are not significantly reduced and persist in the continuous-flow digester.

The production of antibiotic substances by some cellulolytic bacteria in a normal digester flora has been mentioned. It was suggested that while such substances could perhaps regulate the bacterial growth in micro-environments they did not affect the whole digester metabolism. However, antibiotics used in treating outbreaks of disease in farm animals may get into waste streams and so into farm digesters. With present regulations on the use of antibiotics as growth-promoting factors in animal feeds there is less chance than there was of such antibiotics appearing in large amounts in animal excreta. However, ionophores, such as monensin, are used in large amount as feed additives in many countries and these have inhibitory effects on some bacteria and can get into animal excreta. The inhibitory effects of antibiotics are affected by the concentration of antibiotic and the concentration of some may be reduced by microbial degradation in slurry storage tanks and in channels before animal excreta reach a farm digester. Or in some cases the antibiotic may be given to only a part of the animal stock and concentration of any residues is reduced when the waste from these animals is mixed with that from others to form the digester feed. On a pig farm where a digester was being run by one of the authors and his colleagues penicillin, chlortetracycline, aureomycin and sulphadimidine were regularly added to the feed of the baby pigs, but the waste from this section of the piggery was much diluted by the large volumes of waste from the fattening pigs and no adverse effect of the antibiotics was recorded (Hobson, 1988). On the other hand, Fischer *et al.* (1981) found that their farm digester was completely stopped when the pigs on the farm were given 40 g lincomycin per ton of feed for control of swine dysentery. The digester had to be cleaned out and restarted on fresh excreta when feeding lincomycin was stopped.

Poels *et al.* (1984) did extensive experiments with antibiotics added to pig waste at concentrations based on the amount that would occur in the excreta if all of a veterinary prescribed daily dose was excreted, and multiples of this dose. No effects of bacitracin, chlortetracycline, tylosin, erythromycin, chloramphenicol and virginiamycin in normal doses were found. Higher concentrations of bacitracin and virginiamycin did severely inhibit digestion, however. Varel & Hashimoto (1981) found that digestion of waste from cattle fed aureomycin was only slightly (20%) inhibited, but that when waste from cattle given monensin daily in their feed replaced normal cattle waste in a digester feed digestion was completely inhibited. However, in batch digesters run over a period of three months the bacteria adapted to a waste containing monensin and the final methane yield was the same as from the digestion of waste without monensin. Monensin is not degraded by the bacteria of the rumen and digestive tract and so feed additions are largely excreted. Hilpert *et al.* (1982) found monensin in the low concentration of 0·5 mg per litre inhibited digestion of municipal sewage sludge, but lasalocid, another ionophore used as a cattle growth promotor, was inhibitory only at 100 mg per litre. This latter concentration would never be reached in cattle wastes in practice. The antibiotic flavomycin was not inhibitory, but spiramycin, tylosin and virginiamycin were similarly inhibitory to sewage digestion.The chemotherapeutic agents, arsanilic acid, furazolidone, and olaquindox were somewhat inhibitory, while sulphamethazine was virtually non-inhibitory.

The antibiotics and chemical therapeutic agents are usually fed only for short times and possibly to only part of an animal herd, so dilution with uncontaminated wastes and washout from the digester would generally serve to make any inhibition from these sources transient. Monensin is fed continuously and could be a much bigger problem unless the digestion can be adapted to the substance. The effects of monensin on the rumen fermentation and methane production, and the effects on pure cultures of different fermentative and methanogenic bacteria has been reviewed by van Nevel & Demeyer (1988). Antibiotics and substances like monensin may, of course, be present in wastes from fermentations in factories producing these antibiotic substances. Whether these could be treated by anaerobic digestion would be a matter for experiment.

Disinfectants are made to be bactericidal in low concentrations and are used on farms for cleaning of areas such as milking parlours and of milking machinery. In factories, disinfectants may be used in vessels and pipelines and on floors and benches. These disinfectants may enter drains taking wastes that are to be treated by anaerobic digestion. Cleaning is usually intermittent, though, and cleaning waters could be mixed with very much larger volumes of animal or factory wastes and so the disinfectants brought down to concentrations which would not affect digester bacteria. Poels *et al.* (1984) tested disinfectants at concentrations based on washing down of a standard area per pig and the washing water getting into the waste. At concentrations normally recommended, the disinfectants Tego 51, Dettol, sodium hypochlorite and Creolin had no effects on digestion, but higher concentrations of the phenolics, Dettol and Creolin, had inhibitory effects. Pearson *et al.* (1980) found that 500 mg phenol per litre of feedstock could decrease digestion by 50%. Of the disinfectants tested by Hilpert *et al.* (1982) Dekaseptol, a mixture of carbon disulphide, phenols and chloroform, was the only one to inhibit digestion at concentrations that might normally get into digesters. Another disinfectant containing only phenols, Master Mix Environ, had no effect on digestion even in high concentration. Other phenolics from industrial processes may cause digester problems. For example, pentachlorophenol, from a factory waste, at a concentration of 1·4 mg per litre in sewage caused failure of two sewage digesters near London some years ago (Drew & Swanick, 1962). The phenol was a sudden addition to the feedstock of a digester not adapted to phenolics, but phenolics can be degraded anaerobically, particularly if the digester flora is adapted (see later).

Madamwar *et al.* (1991) tested the effects of a number of surfactants added to the feed on the digestion of water hyacinth-cattle dung mixtures in continuous-flow stirred-tanks. The surfactants, Tweens, sodium lauryl sulphate and various Tegoprens, produced maxima and minima in gas production as their concentrations were increased, and, for instance, Tegoprens 43 gave an increase in gas production of 114% at a concentration of 50 mg/l, but gas production declined rapidly at higher concentrations.

Detergents may get into anaerobic digester feedstocks from washing of factory or farm plant or from washing of materials in a factory (e.g. wool processing). Bruce *et al.* (1966) found that primary alcohol sulphate detergents were degraded anaerobically, and in concentrations up to 4·2% of sludge solids produced only an initial slight inhibition of digestion; on prolonged feeding they actually slightly increased gas production. Sodium alkane sulphonates were not degraded and above 2% inhibited gas production. Both 'hard' and 'soft' alkyl benzene sulphonate detergents were non-degradable and at concentrations between 1·5 and 2% of sludge solids caused erratic digestion, while above 2% serious inhibition generally occurred. The ratio of detergent to sludge solids appeared to be more important than the total concentration in the liquid sludge.

The presence of heavy metals in wastes and their discharge to water courses, oceans and the land, has become a topic of much discussion by those concerned with the environment. Heavy metals can contaminate digester feedstocks and could affect the digester bacteria as well as the environment if the digested sludges were discharged. The metals can come from industrial processes and can get into sewage systems from mixing of industrial and domestic wastes, and they also occur in general drainage waters. Copper is also often a component of farm wastes because of its wide use in pig feeds. Heavy metals are toxic to bacteria in general, but they are toxic only when in solution and even then only when above certain concentrations. In low concentrations some of the metals may be growth factors for bacteria; e.g. nickel has previously been mentioned as promoting methane production. The concentrations of dissolved heavy metal ions in digester liquids are generally kept below toxic levels by precipitation of copper, lead, cadmium, zinc and nickel as insoluble sulphides by sulphide produced from sulphur amino acids or sulphate as previously described. Zinc, cadmium, copper and lead can also be precipitated as carbonates at the usual digester pH values. Chromium is precipitated as a hydroxide. Mosey *et al.* (1971) discussed removal of heavy metals from digester liquids and, because of the effects of pH on carbonate precipitation, suggested that one of the methods of recovering a digester poisoned by heavy metals was to raise the pH to about 8 by the addition of sodium carbonate. Lime was not recommended because of the precipitation of calcium carbonate and the associated sludge problems, although calcium carbonate can act as a matrix for granule formation in sludge-blanket and similar digesters (see later): pH adjustment by ammonia carried the risk of producing ammonia toxicity. Sodium sulphide might be added to precipitate the metals as sulphides provided the pH was raised to about 7–7·5 to prevent hydrogen sulphide being given off from the digester contents. (The pH of an inhibited digester often falls as methanogenesis from acids is often reduced before the fermentative production of acids). However, such digester treatment is usually needed only after a shock loading of heavy metals, perhaps from an accidental discharge of factory wastewater to the drains. In the usual concentrations in which they are present metals can be precipitated by the amounts of sulphide and carbonate generally found in digesters. Copper in pig wastes, for instance, has never been found to be a problem as it is precipitated by the sulphide normally found in pig-waste digesters. Organotin compounds, which might be present in some wastes, inhibit methanogens (Boopathy & Daniels, 1991).

McCarty (1964) had earlier suggested sulphide precipitation as a means of detoxifying heavy metals, but he also pointed out that other metals may be stimulatory or inhibitory depending on their concentrations, as previously mentioned. Sodium, potassium, calcium and magnesium, in concentrations of 75 to 400 mg per litre, depending on the metal, stimulated digestion. Magnesium at 3000 mg per litre was inhibitory, while the other metals in concentrations of 1000 to 5000 mg per litre were slightly inhibitory and at 8000 to 12000 mg per litre were strongly inhibitory. Kugelman & Chin (1971) discussed the relationships between stimulatory and toxic concentrations of heavy and light metals and VFA and other digester parameters. Zoetmeyer *et al.* (1982) found phosphate at 1200 mg/l, potassium at 500 mg/l, sodium at 2300 mg/l, chloride at 5000 mg/l and ammonia at 2250 mg/l inhibited the fermentation of glucose at pH 6·0 and 30°C in the acid-forming phase of a two-phase laboratory digestion. Liu & Boone (1991) found that NaCl up to 0·25M had little effect on acid-utilising mixed bacteria, but 0·25M NaCl caused 50% inhibition of lignocellulose-degrading populations. Hydrogen-utilising populations were least affected, even by 1M and higher NaCl. Acetic, propionic and butyric acids also caused product inhibition. Hayes & Theis (1978) discussed the effects of heavy metals on laboratory digestions and the distribution of the metals in digester contents. These phenomena are not confined to digesters, though. Microorganisms in general require small concentrations of many ions and other substances, but in higher concentrations the same substances can inhibit growth. Oleszkiewicz & Sharma (1990) have recently reviewed at length the data on stimulation and inhibition of anaerobic reactions by heavy metals. Gould & Genetelli (1982), following up earlier work (Idem, 1978) showing that heavy metals formed complexes with anaerobically-digesting sludges, found that zinc, nickel and cadmium were solubilised from sludge as the temperature was lowered from 45 to 4°C but that copper was not affected: pH also played a part.

Sewage and other sludges may need to be thickened before digestion (Chapter 5) and chemical coagulants can be used for this. Gossett *et al.* (1978) showed that the coagulants alum and ferric chloride depressed anaerobic digestion of sludges. The insoluble substrates forming complexes with the coagulants were affected, so digestion of proteins, long-chain fatty acids and cellulose was inhibited but digestion of glucose was not (Dentel & Gossett, 1982). Stronach *et al.* (1987) investigated start-up of digestion of a synthetic, completely soluble, waste made up of meat extract, glycerol and salts in a sand expanded-bed digester. Flocculating polymers increased the biomass retention on the bed but did not increase digester performance. The authors suggested that perhaps the wrong types of bacteria were becoming attached to the sand. Salkinoja-Salonen *et al.* (1983) reported successful use of flocculating polymers in digestions, as did Callander & Barford (1984: see Chapter 7).

Sulphide is required as a source of sulphur for production of sulphur amino acids in bacteria, and it also acts as a chemical reducing agent allowing growth of anaerobes. Sulphur requirements of rumen anaerobes have been studied, and there is some information on the requirements of methanogenic and other digester bacteria (Ronnow & Gunnarsson, 1981; Patel, Khan & Roth, 1978; Khan & Trottier, 1978; Wellinger & Wuhrmann, 1977). Speece (1987) calculated that about 11·5 mg unionised H_2S per litre, equivalent to about 23 mg total sulphide per litre and a concentration of H_2S in the gas phase of about 0·5%, was necessary for optimum methanogenesis. This concentration of sulphide in the digester gas is higher than that usually found (0·1–0·2%) and is 5 or 6 times higher than that allowed for operation of engines on biogas (see later). However, sulphide can be toxic to digester bacteria when in too high a concentration. McCarty (1964) and Lawrence & McCarty (1966) suggested that sulphide concentrations of 100 mg per litre were usually toxic to digestion but that 200 mg per litre could be tolerated by an acclimatised digester flora, while concentrations above 200 mg per litre were always toxic. Sulphide is generally produced in digesters from amino acids in feedstocks (e.g. some of the cellulolytic bacteria of Sharma & Hobson (1986) produced H_2S), but most difficulties with sulphide toxicity have been found in digesters with large amounts of sulphate, sulphite and sulphide in the feedstock, such as those treating wastes from pulp and paper-making, olive, palm oil and rubber production, distilling, petrol refining (e.g. Benjamin *et al.*, 1984; Isa *et al.*, 1988; Britz *et al.*, 1983; Welander *et al.*, 1988; Hobson, 1987; Boari *et al.*, 1984; Chin & Wong, 1983). Acclimatisation of the bacteria, removal of sulphate from the feedstock and precipitation of sulphide as metal salts can aid in overcoming sulphide toxicity (Lo *et al.*, 1990). Sulphide can also be removed as gaseous hydrogen sulphide by the gas-stripping action of the methane and carbon dioxide produced by the digestion. Thus it was early recognised that a vigorously-gassing digester could apparently tolerate higher concentrations of sulphide than a poorly-gassing one, at least in the case of sewage and farm-waste types of feedstocks where the concentration of degradable substrates was high and the concentration of sulphides not such as to make the start-up of digestion difficult. Yang *et al.* (1980) showed that sulphide inhibition was reversible: when an inhibitory concentration of sulphide was added to a digester feedstock methane production was reduced in a few hours, but recovery was equally rapid when the sulphide feed was removed and the sulphide concentration was reduced by the normal incoming feed. Molybdate inhibits sulphate-reduction and Hilton & Archer (1988) tested molybdate as a method of overcoming sulphide toxicity in an anaerobic filter digesting a sulphate-rich molasses wastewater. However, methanogenesis was inhibited to some extent as well as sulphate reduction and they concluded that in any case sodium molybdate would be too costly to use in practice, although perhaps some strategy for its use in start-up of digestion of sulphate-containing wastes might be evolved. Rivard *et al.* (1988) found evidence that nitrate was preferentially reduced to nitrite and then to ammonia in digesters fed sulphate and nitrate salts: sulphate remained unchanged. The digesers were acclimated to nitrate by continued feeding of low levels of nitrate. They suggested that this could be a method of removing sulphide toxicity as the ammonia formed was not inhibitory at the maximum concentrations in their digesters (*c.* 2800 mg N/l). Nitrite was very toxic and 5 μmol/ml inhibited 94% of the gas production, but with added nitrate nitrite did not accumulate. Nitrate reduction competed with CO_2 for hydrogen and nitrate was preferentially reduced, thus lowering methane production to some extent.

Dichloromethane (DCM), one of a type of chlorinated solvents, can inhibit methanogenesis, but mixed cultures can acclimate to it (Thiel, 1969; Vargas & Ahlert, 1987). Friedman & Gossett (1991) showed that in a mixed, digester, culture non-methanogens degrade DCM to carbon dioxide and acetate used by methanogens.

Phenolic compounds and other aromatic-ring compounds can result from degradation of lignin in feedstocks containing plant materials, from industrial processing of materials such as leather and wood pulps and from chemical industries, and the possible introduction of phenolic disinfectants to a digester has already been mentioned. Other aromatic-ring compounds can arise from degradation of amino acids.

Indoles and phenols such as p-cresol have been implicated in the offensive odours associated with farm wastes (Spoelstra, 1980; Williams, 1984) and Van Velsen (1977) showed that such compounds in pig waste were destroyed during anaerobic digestion, thus leading to the reduced odour found in digested pig slurry.

Although macro-molecular lignin does not seem to be appreciably depolymerised by anaerobic bacteria, the aromatic alcohols and phenolic acids which are the monomers of lignin-type materials in plants, and, possibly, low molecular weight lignin-type polymers, are degraded by anaerobic bacterial consortia (see earlier). These consortia show another example of inter-species hydrogen transfer in that the aromatic decomposition is always accompanied by removal of hydrogen by a methanogen or sulphate-reducing bacterium, as with the first benzoate-degrading consortium to be investigated, by Mountford & Bryant (1982), and the later work of Barik, Brulla & Bryant (1985) and Tschech & Schink (1986). Phenolic acids, such as ferulic acid, from low molecular weight lignin-type compounds or from the partial degradation of lignin, are inhibitory to bacteria. Sierra-Alvarez & Lettinga (1991) found that lignins from wood-processing industry wastewaters varied in inhibition of methanogenesis and low molecular-weight fractions seemed to be most inhibitory. However, they can probably be detoxified by microbial reduction as in the rumen (Chesson *et al.*, 1982). The degradation of ferulic acid to give methane has also been demonstrated (Healy *et al.*, 1980). Roberts *et al.* (1988) and Boyd *et al.* (1983) are amongst workers reporting anaerobic degradation of cresol and other phenolics. Reports of the actions of phenolics and aromatic-ring compounds in general on the growth and activities of the digester bacteria are somewhat conflicting, examples of inhibition and degradation are given. Benzene-ring compounds (benzene and substituted rings, including chloro compounds) inhibited pure cultures of different methanogenic species to different extents (Winter *et al.*, 1991), but o-cresol was degraded to methane, probably via methyl benzoate (Kaminski *et al.*, 1990; Bisaillon *et al.*, 1991). Evans and Fuchs (1988) reviewed mechanisms of anaerobic microbial degradation of aromatic-ring compounds. However, factory wastewaters containing phenolics can be digested and some practical examples of digestion are mentioned below.

Gollakota & Sarada (1988) digested highly-lignified fibrous tea-processing wastes solids and the tannin-containing liquid wastes in digesters inoculated with bacteria from digestion of de-oiled Sal-seed meal. The bacteria also degraded tannic acid, gallic acid, pyrogallol, phloroglucinol, ferulic acid, caffeic acid, benzoic acid and phenol. Wastes from other food processing, such as olive-oil production, can contain polyphenols, tannins and polyalcohols, but such wastes can be digested (Tsonis & Grigoropoulos, 1988; Balice *et al.*, 1988). Pascik & Henzler (1988) reported on the treatment of wastewater, from a sulphite pulp bleaching mill, containing chlorolignins, chlorophenol and chloroform (which is a specific inhibitor of methanogenesis) and others have reported digestion of similar wastewaters (e.g. Salkonija-Salonen *et al.* 1983; Benjamin *et al.*, 1984; Welander *et al.*, 1988). The anaerobic treatment of chemical-processing wastewaters, difficult to treat aerobically, is possible. For instance, Edeline *et al.*, (1986) could digest coke-oven wastewater so long as the total phenolic content remained below 1250 mg per litre. It was recently reported that lignite could be directly converted to methane by mixed bacteria (Barik *et al.*, 1990). Aromatic nitrogenous compounds are found as amino acids, nucleic acids, animal metabolic products such as uric acid, odourous decomposition products (such as putrescine) in stored wastes, etc. Such compounds can be degraded by anaerobic bacteria and Schink (1986) has reviewed such degradations. Schink (1988) also recently reported the degradation of putrescine, 2-amino benzoate, pyridine and dipicolinic acid to VFA, ammonia, hydrogen and carbon dioxide by pure or mixed cultures of bacteria. The products could then be substrates for methanogenesis. Holliger, Stams & Zehnder (1988) have reviewed the anaerobic degradation of a large number of these aromatic and aliphatic compounds generally regarded as difficult to remove from wastewaters. Bettersby & Wilson (1989) have more recently surveyed the degradability in digesters of 77 organic compounds (phenols, benzoates, pesticides, other aromatic-ring compounds, etc.) Some were completely degraded, others were undegradable or toxic.

Since compounds containing aromatic structures can be inhibitory to digester bacteria, in some cases the concentration of the compound must be kept below a certain limit, if necessary by diluting the waste or by chemical or other treatment. In other cases the digester bacteria can adapt to the compound when the concentration in the digester feed is gradually raised or a partially-inhibited digester is allowed to function for, perhaps, some months. Field *et al.* (1987) reported that the aromatic amino acid tyrosine was not toxic to digester bacteria and could be degraded, through the intermediates phenol and cresols, after the digester bacteria had been exposed to tyrosine in the feedstock for about 8 weeks. Tyrosine is contained in large amount in potato wastes and would not prevent the digestion of these wastes. However, during storage of the wastes prior to digestion the tyrosine could be converted by potato enzymes to the rather similar amino acid L-dopa, and this was toxic to digestion and little degraded even after long exposure of the bacteria to the compound. Field *et al.* (1989) found that the inhibitory effects of phenolic compounds (tannins) to digestion depended on their molecular weights. The phenolics could be polymerised to coloured compounds by, for instance, autooxidation of wastewaters. Monomeric phenolics were not as inhibitory as oligomeric compounds, while the polymers produced on prolonged oxidation were practically non-inhibitory. On the basis of theories about the action of tannins in tanning leathers, it was suggested that the oligomers were more effective in forming hydrogen bonds with bacterial enzymes than were the monomers, while the polymers were too large to penetrate the bacterial cell. High molecular-weight phenolic compounds such as soil or peat humic substances have not been found toxic to digester bacteria (Brons *et al.*, 1985). The effects of oxidative colouration and changes in molecular size on the toxicity and degradability of a number of phenolic compounds have been investigated by Field & Lettinga (1989).

Sierra-Alvarez & Lettinga (1990) showed that many of the constituents of wood resins (long-chain fatty acids, phenols, volatile terpenes but not triterpenes, and resin acids) inhibited methanogenesis by mixed digester bacteria. These compounds can be found in wood-processing wastes.

The retained-biomass type of digester may not be so greatly affected by toxic compounds in feedstocks as the stirred-tank digester, or it may adapt better to such feedstocks. This can be explained by diffusion of the toxic substances through films of attached or granulated bacteria being less than that to free bacteria and so the concentration of toxic substance at the surfaces of the main mass of bacteria is low. The high concentration of bacteria in films and granules and their long residence-times allow of mutants and adapted varieties appearing before all the bacteria are killed by the toxin (Canovas-Dias & Howell, 1987; Howell, 1978). Thiele *et al.* (1990) discussed experimental improvements of granular sludges as starter cultures or for treatment of potentially toxic wastes.

Olson *et al.* (1991) found that growth of methanogens was inhibited by visible light. This is unlikely to affect large-scale digesters but it might affect digestion in the glass vials and bottles sometimes used as laboratory digesters (see Chapter 6).

Startup of digesters

In this Chapter only an overview of the digester bacteria and the degradable and toxic constituents of digester feedstocks has been possible: space precludes the mention of all papers and all aspects of each topic. However, enough has been said to show that anaerobic digestion, like the reactions of most natural microbial systems, is the product of the actions, and especially the interactions, of consortia of large numbers of bacterial species.

Fortunately for the digester engineer, although the reactions of these bacteria are complicated, the bacteria adjust themselves to the degradation of the feedstock without complex external control systems. There may be some problems, but the startup of a digester is generally relatively easy, with the bacteria of the inoculum changing in relative numbers, mutating or expressing different enzymic activities as they adapt to the feedstock. It has previously been mentioned that laboratory experiments have been carried out in which a defined culture of three or so bacteria have converted cellulose or a sugar or an acid to methane and carbon dioxide; they have carried out an anaerobic digestion of the substrate. In theory, then, it should be possible to inoculate a simple, defined digester feedstock with known bacteria and obtain digestion. However, the laboratory experiments are carried out in sterile conditions in apparatus kept anaerobic by chemical and physical means. The factory digester has to operate under non-sterile conditions and under imperfect anaerobic conditions. Some oxygen-using bacteria will most likely be needed to allow the defined methanogenic culture to operate. It was previously said that there seems to be continual change in natural mixed cultures such as the digester bacteria. The defined culture may mutate or not be stable against inoculation from outside sources and may be replaced by another flora. Although this may carry out the reactions of the original flora, it is possible that it may not do so as efficiently, and it is possible that bacteria producing different end-products or different proportions of end-products might take over. As has been previously mentioned, it is possible that a complex, mixed bacterial population is necessary to ensure the stability and proper function of the principal bacteria and that a simple mixed culture may be unstable in the long run: it should be remembered that once a digester is commissioned it is is expected to run for years with no further attention, so far as the bacteria are concerned. There is also the problem that bacteria may change in subtle ways after isolation and continued culture in the laboratory and may not be capable of life in the original natural habitat. A further problem is the purely mechanical one of producing enough of bacteria which are difficult to grow in the laboratory to inoculate a digester of many hundreds or thousands of cubic metres capacity. But a major problem in starting digesters is that in most cases the composition of the feedstock is not known with any accuracy, so it is not possible to select pure cultures which could properly digest the waste. It wll also have been gathered from the preceding paragraphs that our knowledge of the bacteria involved in the digestion of complex wastes is rather small.

For all these reasons digesters are inoculated with undefined mixed cultures. Some practical aspects of the start-up of particular digesters will be considered later, a few general points are described here. It may be possible to inoculate a digester from one which is actively digesting a similar feedstock. This is generally done in the case of sewage digesters where many hundreds of gallons of sludge from one digester can be relatively easily conveyed to a new digester, and sewage sludge feedstocks are similar wherever they occur. Since farm-animal wastes are similar to human faecal wastes it is possible to start-up a farm digester with an inoculum from a sewage digester. The bacterial composition may change to some extent to allow for digestion of the higher, and different, fibre content of the farm waste (see e.g. Hobson & Shaw, 1976), but the essential bacteria are in the inoculum. However, in this case the feedstock itself contains bacteria which can form a digester flora, although they are not in the correct proportions, and so by allowing the right conditions of operation of the digester an active flora may be built up from the bacteria in a farm waste. This was also shown in the experiments of Hobson & Shaw (1976). So, in the case of the farm-waste digester started with an inoculum of digesting sewage both inoculum and feedstock may contribute to the final active flora. Digesting sewage sludge is also used to inoculate digesters for factory wastes. Again there will generally be a shift in the components of the microbial population to encouraging the growth of bacteria able to degrade the main constituents of the factory waste. Digesters may also be inoculated from some of the feedstock which has been allowed to stand for some time (weeks or months) in tanks or lagoons and which has begun to degrade anaerobically. The bacteria in this case may have come from the feedstock or may have come from the air and soil. The stored feedstock will then have at least small numbers of all the types of bacteria needed in the digester and these can be developed when the waste is transferred to the digester. However, the most certain way of inoculating a digester is to use organisms from a long-running and active digester treating the same waste. In particular, the attainment of a granular sludge in UASB digesters may be promoted by an inoculum of a well-granulated sludge from a working digester. Sales of surplus sludge for inocula can contribute towards the costs of running a factory digester.

A suitable inoculum, at least 10% of the digester volume, of bacteria having been obtained, the start-up procedure then follows similar lines for all inocula in that a continuous feed of dilute waste at a flow rate lower than the design is started to fill up the digester tank if necessary and then to start off the continuous digestion. One of the greatest chances of failure of digestion occurs during start-up, as the methanogenic flora generally takes the longest time to develop and so there is a possibility of production of acids which are not metabolised to methane and cause a low pH and 'souring' of the digester contents. The low pH then inhibits methanogenesis and so the situation becomes worse. Loading at low rate is designed to prevent fermentation of large amounts of the primary substrate and overproduction of acids. More regular monitoring of the digestion, by measurement of gas production and flammability (or, where possible, constituents) and (again, if possible) of acids and pH, is required during start-up than later on. If signs of an unbalanced digestion, in a gas largely of CO_2, VFA in thousands rather than a few hundred mg per litre and a pH dropping to 6 or below, are found, then remedial action must be taken. This is generally decreasing the loading rate or stopping loading altogether. The digestion should then correct itself over some days and loading can be continued. If the pH has fallen drastically to, say, 5, with high acid concentration and methane production has stopped, the digestion may not recover and partial or complete emptying of the digester and reinoculation may be needed. Addition of large amounts of alkali, usually as lime, to attempt to recover a sour digester can cause more trouble and this procedure is generally not recommended. Depending on the inoculum and the feedstock, a stirred-tank digester will take some weeks or, perhaps months, to attain a stable digestion and then the loading rate can be steadily increased to the design. The special problems of obtaining good bacterial granules in a sludge-bed digester and of getting a good film of bacteria on a filter matrix or fluidised-bed will be described in more detail later, and these processes may take many months.

The 'rumen' digester

The forgoing discussion has been about the bacterial populations of digesters inoculated with the bacteria from a decomposing waste or where the population has been allowed to develop from faecal bacteria in the feedstock under the digester conditions of operation. Although much work has been done on farm animal waste digesters degrading faecal wastes from ruminants, from these, and other digesters, few, if any, rumen bacteria have been isolated. The populations seem to differ in general from those of the rumen and some tests have suggested that bacteria cultured from the rumen have a limited life when introduced into digesters (Summers, 1978; Sharma & Hobson, 1986). It is possible that the digester bacteria previously mentioned as being similar to rumen types may have adapted in some way to the digester conditions and are not quite like those in the rumen. Although the temperature in many digesters and the anaerobic conditions are similar to those in the rumen, the fibres of animal- and human-waste feedstocks are in general more difficult to degrade than those in ruminant feeds and the retention times of the substrates and bacteria in the systems are much longer than the rumen retention times. Other parameters, such as ammonia and VFA concentrations differ from those of the rumen. Since these anaerobic mixed cultures are well adapted to their habitats, it may not be surprising that the digester and rumen poulations seem to differ. However, Gijzen and colleagues have recently suggested that rumen bacteria can be used as digester floras (Gijzen, 1987; Gijzen *et al.*, 1986, 1987, 1988*a,b*). An inoculum of rumen contents (bacteria and protozoa) is used in the first, hydrolytic and fermentative, phase of a two-phase digester for cellulosic wastes. The acids produced are then converted to methane by a different population in a second, retained-biomass digester. The cellulose degradation in the first phase seems good. However, the system has been tested only on a laboratory scale over comparatively short time periods and the feedstocks have been relatively bacteria-free and not composed of very highly lignified cellulose; grass-grain, municipal waste fractions and paper-mill sludge for instance. A buffer salts medium is also continuously added with the feed. Whether the system could be used on a full-scale and whether the microbial population would be stable under working conditions remains to be seen. It would seem likely that the microbial population would be stable only in the absence of large external contamination such as that from faecal wastes.

CONCLUSION

The biochemical pathways in the anaerobic digestion of many different types of feedstocks have been well described, and only, perhaps, some details remain to be filled in. Bacteria which will carry out these reactions have been cultured. However, many of the isolates mentioned have come from digesters with different feedstocks and little systematic culture of a range of bacteria carrying out the reaction pathways in one digester has been done. Also, much remains to be done in analysis of the long-term stability of digester flora. Similarly, little is known about whether all digesters treating the same, or very similar, waste have the same microbial populations, or whether the population depends on the starting inoculum or other factors.

Culture and identification of hundreds of bacteria is, of course, immensely consuming of time, manpower and apparatus: a whole laboratory would be needed for even an incomplete survey. Methods using antibodies, nucleic acid compositions or cell constituents may enable more rapid scans of bacterial populations to be made, but even here bacteria must be isolated to define the 'probes' being used.

To some extent, the study of the microbiology of digesters is academic: all the digesters now commercially running or being tested on pilot or laboratory scale have been started without knowledge of the bacteria involved but using relatively simple processes which are known to allow stable digestion to be attained, and so the inoculun bacteria to adjust in relative numbers or adapt to the digestion conditions. Again, two-phase, retained-biomass and other types of digester have been designed without detailed knowledge of the bacterial populations but on some 'rules' found from observation of the behaviour of digesters or the overall properties of some metabolic groups of bacteria; hydrolytic, methanogenic, etc. Nevertheless, more knowledge of the digester flora is worthwhile not only from the academic view of knowledge of the bacteria in complex, mixed, populations (and such knowledge has relevance to the behaviour of bacteria in other habitats), but this knowledge will be put to use in improving anaerobic digestion in the future as greater efficiency and digestion of more recalcitrant feedstocks is demanded.

REFERENCES

Ashley, N.V., Davies, M. & Hurst, T.J. (1982). *Water Res.* **16**, 963.

Ashley, N.V., Hurst, T.J. (1981). *Water Res* **15**, 633.

Andrews, J.F. (1969). *J. Sanitary Engin. Div.* **95**, 95.

Attal, A., Ehlinger, F., Audic, J.M. & Faup, G.M. (1988). In *Anaerobic Digestion 1988*, ed. E.R. Hall & P.N. Hobson, Pergammon Press, Oxford. p.71.

Babich, H. & Stotzky, G. (1983). *Appl. Environ. Microbiol.* **29**, 195.

Balch, W.E., Fox, G.E., Magrum, L.J., Woese, C.R. & Wolfe, R.S. (1979). *Microbiol. Rev.* **43**, 260.

Balice, V., Carrieri, C., Cera, O. & Rindone, B. (1988). In *Anaerobic digestion 1988*, ed. E.R. Hall & P.N. Hobson, Pergamon Press, Oxford. p.275.

Barik, S., Brulla, W.J. & Bryant, M.P. (1985). *Appl. Environ. Microbiol.* **50**, 304.

Barik, S., Tiemens, K., Harding, R., Hawley, B. & Isbister, J. (1990). Proceedings: 1990 First International Symposium on the Biological Processing of Coal. Sect. 5, **20**.

Belay, N., Sparling, R. & Daniels, L. (1986). *Appl. Environ. Microbiol.* **52**, 1080.

Benjamin, M.M., Woods, S.L., Sandra, L. & Ferguson, J.F. (1984). *Water Res.* **18**, 601.

Benner, R., Maccubin, A.E. & Hodson, R.E. (1984). *Appl. Environ. Mic.* **47**, 998.

Bettersby, N.S. & Wilson, V. (1989). *Appl. Environ. Microbiol.* **55**, 433.

Bhadra, A., Scharer, J.M. & Moo-Young, M. (1987). *Biotech. Bioeng.* **30**, 314.

Bisaillon, J-G., Lepine, L., Beaudet, R. & Sylvestre, M. (1991). *Appl. Environ. Microbiol.* **57**, 2131.

Boari, G., Brumetti, A., Passino, R. & Rozzi, A. (1984). *Agric. Wastes*, **10**, 161.

Bogdahn, M. & Kleiner, D. (1986). *Arch. Microbiol.* **145**, 159.

Bolle, W.L., van Breugel, J., van Eybergen, G.C. & van Gils, W. (1983). In *Proc. Europ. Symp. on Anaerobic Wastewater Treatment*, ed. W.J. van den Brink, TNO Corporate Communication Dept., The Hague. p. 100.

Boone, D.R., Johnson, R.L. & Liu, Y. (1989). *Appl. Environ. Microbiol.* **55**, 1735.
Boopathy, R. & Daniels, L. (1991). *Appl. Environ. Microbiol.* **57**, 1189.
Bousfield, S., Hobson, P.N. & Summers, R. (1979). *Agric. Wastes*, **1**, 161.
Boyd, S.A., Shelton, D.R., Berry, D. & Tiedje, J.M. (1983). *Appl. Environ. Microbiol.* **46**, 50.
Breure, A.M. & Van Andel, J.G. (1988). In *5th. Int. Symp. An. Dig. Poster Papers*, ed. A. Tilche & A. Rozzi, Monduzzi Editore, Bologna. p. 463.
Breure, A.M., Buftink, H.H., Verkuijlen, J. & Van Andel, J.G. (1986*a*). *Appl. Microbiol. Biotechnol.* **23**, 245.
Breure, A.M., Mooijman, K.A. & Van Andel, J.G. (1986*b*). *Appl. Microbiol. Biotechnol.* **24**, 426.
Briand, X. & Morand, P. (1987). In *Biomass for energy and industry*, 4th. EC conference, ed. G. Grassi, D. Pirrwitz & H. Zibetta, Elsevier Applied Science, London. p. 834.
Britz, T.J., Meyer, L.C. & Botes, P.J. (1983). *Biotechnol. Lett.* **5**, 113.
Brons, H.J., Field, J.A., Lexmond, W.A.C. & Lettinga, G. (1985). *Agric. Wastes*, **13**, 321.
Bruce, A.M., Swanwick, J.D. & Ownsworth, R.A. (1966). *J. Inst. Sew. Purif.* **5**, 3.
Buchholz, K., Arntz, H.J., Pellegrini, A. & Stoppok, E. (1986). *Zuckerindustrie*, **111**, 837.
Buisson, M.N. (1988). In *Granular anaerobic sludge: microbiolgy and technology*, ed. G. Lettinga et al., Pudoc, Wageningen. p. 55.
Buswell, A.M. & Neave, S.L. (1930). Ill. *State Wat. Survey Bull.* **30**, 1.
Cail, R.G. & Barford, J.P. (1985). *Agric. Wastes*, **13**, 295.
Callander, I.J. & Barford, J.P. (1984). *Agric. Wastes*, **11**, 1.
Canovas-Diaz, M. & Howell, J.A. (1986). *Biotech. Lett.* **8**, 287.
Chandler, J.A., Jewell, W.J., Gossett, J.M., Van Soest, P.J. & Robertson, J.B. (1980). *Biotech. Bioeng. Symp.* **10**, 93.
Canovas-Diaz, M. & Howell, J.A. (1987). *Biotech. Lett.* **8**, 379.
Chen, Y.R. (1983). *Agric. Wastes*, **8**, 65.
Chen, M. & Wolin, M.J. (1977). *Appl. Environ. Microbiol.* **34**, 756.
Chesson, A., Stewart, C.S. & Wallace, R.J. (1982). *Appl. Environ. Microbiol.* **44**, 597.
Chin, K.K. & Wong, K.K. (1983). *Water Res.* **17**, 993.
Chung, K.T. (1972). *Abs. Am. Soc. Microbiol.* p. 64.
Chynoweth, D.P. & Mah, R.A. (1971). *Anaerobic Biological Treatment Processes*, Ad. Chem. Series, **105**, 41.
Cohen, A., Zoetemeyer, R.J., Van Deursen, A. & Van Andel, J.G. (1979). *Water Res.* **13**, 571.
Colberg, P.J. & Young, L.Y. (1985). *Appl. Environ. Microbiol.* **49**, 345.
Colin, F., Ferrero, G.L., Gerletti, N., Hobson, P.N., L'Hermite, P., Naveau, H. & Nyns, E.J. (1983). *Agric. Wastes*, **7**, 183.
Cox, D.J. (1978). *J. Appl. Bact.* **45**, 259.
Dar, G.H. & Tandon, S.M. (1987). *Biol. Wastes*, **22**, 261.
Demeyer, D.I. & Henderickx, H.K. (1967). *Biochim. Biophys. Acta*, **137**, 484.
Dentel, S.K. & Gossett, J.M. (1982). *Water Res.* **16**, 707.
Dickert, G., Klee, B. & Thauer, R.U. (1980). *Arch. Microbiol.* **124**, 103.
Drew, E.A. & Swanwick, J.D. (1962). Publ. Wks. Mun. Serv. Cong. p. 1.
Doddema, H.J. & Vogels, G.D. (1978). *Appl. Environ. Microbiol.* **36**, 752.
Dolfing, J. & Mulder, J-W. (1985). *Appl. Environ. Microbiol.* **49**, 1142.
Dubourguier, H.C., Archer, D.B., Albagnac, G. & Prensier, G. (1988). In *Anaerobic digestion 1988*, ed. E.R. Hall & P.N. Hobson, Pergamon Press, Oxford. p.13.
Dubourguier, H.C., Prensier, G. & Albagnac, G. (1988*a*). In *Granular anaerobic sludge: microbiology and technology*, ed. G. Lettinga, A.J.B. Zehnder, J.T.C. Grotenhuis & L.W. Hulshoff. Pol. Pudoc, Wageningen, The Netherlands. p. 18.
Edeline, F., Lambert, G. & Fatticioni, H. (1986). *Process Biochem.* **21**, 58.
Ekwenchi, M. M., Akunwanne, B. U., Okeke, N.R. & Ekpenyong, K.I. (1990). *Fuel* **69**, 1569.
Evans, W.C. & Fuchs, G. (1988). *Ann. Rev. Microbiol.* **42**, 289.

Feilden, N.I.H. (1981). *Agric. Wastes*, **3**, 75.
Ferguson, T. & Mah, R. (1983). *Appl. Environ. Microbiol.* **45**, 265.
Ferguson, T & Mah, R. (1987). In *Anaerobic digestion of biomass*, ed. D.P. Chynoweth & R. Isaacson, Elsevier Applied Science, London. p. 49.
Field, J.A. & Lettinga, G. (1989). *Biol. Wastes*, **29**, 161.
Field, J.R., Kortekans, S. & Lettinga, G. (1989). *Biol. Wastes*, **30**, 111.
Field, J.R., Letinga, G. & Geurts, M. (1987). *Biol. Wastes*, **21**, 313.
Fischer, J.R., Iannotti, E.L. & Sievers, D.M. (1981). *Agric. Wastes*, **3**, 201.
Forro, J.R. & Lescinski, D.M. (1985). Proc. int. gas research conf., Washington 1984, Government Institutes Inc., PO Box 1096, Rockville, MD20850, USA. p. 488.
Friedman, D.L. & Gossett, J.M. (1991). *Appl. Environ. Microbiol.* **57**, 2847.
Fukuzaki, S., Nishio, N. & Nagai, S. (1990*a*). *Appl. Environ. Microbiol.* **56**, 3158.
Fukuzaki, S., Nishio, N., Shobayashi, M. & Nagai, S. (1990*b*). *Appl. Environ. Microbiol.* **56**, 719.
Fynn, G.H. & Ibba, M. (1988). In *Fifth int. symp. anaerobic digestion.* Poster papers, ed. A. Tilche & A. Rozzi, Mondizzi Editore, Bologna. p. 39.
Galbraith, H., Miller, T.B., Paton, A. & Thompson, J.K. (1971). *J. Appl. Bact.* **34**, 803.
Georgacakis, D., Sievers, D.M. & Iannotti, E.L. (1982). *Agric. Wastes*, **4**, 427.
Ghosh, S. & Klass, D.L. (1978). *Process Biochem.* **13**, 15.
Gijzen, H.J. (1987). *Anaerobic digestion of wastes by a rumen-derived process.* Thesis, University of Nijmegen, The Netherlands.
Gijzen, H.J., Zwart, K.B., van Gelder, P.T. & Vogels, G.D. (1986). *Appl. Microbiol. Biotechnol.* **25**, 155.
Gijzen, H.J., Lubberding, H.J., Verhagen, F.J., Zwart, K.B. & Vogels, G.D. (1987). *Biol. Wastes*, **22**, 81.
Gijzen, H.J., Zwart, K.B., Teunissen, M.J. & Vogels, G.D. (1988*a*). *Biotech. Bioeng.* **32**, 749.
Gijzen, H.J., Zwart, K.B., Verhagen, F.J. & Vogels, G.D. (1988*b*). *Biotech. Bioeng.* **31**, 418.
Gollakota, K.G. & Sarada, R. (1988). In *Anaerobic digestion 1988*, ed. E.R. Hall & P.N. Hobson, Pergamon Press, Oxford. p. 281.
Gould, M.S. & Genetelli, E.J. (1978). *Water Res.* **12**, 505.
Gould, M.S. & Genetelli, E.J. (1982). *Water Res.* **16**, 749.
Gossett, J.M., Wilson, J.C., Evans, D.S. & McCarty, P.L. (1978). *J. Water Pollut. Cont. Fed.* **50**, 533.
Grotenhuis, J.Y.C., Smit, M., Plugge, C. M., Xu Yuansheng, van Lammeren, A.A.M., Stams, A.J.M. & Zehnder, A.J.B. (1991). *Appl. Environ. Microbioil.* **57**, 1942.
Grbic-Galic, D. (1983). *Appl. Environ. Microbiol.* **46**, 1442.
Gujer, W. & Zehnder, A.J.B. (1983). *Wat. Sci. & Technol.* **15**, 127.
Guta, G.S., Raghavendra, S. & Reddy, T.K.R. (1986). *Agric. Wastes*, **17**, 153.
Hanaki,K., Matsuo, T. & Nagase, M. (1981). *Biotech. Bioeng.* **23**, 1591.
Hanssen, J.F., Indergaard, M., Ostgaard, K., Baevre, O.A., Pedersen, T.A. & Jensen, A. (1987). *Biomass*, **14**, 1.
Hansson, G. (1979). *Europ. J. Appl. Microbiol. Biotechnol.* **6**, 351.
Hansson, G. & Molin, N. (1981*a*). *Europ. J. Appl. Microbiol. Biotechnol.* **13**, 242.
Hansson, G. & Molin, N. (1981*b*). *Europ. J. Appl. Microbiol. Biotechnol.* **13**, 236.
Harkness, N. (1966). *J. Proc. Inst. Sew. Purif.* p. 542.
Hashimoto, A.G. (1983). *Agric. Wastes*, **6**, 175.
Hashimoto, A.G. (1984). *Agric. Wastes*, **9**, 299.
Hashimoto, A.G., Varel, V.H. & Chen, Y.R. (1981). *Agric. Wastes*, **3**, 241.
Hayes, T.D. & Theis, T.L. (1978). *J. Wat. Pollut. Cont. Fed.* **50**, 61.
Healy, J.B. & Young, L.Y. (1979). *Appl. Environ. Microbiol.* **38**, 84.
Healy, J.B., Young, L.Y. & Reinhard, (1980) *Appl. Environment Microbiol.*, **39**, 463.
Henderson, C. (1973). *J. Agric. Sci. (Camb)* **81**, 107.

Henderson, C. (1980). *J. Gen. Microbiol.* **119**, 485.
Henson, J.M., Smith, P.H. & White, D.C. (1985). *Appl. Environ. Microbiol.* **50**, 1428.
Heukelekian, H. & Mueller, P. (1958). *Sew. Ind. Wastes*, **30**, 1108.
Heyes, R.H. & Hall, R.J. (1981). *Biotech. Lett.* **3**, 431.
Hickey, R.F., Vandenveilen, J. & Switzenbaum, M.S. (1987). *Biotechnol. Lett.* **9**, 63.
Hill, D.T. & Bolte, J.P. (1989). *Biol. Wastes*, **28**, 33.
Hill, D.T., Cobb, S.A. & Bolte, J.P. (1987). *Trans, Asae*, **30**, 496.
Hilpert, R., Winter, J. & Kandler, O. (1982). In *Energy from biomass E.C. Conference*, Applied Science Publishers, London. p. 552.
Hilton, M.G. & Archer, D.B. (1988). *Biotech. Bioeng.* **31**, 885.
Hobson, P.N. (1969). In *Microbial growth*, ed. S.J. Pirt, University Press, Cambridge, England. p. 43.
Hobson, P.N. (1976). *Rep. Rowett Res. Inst.* **32**, 109.
Hobson, P.N. (1971). *Prog. Ind. Microbiol.* **9**, 41.
Hobson, P.N. (1979). In *Straw decay and its effects on disposal and utilisation*, ed. E. Grossbard, John Wiley & Sons, Chichester, England. p. 217.
Hobson, P.N. (1983). *J. Chem. Tech. Biotechnol.* **33B**, 1.
Hobson, P.N. (1985). *Agric. Wastes*, **14**, 255.
Hobson, P.N. (1988*a*). *The rumen microbial ecosystem*. Elsevier Applied Science, London.
Hobson, P.N. (1988). In *Anaerobic digestion 1988*, ed. E.R. Hall & P.N. Hobson, Pergamon Press, Oxford. p.425.
Hobson, P.N. (1988). In *Agriculutural waste management and environmental protection* ed. E. Welte & I. Szabolcs, CIEC and FAL, Braunschweig, F.D.R. **2**, 27.
Hobson, P.N., Bousfield, S. & Summers, R. (1974). *Crit. Rev. Environ. Cont.* **4**,131.
Hobson, P.N., Bousfield, S., Summers, R. & Mills, P.J. (1980). In *Anaerobic digestion*, ed. B.A. Stafford, B.I. Wheatley & D.E. Hughes, Applied Science Publishers, London. p. 237.
Hobson, P.N., Mackay, E.S.M. & Mann, S.O. (1955). *Res. Corres.* **8**, 30.
Hobson, P.N. & Mann, S.O. (1957). *J. Gen. Microbiol.* **16**, 463.
Hobson, P.N. & Mann, S.O. (1961). *J. Gen. Microbiol.* **25**, 227.
Hobson, P.N., Mann, S.O. & Oxford, A.E. (1958). *J. Gen. Microbiol.* **19**, 462.
Hobson, P.N., Mann, S.O. & Stewart, C.S. (1981). *J. Gen. Microbiol.* **126**, 219.
Hobson, P.N. & Richardson, A.J. (1985). In *Anaerobic digestion of farm waste, Tech. bull. 7* ed. B.F. Pain & R.Q. Hepherd, Food Research Institute, Shinfield, Reading, England. p. 15.
Hobson, P.N. & Shaw, B.G. (1971). In *Microbial aspects of pollution* ed. G. Sykes & F.A. Skinner, Academic Press, London. p. 103.
Hobson, P.N. & Shaw, B.G. (1973). *Water Res.* **7**, 437.
Hobson, P.N. & Shaw, B.G. (1974). *Water Res.* **8**, 507.
Hobson, P.N. & Shaw, B.G. (1976). *Water Res.* **10**, 849.
Hobson, P.N. & Wallace, R.J. (1982). *Critic. Rev. Microbiol.* **9**, 165 & 253.
Hollinger, C., Stams, A.J.M. & Zehnder, A.J.B. (1988). In *Anaerobic digestion 1988*, ed. E.R. Hall & P.N. Hobson, Pergamon Press, Oxford. p. 211.
Howell, J.A. (1978). In *Oil industry and microbial ecosystems*, ed. Chater & Someville, Heyden, p. 199.
Hungate, R.E. (1950). *Bact. Rev.* **14**, 1.
Hungate, R.E. (1965). *The rumen and its microbes*. Academic Press, New York.
Hungate, R.E., Smith, W., Bauchop, T., Yu, I. & Rabinowitz, J.C. (1970). *J. Bact.* **102**, 389.
Hunik, J.H., Hamelers, H.V.M. & Koster, I.W. (1990). *Biol. Wastes*, **32**, 285.
Iannotti, E.L., Porter, J.H., Fischer, J.R. & Sievers, D.M. (1979). *Developments in Ind. Microbiol.* **20**, 519.
Iannotti, E.L., Fischer, J.R. & Sievers, D.M. (1982). *Appl. Environ. Microbiol.* **43**, 136.
Isa, Z., Subbiah, R.M. & Ombregt, J.P. (1988). In *Fifth Int. Symp. An. Digestion, Poster Papers*, ed. A. Tilche & A. Rozzi, Monduzzi Editore, Bologna. p. 657.
Jack-Jones, W.P., Guyot, J.P. & Wolfe, R.S. (1984). *Appl. Environ. Microbiol.* **47**, 1.

Jain, M.K., Singh, R. & Kalsy, J.S. (1986). *Environ. Ecol.* **4**, 30.
Jain, M.K., Bhatnagar, L. & Zeikus, J.G. (1988). *Indian J. Microbiol.* **28**(3), 143.
Jain, M.K. & Zeikus, J.G. (1985). In *1st. symp. on Biotechnological advances in processing municipal wastes for fuels and chemicals*, ed. A.A. Antonopoulus, Academic Press, New York.
Jain, M.K. & Zeikus, J.G. (1989). *Appl. Environ. Microbiol.* **55**, 366.
Jimeno, A., Bermudez, J.J., Canovas-Diaz, M., Manjon, A. & Iborra, J.L. (1990). *Biol. Wastes* **34**, 241.
Kaminski, U., Kuschk, P. & Janke, D. (1990). *J. Basic Microbiol.* **30**, 259.
Kelly, C.R. & Switzenbaum, M.S. (1984). *Agric. Wastes*, **10**, 135.
Kemp, H.A., Morgan, M.R.A. & Archer, D.B. (1986). In *Anaerobic treatment. A grown-up technology*. Aquatech '86, Amsterdam. p. 39.
Kenealy, W.R. & Zeikus, J.G. (1982). *J. Bact.* **151**, 932.
Khan, A.W. (1980). *Febs. Micro. Lett.* **9**, 233.
Khan, A.W. & Trottier, T.M. (1978). *Appl. Environ. Microbiol.* **35**, 1027.
Khan, A.W., Wall, D. & van den Berg, L. (1981). *Appl. Environ. Microbiol.* **41**, 1214.
King, G.M., Guist, G.G. & Lauterbach, G.E. (1985). *Appl. Environ. Microbiol.* **49**, 588.
Kirsch, E.J. (1969). *Dav. Ind. Microbiol.* **10**, 170.
Koster, I.W. (1987). *Biol. Wastes*, **22**, 295.
Koster, I.W. & Kooman, E. (1988). *Appl. Microbiol. Biotechnol.* **28**, 500.
Koster, I.W. & Lettinga, G. (1984). *Agric. Wastes*, **9**, 205.
Koster, I.W. & Lettinga, G. (1988). *Biol. Wastes*, **25**, 51.
Kotze, J.P., Thiel, P.G., Torien, D.F., Hattingh, W.H.J. & Siebert, M.L. (1968). *Water Res.* **2**, 198.
Kristjansson, J.K., Schonheit, P. & Thauer, R.K. (1982). *Arch. Microbiol.* **131**, 278.
Kugelman, I.J. & Chin, K.K. (1971). *Adv. Chem. Series*, **105**, 55.
Kugelman, I.J. & Drake, H.L. (1990). *Appl. Environ. Microbiol.* **56**, 81.
Labat, M & Garcia, J.L. (1986). *Appl. Microbiol. Biotechnol.* **25**, 163.
Laube, V.M. & Martin, S.M. (1981). *Appl. Environ. Microbiol.* **42**, 413.
Lawrence, A.W. & McCarty, P.L. (1966). *Int. J. Air and Wat. Poll.* **10**, 207.
Lequerica, J.L. & Toldra', F. (1986). *Rev. Agriquim. Tecnol. Aliment.* **26**, 90.
Leschine, S.B., Holwell, K. & Canale-Parola, E. (1988). *Science*, **242**, 1157.
Lin Chou, W., Speece, R.E., Siddiqui, R.H. & McKeon, K. (1978). *Prog. in Water Technol.* **10**, 545.
Liu, Y. & Boone, D.R. (1991). *Biores. Technol.* **35**, 271.
Lorowitz, W.A. & Bryant, M.P. (1984). *Appl. Environ. Microbiol.* **47**, 961.
Lo, K.V. Chen, A. & Liao, P.H. (1990). *Biomass* **23**, 25.
Madamwar, D., Patel, A. & Patel, V. (1991). *Biores. Technol.* **37**, 157.
McCarty, P.L. (1964). *Public Wks.* **95**, 91.
McCarty, P.L. (1982). In *Anaerobic digestion 1981*, ed. D.E. Hughes et al., Elsevier Biomedical Press, Amsterdam. p. 3.
McCarty, P.L. & McKinney, R.E. (1961). *J. Water Poll. Cont.* **33**, 399.
McInerney, M.J. & Bryant, M.P. (1981). *Appl. Environ. Microbiol.* **41**, 346.
McInerney, M.J., Bryant, M.P., Hespell, R.B. & Costerton, J.W. (1981). *Appl. Environ. Microbiol.* **41**, 1029.
McInerney, M.J., Bryant, M.P. & Pfennig, N. (1979). *Arch. Microbiol.* **122**, 129.
Mackie, R.I. & Bryant, M.P. (1981). *Appl. Environ. Microbiol.* **41**, 1363.
Mackie, R.I. & Bryant, M.P. (1990). *Appl. Environ. Microbiol.* **56**, 87.
Macario, A.J.L. & Conway de Macario, E. (1983). *Syst. Appl. Microbiol.* **4**, 451.
Macario, A.J.L. & Conway de Macario, E. (1988). *Appl. Environ. Microbiol.* **54**, 79.
Macario, A.J.L., Conway de Macario, E., Ney, U., Schoberth, S.M. & Sahm, H. (1989). *Appl. Environ. Microbiol.* **55**, 1996.
Macario, A.J.L., Earle, J.F.K., Chynoweth, D.P. & Conway de Macario, E. (1989). *Syst. Appl. Microbiol.* **12**, 216.

Mah, R.A. & Susman, C. (1986). *Appl. Microbiol.* 16, 358.
Märkel, H., Mather, M. & Witty, W. (1984). In *Proc. Third Cong. on Biotechnol.*, München. Verlag Chemie, Weinheim, FRG. Vol. III, p. 137.
Maki, L.W. (1954). *Antonie van Leeuwenhoek*, **20**, 185.
Melbinger, N.R. & Donnellon, J. (1971). *J Water Poll. Cont. Fed.* **43**, 1658.
Minato, H. & Suto, T. (1981). *J. Gen. Appl. Microbiol.* **27**, 21.
Morelli, A., Rindone, B., Andreoni,V., Villa, M., Sorlini, C. & Balice, V. (1990). *Biol. Wastes*, **32**, 253.
Mosey, F.E., Swanwick, J.D. & Hughes, D.A. (1971). *J. Wat. Pollut. Cont. Fed.* **6**, 2.
Mountford, D.O. & Bryant, M.P. (1982). *Arch. Microbiol.* **145**, 396.
Mountford, D.O., Brulla, W.J., Krumholz, L.R. & Bryant, M.P. (1984). *Int. J. Syst. Bact.* **34**, 216.
Mylroie, R.L. (1954). *Canad. J. Microbiol.* **1**, 55.
van Nevel, C.J. & Demeyer, D.I. (1988). In *The rumen microbial ecosystem*, ed. P.N. Hobson, Elsevier Applied Science, London. p. 387.
Ng., T.K., Weimer, P.J. & Zeikus, J.G. (1977) *Arch. Microbiol.* **114**, 1.
Novaes, R.F.V., Rech, C.M., Figueirdo, M.G. & Giaj-Levra, L.A. (1988). In *Fifth Int. Symp. An. Digestion, Poster papers*, ed. A. Tilche & A. Rozzi, Monduzzi Editore, Bologna. P. 61.
Ohwaki, K. & Hungate, R.E. (1977). *Appl. Environ. Microbiol.* **33**, 1270.
Oleszkiewicz, J.A. & Sharma, V.K. (1990). *Biol. Wastes*, **31**, 45.
Ollivier, B., Smiti, N., Mah, R.A. & Garcia, J.L. (1986). *Appl. Microbiol. Biotechnol.* **24**, 79.
Olson, K.D., McMahon, C.W. & Wolfe, R.S. (1991). *Appl. Environ. Microbiol.* **57**, 2683.
Pascik, I & Henzler, H-J. (1988). In *Fifth Int. Symp. An. Digestion, Poster papers*, ed. A. Tilche & A. Rozzi, Monduzzi Editore, Bologna. p. 491.
Patel, G.B., Khan, A.W., Agnew, B.J. & Calvin, J.R. (1980). *Int. J. Syst. Bact.* **30**, 179.
Patel, G.B., Khan, A.W. & Roth, C.A. (1978). *J. Appl. Bact.* **45**, 347.
Patel, G.B. & Roth, L.A. (1978). *Can. J. Microbiol.* **24**, 1007.
Pearson, F., Shiun-Chung, C. & Gautier, M. (1980). *J. Wat. Pollut. Cont. Fed.* **52**, 472.
Pechan, Z., Knappova, O., Petrovicova, B. & Adamec, O. (1987). *Biol. Wastes*, **20**, 117.
Peek, M.W. & Chynoweth, D.P. (1990). *Biotech. Lett.* **12**, 17.
Perski, H.J., Schönheit, R. & Thauer, R.K. (1982). *FEBS Lett.* **143**, 323.
Petersen, S.P. & Ahring, B.K. (1988). In *Fifth Int. Symp. An. Digestion, Poster papers*, ed. A. Tilche & A. Rozzi, Monduzzi Editore, Bologna. p. 65.
Pfeffer, J.T. (1978). *Process Biochem.* **13**, 6.
Phelps, T.J., Conrad, R. & Zeikus, J.G. (1895). *Appl. Environ. Microbiol.* **50**, 589.
Pirt, S.J. (1975). *Principles of Microbe and Cell Cultivation*, Blackwells, Oxford.
Pirt, S.J. & Lee, Y.K. (1983). *FEMS Microbiol. Lett.* **18**, 61.
Poels, J., van Assche, P. & Verstraete, W. (1984). *Agric. Wastes*, **9**, 239.
Pohland, F.C. (1962). *Eng. Ext. Bull. Purdue Univ. USA,* **20**, 583.
Prensier, G. Dubourguier, H.C., Thomas, I.,, Albagnac, G. & Buisson, M.N. (1988). In *Granular anaerobic sludge: microbiology and technology*, ed. G. Lettinga et al., Pudoc, Wageningen. p. 55.
Rajasekaran, P., Murugesan, R. & Palanisamy, A. (1986). *Agric. Wastes*, **17**, 83.
Richardson, A.J., Hobson, P.N. & Campbell, G.P. (1987). *Lett. Appl. Microbiol.* **5**, 119.
Rivard, C.J., Bordeaux, F., Henson, J.M. & Smith, P.H. (1988). *Appl. Biochem. Biotech.* **17**, 245.
Robbins, J.E., Arnold, M.T. & Lacher, S.L. (1979). *Appl. Environ. Microbiol.* **38**, 175.
Robbins, J.E., Gerhardt, S.A. & Kappel, T.J. (1989). *Biol. Wastes*, **27**, 1.
Roberts, D.J., Fedorak, P.M. & Hrudey, S.E. (1990). *Appl. Environ. Microbiol.* **56**, 472.
Robinson, J.P. & Hungate, R.E. (1973). *Int. J. Syst. Bact.* **23**, 171.
Ronnow, P.H. & Gunnarsson, L.A.H. (1981). *Appl. Environ. Microbiol.* **42**, 580.
Rose, C.S. & Pirt, S.J. (1981). *J. Bact.* **147**, 248.

Roy, F., Albagnac, G. & Samain, E. (1985). *Appl. Environ. Microbiol.* **49**, 702.
Salkinoja-Salonen, M.,Nuys, E.J., Sutton, P.M., van den Berg, L. & Wheatley, A.D. (1983). *Water Sci. Technol.* **15**, 305.
Salkinoja-Salonen, M., Valo, R., Apajalahti, J., Hakulinen, R. & Silakoski, L. (1983). In *Third Int. Symposium Anaerobic Digestion*, 99 Erie Street, Cambridge, MA, USA. p. 107.
Scheifinger, C.C., Latham, M.J. & Wolin, M.J. (1975). *Appl. Microbiol.* **30**, 961.
Schink, B. (1986). In *Biology of anaerobic bacteria*, ed. H.C. Dubourgier et al., Elsevier, Amsterdam. p. 2.
Schink, B. (1988). In *Fifth int. symp. an. digestion*, ed. A. Tilche & A Rozzi, Monduzzi Editore, Bologna. p. 459.
Segretain, C. & Bories, A. (1987). In *Biomass for energy and industry*, 4th EC conference. Elsevier Science Publishers, New York. p. 808.
Sewell, G.W., Aldrich, D., Williams, D., Mannarelli, B., Wilkie, A., Hespell, R.B., Smith, P.H. & Ingram, L.O. (1988). *Appl. Environ. Microbiol.* **54**, 1085.
Sharma, V.K. (1983). Ph.D. Thesis, Aberdeen University.
Sharma, V.K. & Hobson, P.N. (1985). *Agric. Wastes*, **14**, 173.
Sharma, V.K. & Hobson, P.N. (1986*a*). *J. Appl. Bact.* **61**, 257.
Sharma, V.K. & Hobson, P.N. (1986*b*). *Mic. Ecol.* **12**, 232.
Sharma, V.K. & Hobson, P.N. (1990). *Biol. Wastes* **33**, 81.
Sharma, V.K. & Hobson, P.N. (1991). *Proc. ASM Annual Meeting*, 1991. xx.
Shaw, B.G. (1971). Ph.D. Thesis, Aberdeen University.
Shelef, G., Kimchie, S. & Grynberg, H. (1980). *Biotech. Bioeng. Symp.* **10**, 341.
Shelton, D.R. & Tiedje, J.M. (1984). *Appl. Environ. Microbiol.* **48**, 840.
Siebert, M.L. & Torien, D.F. (1969). *Water Res.* **3**, 241.
Sierra-Alvarez, R. & Lettinga, G. (1990). *Biol. Wastes,* 33, 211.
Sierra-Alvarez, R. & Lettinga, G. (1991). *J. Chem. Technol. Biotechnol.* **50**, 443.
Singh, R., Kalsey, J.S. & Jain, M.K. (1986). *Environ. & Ecol.* **4**, 64.
Sleat, R. & Mah, R.A. (1985). *Int. J. Syst. Bact.* **35**, 160.
Sleat, R., Mah, R.A. & Robinson, R. (1984). *Appl. Environ. Microbiol.* **48**, 88.
Smith, D.P. & McCarty, P.L. (1989). *Biotech. Bioeng.* **34**, 885.
Smith, P.H. & Mah, R.A. (1966). *Appl. Microbiol.* **14**, 368.
Sowers, K.R. & Ferry, J.G. (1985). Proc. int. gas res. conf., Washington, 1984. Government Institutes Inc., PO Box 1096, Rockville, MD 20850, USA. p. 527.
Speece, R.E. (1987). In *Anaerobic digestion of biomass*, ed. D.P. Chynoweth & R. Isaacson, Elsevier Applied Science, London. p. 109.
Spoelstra, S.F. (1980). *Microbial Aspects of the Formation of Malodorous Compounds in Anaerobically Stored Piggery Wastes.* Thesis, Wageningen, The Netherlands.
Sprott, G.D. & Patel, G.B. (1986). *System. Appl. Microbiol.* **7**, 358.
Sprott, G.D., Shaw, K.M. & Jarell, K.F. (1984). *J. Biol. Chem.* **1259**, 12602.
Sprott, G.D., Shaw, K.M. & Jarell, K.F. (1985). *J. Biol. Chem.* **1260**, 9244.
Stanogias, G., Tjandraatmadja, M. & Pearce, G.R. (1985). *Agric. Wastes*, **12**, 37.
Stronach, S.M., Diaz-Biaz, M.C., Rudd, T. & Lester, J.N. (1987). *Biotech. Bioeng.* **30**, 611.
Summers, R. (1978). In *The microbial breakdown of straw in the rumen and breakdown of straw and manure for methane production.* MAFF, Agricultural Development and Advisory Services, Oxford. p. 84.
Summers, R. & Bousfield, S. (1980). *Agric. Wastes*, **2**, 61.
Summers, R., Hobson, P.N., Harries, C.R. & Richardson, A.J. (1987). *Biol. Wastes*, **20**, 43.
Tatton, M.J., Archer, D.B., Powell, G.E. & Parker, M.L. (1989). *Appl. Environ. Microbiol.* **55**, 440.
Temper, U., Winter, J. & Kandler, O. (1985). An. Digestion, MIRCEN, Stockholm, **2**, 70.
Thiel, P.G. (1969). *Water Res.* **3**, 215.

Thiele, J.H. & Zeikus, J.G. (1988). *Appl. Environ. Microbiol.* **54**, 20.
Thiele, J.H., Wu, W-M., Jain, M.K. & Zeikus, J.G. (1990). *Biotech. Bioeng.* **35**, 990.
Torien, D.F. (1967). *Water Res.* **1**, 147.
Torien, D.F., Thiel, P.G. & Hattingh, W.H.J. (1968). *Water Res.* **2**, 505.
Tscheck, A. & Schink, B. (1986). *Arch. Microbiol.* **145**, 396.
Tsonis, S.P. & Grigoropolous, S.G. (1988). In *Anaerobic digestion 1988*, ed. E.R. Hall & P.N. Hobson, Pergamon Press, Oxford. p.115.
Ueki, A, Miyagawa, E., Minato, H., Azuma, R. & Suto, T. (1978). *J. Gen. Appl. Microbiol.* **24**, 317.
Varel, V.H., Isaason, H.R. & Bryant, M.P. (1977). *Appl. Environ. Microbiol.* **33**, 298.
Varel, V.H. & Hashimoto, A.G. (1981). *Appl. Environ. Microbiol.* **41**, 29.
Vargas, C. & Ahlert, R.C. (1987). *J. Wat. Pollut. Cont. Fed.* **59**, 964.
van Velsen, A.F.M. (1977). *Neth. J. Agric. Sci.* **25**, 151.
van Velsen, A.F.M. (1979). *Water Res.*, **13**, 995.
Visser, F.A., van Lier, J.B., Macario, A.J.L. & Conway de Macario, E. (1991). *Appl. Environ. Microbiol.* **57**, 1728.
Waite, R. & Gorrod, A.R.M. (1959). *J. Sci. Food Technol.* **10**, 317.
Webb, A.R. & Hawkes, F.R. (1985). *Agric. Wastes*, **14**, 135.
Weimer, P.J. & Zeikus, J.G. (1977). *Appl. Environ. Microbiol.* **33**, 289.
Welander, T., Malmqvist, A. & Yu, P. (1988). In *Anaerobic digestion* 1988 ed. E.R. Hall & P.N. Hobson, Pergamon Press, Oxford. p. 267.
Wellinger, A. & Wuhrmann, K. (1977). *Arch. Microbiol.* **115**, 13.
Westerman, P., Ahring, B. K. & Mah, R. A. (1989). *Appl. Environ. Microbiol.* **55**, 2257.
Wiegant, W.M. & Zeeman, G. (1986). *Agric. Wastes*, **16**, 243.
Wilkie, A., Goto, M., Bordeaux, F.M. & Smith, P.H. (1986). *Biomass*, **11**, 135.
Williams, A.G. (1984). *Agric. Wastes*, **10**, 15.
Winfrey, M. R. & Zeikus, J.G. (1977). *Appl. Environ. Microbiol.* **33**, 275.
Winter, J.U. & Wolfe, R.S. (1979). *Arch. Microbiol.* **121**, 97.
Winter, J.U. & Wolfe, R.S. (1980). *Arch. Microbiol.* **124**, 73.
Winter, B., Fiechter, A. & Zimmerman, W. (1991). *Appl. Environ. Microbiol.* **57**, 2858.
Woese, C.R., Magrum, L.J. & Fox, G.E. (1978). *J. Mol. Evolut.* **11**, 245.
Yang, J.C., Chynoweth, D.P., Williams, D.S. & Li, A. (1990). *Int. J. Syst. Bact.* **40**, 268.
Yang, S.H. & Drake, H.L. (1990). *Appl. Environ. Microbiol.* **56**, 81.
Yang, J., Speece, R.E., Parkin, G.F., Gossett, J. & Kocher, W. (1980). *Prog. Water Technol.* **12**, 977.
Yang, S-T. & Okos, M.R. (1987). *Biotech. Bioeng.* **30**, 661.
Yang, S-T., Tang, I-C. & Okos, M.R. (1988). *Biotech. Bioeng.* **32**, 28.
Zeeman, G., Wiegant, N.M., Koster-Treffers, M.E. & Lettinga, G. (1985). *Agric. Wastes*, **14**, 19.
de Zeeuw, W.J. (1984). *Acclimatization of anaerobic sludge for UASB reactor start-up.* Ph.D. Thesis, Agricultural University, Wageningen.
Zehnder, A.J.B., Ingvorsen, K. & Marti, T. (1981). In *Anaerobic digestion 1981*, ed. D. Hughes *et al.*, Elsevier Biomedical, Amsterdam. p. 45.
Zinder, S.H. (1988). In *Anaerobic digestion 1988*, ed. E.R. Hall & P.N. Hobson, Pergamon Press, Oxford. p. 1.
Zinder, S.H., Anguish, T. & Cardwell, S.C. (1984). *Appl. Environ. Microbiol.* **47**, 808.
Zinder, S.H. & Mah, R.A. (1979). *Appl. Environ. Microbiol.* **38**, 996.
Ziomek, E.Z. & Williams, R.E.W. (1989). *Appl. Environ. Microbiol.* **55**, 2262.
Zoetmeyer, R.J., Malthusen, A.J.C.M., Cohen, A. & Boelhouever, C. (1982). *Water Res.* **16**, 633.

Chapter 3

MATHEMATICAL MODELS

INTRODUCTION

As with all fermentation processes, the design of a large-scale digester is undertaken only when laboratory and pilot-plant experiments have shown that the digestion is feasible and some parameters such as retention time, loading rate and possible type of digester have been determined. The type of digester is basically determined by the feedstock in that, for instance, high suspended-solids wastes would usually call for a stirred-tank digester, while wastewaters with low solids would need a retained biomass digester. However, there may be other possibilities. It might be suggested that a high-solids feedstock could be treated by removing some, or all, of the solids for digestion in a 'solid state' tank digester while the liquid was treated in a retained-biomass digester. A retained-biomass system may be a filter or some other type of digester. In many cases there can exist a number of possibilities for design of the digester plant. However, if a full-scale plant for a feedstock already exists then design of the digester plant may be only a case of altering some dimensions to deal with the different flow of feedstock at the new site, the type of digester and ancillary plant remaining as before. This is generally the case with sewage works where a large number of digesters have been made and the feedstock is similar at most sites. The requirements of the digestion in purification of the feedstock and the uses of the biogas and of the digested sludge are similar at many works. The financial basis used for new plants and the costs of sewage treatment are governed by local taxes and similar factors in most cases.

However, in the case of factory and farm wastes the situation can be quite different. A digester to treat the particular waste, or combination of wastes, may not have been built before. The situation on the site with regard to the degree of pollution control required, the disposal or further treatment of the digested waste, and the use of the biogas, may be different from that on other sites even if the wastes are similar. It may be necessary to take into account different types of secondary treatment of the digester effluent which may be possible and which may affect the type of digester or its running. There may be a use for the solids of the waste which would make initial solids separation and use of a retained-biomass- instead of a stirred-tank-digester a practical proposition: or better use might be made of the solids separated after digestion.

In these cases planning of the digester and associated systems involves the collection and assessment of data. Digesters may broadly be said to be installed in three types of system. In one, the digester is to be a whole, or part, of a pollution-control system and the production of biogas is ignored (it could be flared off). In a second, the digester is part of a pollution- control system but the biogas is counted on to repay part of the costs of the system or is needed as a fuel because of lack of other fuels on the site. The first of these might be exemplified by a factory digester where other fuels are available; the second could be a digester in a developing country where the wood for fuel is rapidly disappearing. In the third case the digester is required only for the production of fuel gas: this is the caase in 'alternative-energy' schemes.

In each case there is need for detailed assessment of the digester feedstock; how much and of what composition?; might it contain some toxic compound? This type of data can suggest the type of digester needed and the possible size and running parameters based on comparison with published data or the results of pilot-plant experiments, and the need to optimise pollution control or other factors. Such calculations will also include gas production and from that its use. And, as said before, there may be more than one option to consider. The calculated, or experimental, data will also give effluent composition and so be a basis for design of further treatment- or disposal-systems. The general scheme having been worked out, more detailed engineering and construction data can be considered and the costs/benefits of the scheme determined; and so on.

In all this work, even if no equations are being written or formal statements made, a form of mathematical modelling is being used, because mathematical modelling is concerned with flows and rates: flow of the feedstock through the plant, flow of biogas from digester through use; rates of input and degradation of feed and output of products; rates of production and use of energy; and flow of cash in costs and returns. This latter involves the mechanical and civil engineering, and running, costs, and the returns from the primary and secondary products. This could involve direct returns in, say, biogas energy, and indirect returns in reduction of costs involved in other forms of waste disposal, or even of the value of permission to keep an industry going because pollution has been decreased. The 'modelling' aspect may be only implied, and these concepts are widely used in planning any large chemical or other plant. The mathematical model attempts to formalise the concepts in flow sheets and equations which relate different parts of the proposed digester system to each other and to the whole and to possible alternative systems. Formerly, the amount of modelling that could be done was limited by the limits of 'hand' calculations as well as lack of data and concepts. The advent of the computer has brought almost no limit to the possibilities of calculations, but lack of data remains a major problem.

There are two types of model used in anaerobic digestion technology. The first might be termed the 'engineering' or 'economic' model; the second the 'biological' model. The first is concerned with the design and construction of the digester and its ancillary plant and the provision of feedstock and removal or use of digester effluent and gases. The second is concerned with the metabolic activities of the bacteria in the digester and not with the actual design of the digester (except insofar as the model will assume a certain digester configuration; say, a stirred-tank with theoretically optimised feeding and mixing). The first model has to rely on data about the effects of bacterial metabolism, in that the specific gas production, reduction in polluting factors in the feedstock and so on are required, but not on the details of this metabolism. These data may be obtained from experimental observations or from a biological model, so the two types of model may be linked.

One objective of the biological model has been to provide data from which an optimised digester may be built. Examples of this discussed later are the models of animal-waste digestions of Hill (1983*a,b,c,d*). These models set out to determine the optimum and minimum (in terms of solids reduction and gas production) retention times, optimum solids concentrations in the feeds, and the best start-up procedures for digestions at mesophilic and thermophilic temperatures: they did not consider these parameters in terms of cost of digester and use of gas or other economic factors, but the conclusions from the biological model could be transferred to an economic model. The two types of model may, then, be related, but, in general, the biological model has been conceived as an exercise in explaining and defining bacterial metabolism rather than as an adjunct to digester design.

Some examples of both types of model will be considered below.

The economic and engineering model

There are fewer papers on the economic and engineering modelling of digesters than on biological modelling. For this reason, and because the engineering and economic model as such is probably less used now than formerly, this section of the chapter will be briefer than the section on the biological model.

As experimental data on small and, particularly, full-scale digester plants have increased, so the need for modelling in design of digester plants has decreased. The degree of degradation, gas production, etc., for the particular (or very similar) feedstock is known: if it is not known then data must generally be collected from experiment not from theory. The type of digester will be decided from the nature of the feedstock, but if a retained-biomass digester is decided on, the design—filter, UASB, etc. — will probably be determined by the fact that the firm building the plant specialises in one particular design rather than any theoretical concept. Similarly, the materials of construction will be largely dictated by the experience of the construction firm rather than modelling and comparison of the properties of a number of materials. The costs and performance of ancillary equipment, engines, boilers, etc., will be known and they will probably be supplied as package deals with the installation costs determined by the costings usual for any civil engineering job. Thus, as said before, while there is calculation involved there is probably little reliance on formal models as the economic or engineering model was conceived some years ago.

At the time of the first international symposium on anaerobic digestion (1979) there were comparatively few data on digestion of materials other than sewage and some farm-animal sludges and the main type of digester in use was the stirred-tank. In the proceedings of the symposium is a paper by Hawkes (1980) in which the major operating parameters affecting digester productivity are discussed and relatively simple computer models show the effects of varying these parameters (retention time, solids concentration, etc.) on maximising net gas production. In a similar paper Horton (1980) also considers the interaction of operating parameters and concludes, for instance, that for pollution control an operating temperature of 35°C, where gas production in the mesophilic range is often at a maximum, would be chosen. However, if gas utilisation for digester heating, and ambient temperatures, are taken into account then it would be better to run the digester at 25°C, even though the gas production is lower than at 35°C. The model of Feilden (1981) for determining the best combination of feed solids and digestion temperature to obtain maximum net gas production has been mentioned elsewhere.

Hawkes & Rosser (1983) discussed a much more complex computer model based on the previous conceptions. This model, named 'ADMODL', considered not only the parameters, such as those mentioned above, affecting the digestion itself but also value of the gas utilisation and the value of the digester effluent as fertilizer, animal feed, etc., set against all the running and capital costs of the digester and ancillary equipment and taking into account Government grants available, taxes, and so on. Some similar models have been constructed but it seems doubtful if they have ever been used in practical situations.

Bala & Satter (1990) proposed a model based on Contois or Adams-Eckenfeder models for predicting solids reduction and biogas formation in Indian digesters and the procedure described by Audsley & Wheeler (1978) for calculating the cost of machinery to take into account the economics of the digestions, assuming, for instance different life-spans of the digester. They found that the annual cost of digesters was minimum for 15 days RT and the slurry from 2 or 4 cows, and for 20 day RT with slurry from 6 to 8 cows. Although the cost was reduced considerably for over 4 cows the cost of biogas for cooking was higher than that of fuel wood, though assigning a value for pollution control and fertilzer value of the digested slurry did make some improvement. Relatively large increases in the price of fuel wood or reductions in the cost of digesters were needed to make gas usage economic. However, against this must be put the fact that the present use of fuel wood is rapidly devastating woodlands and wood is becoming scarcer in countries like India.

Lusk (1991) examined the economics of a psychrophylic covered-lagoon type of digester and a mesophilic stirred-tank farm digester and found in favour of the psychrophylic digester because of lower capital and operating costs. This study did not include a factor for pollution control and the desirability of including some values for pollution control and environmental protection in the costs of digesters has been mentioned elsewhere. Videla *et al.* (1990) applied a modelling technique to the design and operation of a digester treating landfill leachate. The models were to reduce costs to a minimum and to maximise energy production with at least 75% COD removal as a pollution-control benefit. Likewise, the value of products other than the biogas should be taken into account if possible. Hashimoto and Chen (1981) modelled digester systems treating cattle waste and crop residues to maximise the net energy production per unit digester cost. They found, as have others, that only the largest farm digesters could be economically viable on energy production, but if the effluent could be valued as a feedstuff then the economic size of plant decreased considerably. Although he did not make a formal model, Marchaim (1983) argued from the available data that the use of digester effluents was a key factor in the profitability of digester plants.

Hamad *et al.* (1989) planned a model village for Egyptian desert communities. The design and layout of the houses and the adjacent animal sheds (for 4 or 5 cattle) was such as to facilitate each house, or each two houses, having a small digester to take animal and human excreta to provide gas for lighting and cooking. This biogas could be augmented by gas from large digesters on the edge of the village which ran on crop residues and slurry from big animal sheds (for 400 cattle) and poultry units run by the bigger farmers or government agencies. The village road system provided for movement of animals, and removal of digested slurry, behind the houses. Such a village layout made the economics and the practicalities of village digesters much better than the ad hoc building of digesters in existing villages.

Tasdemiroglu (1991) used a model to compare the economics of biogas and traditional methods of house heating in rural Turkey and decided biogas was the best system. However, the various assumptions made about construction and use of biogas plants might not have been entirely valid in all villages and the biogas plants might not be run as efficiently as the experimental plants from which some data were obtained.

Other papers have described models similar to these briefly mentioned here and many of the papers on digestion of particular feedstocks or on particular digester types mentioned in other chapters include some calculations on the economics of use of the feedstock or digester. But, for example, two papers devoted to the economics of cattle-waste digestion are those by Oliver *et al.* (1986) and Parsons (1986). Oliver *et al.* used a computer model to assess the effects of changing various running parameters on the relative costs of digesting dairy-cattle slurry. Relative costs were used as energy was the only output value given to the digestion and it was difficult to assess the possible usage of energy on an actual farm; complete use of energy was supposed. The design and performance data used came from the operation of two 125 cubic metre digesters on a dairy unit (see Chapter 6) and costs of construction and ancillary equipment came from manufacturers' data. With hot water or electricity production as the sole credit digestion made an operating loss, but this could be reduced by reducing the RT, increasing the feed solids (about 8% TS was thought to be the limit of pumping capacity) and decreasing the operating temperature from 35°C to 20°C. Big changes in capital costs, interest rates and the prices of electricity or hot water generated were needed to make reasonable reductions in the operating loss. However, the digester was run only for six months of the year while the cattle were housed in the dairy unit and running the digester for a whole year by using other feedstocks improved the economics. Digesting whole rather than separated slurry (Chapter 6) was, on an energy production basis, better economically despite the lower costs of digesting separated slurry. Giving a value to pollution control and marketing digested slurry much improved the economics. The calculations of Parsons, also related to dairy farms, gave much the same results as those just quoted.

These models were for plants whose feedstocks were wastes and whose primary purpose was treatment of wastes. Mishoe *et al.* (1988) described a model for use in generation of 'alternate' energy from crops grown as feedstocks. This BIOMET model considers the growth, harvesting, transport and preparation of the crop and its digestion, and includes energy required and used and plant costs. Variations in one or other parameter (say transport distances) can be introduced to see the effects on the production and costs of the biogas. Thus, for instance, minimum costs of gas can be compared with present or future (projected) costs of other fuels to see when, or under what conditions, the alternative fuel can compete with conventional fuels.

The various economic models have given broadly similar results for many aspects of digestion. However, costs of plant and the methods of building digesters change from year to year, thus altering the economic calculations: a major example is the fall in oil prices between the 70's and the 80's and the effects of this on the value of digesters as energy producers. The importance of being able to apply some economic value to the digester as a means of pollution control is stressed by the models, but pollution control is difficult to value and there are many related factors, such as prevention of the destruction of trees mentioned above, which are unquantifiable in strict monetary and modelling terms and it is these which will probably have most influence on the building of digesters in the future.

The biological model

General considerations

Biological modelling of digestion was attempted before much was known about the actual bacteria of digesters and was based on the overall reactions of digestion determined by biochemical analysis. Although more is now known about the digester bacteria most modelling is still based on the biochemical reactions and not on any detailed data on individual bacteria or groups of bacteria derived from experiments on culture of isolated bacteria.

A bacterium in pure culture and growing on a defined substrate in a defined, or mainly defined, medium with dissolved constituents behaves according to certain 'rules' and its performance under changed conditions of growth in batch- or continuous-culture can be reasonably well predicted from models. This is possible because a very large amount of data on the growth of bacteria has been obtained and such data have been used to derive models and to test their validity against the experimental results from growth of different bacteria under different conditions of culture. But even here exceptions to the 'rules' are found. These may be due to previously unconsidered constituents of the culture medium or to bacterial reactions which come into effect only in relatively extreme culture conditions. This latter may limit the model to certain ranges of growth rate or other parameter.

Although a bacterium in pure culture may behave according to relatively simple rules, the behaviour of two bacteria competing for the same substrate, or where one is growing on the products of another, becomes much more complicated. Cultures with more than two bacteria become almost impossible to analyse in detail microbiologically and biochemically, and modelling becomes equally, or more, complicated. In a digester we have, in general, a substrate (feedstock) whose composition is only partially known, and then often in terms which have no precise chemical meaning, and a mixed bacterial population of largely unknown bacteria in undefined numbers, but all interacting. Even for those bacteria which have been isolated few data suitable for incorporating in models exist. The biological model is, then, still constructed largely in terms of the overall biochemical reactions taking place in the digester and not on the detailed behaviour of individual bacteria or groups of bacteria concerned in all the steps in the pathways from feedstock to methane.

The digestion process thus has to be simplified. The substrate may be simplified as shown later. The previous chapter showed that many different types of bacteria are involved in each of the main steps in digestion; for instance, in the breakdown and fermentation of fibre polysaccharides. Since there are no data available on the contribution of each of these bacteria to the overall reaction, another simplification is made by assuming that the consortium of bacteria can be modelled as one bacterium converting the initial substrate to the final products. A model also needs data on which to base the various numerical factors involved in describing the rates and extents of microbial reactions, and data from similar, but somewhat different, digestions with which to test the validity of predictions made from the model. There are two main goals in model making: one is to describe accurately a particular situation, the other is to predict what will happen to the system when it is operated under different conditions. In theory, models could do away with the need for experiment in digestions and other microbial systems. But before a model can be relied upon to predict the unknown, some predictions have to be tested against a known system. For a number of reasons a modeller may have only limited data from his own experiments and this may be used to construct the model. The data for testing of the model will have to come from the experiments of others reported in the literature. It will very often be found that others doing experiments on similar feedstocks have not made the same analyses as the modeller and so it is difficult to test the model on the new data. Or, the workers may both have analysed for the complex substrate 'Y' but used different methods of analysis and it is not certain if exactly the same substrate degradation is being carried out in each case. However, a model which cannot be fully tested because of lack of data may serve a useful purpose in showing where experiment is needed to obtain more knowledge of the substrates or bacteria, and this can be one important aspect of the use of models.

Since most digesters are continuous-cultures, the biological model is, then, concerned with the flow of feedstock through the digester and the kinetics of the breakdown of microbial substrates and formation of endproducts of intermediate and final reactions. The other consideration is the growth of bacteria and the flow of these out of the digester (it is generally assumed that the feedstock is sterile and there is no flow of bacteria into the digester). The kinetics of these two flows may be directly, or only indirectly, related. A third aspect, which may or may not be considered, is the effects of the end-products on physical aspects of the digester contents (e.g., acid concentration and pH changes) and on the bacterial activities.

In most cases primary substrates are considered not as a mixture of known compounds but as undefined entities such as 'pig waste' or 'fibre', or even 'COD'. Only in the methanogenic phase of the digestion are chemically-defined substrates, acetic acid, hydrogen and carbon dioxide, usually mentioned. However, the actual reactions of methanogenesis may be omitted. For instance, one might define the digestion of pig-waste in a simple model by collecting data from a large number of digestions and, by averaging all the data, formulate the equation:

$$1 \text{ unit weight pig waste TS} \longrightarrow X \text{ litres biogas (at } 35°\text{C)}$$

(The pig waste being optimally degraded and the biogas having a certain composition).

From such data it would be possible to calculate the biogas yield from different numbers of pigs or different-sized digesters, etc. However, the ash content of pig wastes may vary. As the ash does not contribute to gas production it could be more accurate to use a figure for gas production per unit weight of Volatile Solids if the ash content of the pig-waste feedstock being modelled were known. (The Volatile Solids, being the organic matter in the waste, contains the material which will be converted to gas, but in feedstocks such as pig wastes not all the organic matter is biodegradable. 'Complete degradation' then means the conversion of all the potentially-degradable material to gas, not the conversion of the entire organic matter; Fig. 6). This simple equation may be extended to become part of a model predicting, say, gas production at different RT in a continuous stirred-tank digester. However, it ignores all reactions between feed and biogas and so the data could not be used if concentrations of VFA in the modelled digestion had to be predicted. In this case, since it is experimentally difficult, or impossible to obtain data on the amount of VFA produced by complete degradation of pig-waste it is usual to assume that the degradable fractions of the waste are composed of hexose polymers and the amount of acetic acid and hydrogen generated by complete fermentation of this material can then be calculated (formation of other acids and their conversion to acetic is assumed). If the analyses are available it may be possible to divide the degradable material into carbohydrates (assumed hexose) and lipids (assumed to be one representative triglyceride) and so get a, perhaps, more exact figure for acid production. The acetic acid then becomes the substrate for the methanogenic bacteria and the kinetics of growth and metabolism of these, the degradation of the acid to gas, and so the residual acids, can be considered as a separate part of a more detailed model.

These general considerations apply to models of all types of digester systems, but the different types of digester have to be considered individually. The most common type of laboratory continuous-culture is the stirred-tank or chemostat, with no feed-back of medium or bacteria, and most modelling of microbial growth and substrate utilisations in continuous-cultures has been done on chemostat systems. The earlier anaerobic digesters were all stirred-tanks, so modelling began with equating the stirred-tank digester and the laboratory continuous-culture of one bacterium. Most subsequent modelling has been concerned with the stirred-tank digester and so this will be discussed first.

The stirred-tank digester

Steady-state models

Since so much work has been done on modelling of stirred-tank digesters it is impossible in a short chapter to mention all models, only some representative account can be given. It will be remembered that in almost all cases the feedstock of a stirred-tank digester contains mainly particulate substrates with a small amount of dissolved substrates.

There are basically two types of models of continuous-cultures, the 'steady-state' and the 'dynamic' model, and each was originally developed to refer to cultures with a dissolved substrate in a liquid medium, not solid-substrate cultures. When a chemostat culture of one bacterium has been started up and fresh medium is being added and used medium is being withdrawn at a constant rate, then the culture will come to a steady-state in which the numbers of bacteria in the culture vessel remain constant and the degradation of substrate and production of end-products is at a constant rate (Chapter 2 and Fig. 4). In the same way, when a digester has been started up and is being fed with some feedstock at regular intervals and digester contents are being withdrawn at the same rate the system will eventually come to a steady-state in which, in theory, the contents of the digester will be constant with time in concentration and chemical composition, the number of bacteria will be constant, and the production of gas will be at a constant rate. The digester is then analagous to the pure-culture chemostat. One type of model describes the behaviour of the chemostat or digester under such steady-state conditions with different running parameters, but it cannot describe the temporal changes in bacteria and products during the transition from one steady-state to another or the initial start-up of the system. The steady-state values for products and residuals formed are assumed to be constant with time. The 'dynamic' model attempts to describe the digester from day to day or hour to hour during the start-up of digestion and the changes from one steady-state to another, or during the steady-state running. It should be noted, though, that all digester models 'idealise' the system. Because it has a complex microbial population, because of unavoidable irregularities in mechanical control (small fluctuations in pumping rates of sludges and in temperature, for instance) the actual digester never attains the invariant steady-state of the model (or the ideal pure culture), but averaged over some time the digester approximates to the model.

A factor in digester operation which is constant is the digester working volume. The rate of flow of feedstock into this volume is controlled and is constant over long periods of time, although it may be possible to alter it. The concentrations of microbial substrates and undegradable material in the feedstock, and the working temperature of the digester, are generally constant, but, again, may be varied in a controlled manner. From the digester volume (V; volume units) and feed flow rate (f; volume units per unit time), the Retention Time (RT or HRT [Hydraulic RT], sometimes Θ-; V/f, the mean time for which a volume or particle of feed, or a bacterium suspended in the digester liquid, or the liquid itself [HRT], remains in the digester) and the Dilution Rate (D; f/V, the rate at which feed and bacteria flow through the digester [reciprocal time units]), can be obtained. From the concentration of material (dissolved, suspended; substrate or inert) in the feed (C; weight units per volume unit) and the flow rate, the Volumetric Loading Rate (B_v; weight units per digester volume unit per unit time) can be obtained. Since the other parameters are often defined by the situation of the digester, probably the most common use of a model in practical digester design is to determine an optimum RT for maximum feed degradation (Fig. 6). This can be obtained in practice by altering feed flow-rate with a fixed-size digester or by altering the digester size to optimise treatment of a fixed flow of feedstock.

Monod (1942) found that, over certain concentration ranges, the growth rate of a bacterium growing in a medium in which all the bacterial nutrients except one were in excess was related to the concentration of this one 'limiting' substrate. There was, though, a concentration above which growth rate no longer increased. This 'maximum growth rate' differed with different bacteria and substrates and it is probably defined by a factor such as the maximum rate at which nutrients can be transported into the cell or some cell constituent can be synthesised. This theory was used as the basis of one of the first models of chemostat continuous-cultures (Herbert, Elsworth & Telling, 1956) and it determined the concentration of residual substrate (and so of the substrate used) during the steady-state running of a culture at different values of RT or D. The authors showed that at steady-state the bacteria grew at the Dilution Rate of the culture and that the bacteria adjusted the concentration of the limiting substrate to that at which this growth rate was sustained. The limiting substrate in this test culture and the model based on it was the energy and carbon source for aerobic growth, and in most subsequent uses of the model the energy source has been the limiting substrate; nitrogen, trace elements, etc., being in excess. In some anaerobic fermentations almost 100% of the energy source is converted to fermentation products, in others a, usually small, part of the energy source is used in cell synthesis (Hobson, 1965; Hobson & Summers, 1967). This Monod

Herbert model of continuous-cultures has been the basis of, probably, the majority of models of anaerobic digestions. The limiting substrates being taken to be the bacterial energy sources; carbohydrates, acetic acid and hydrogen-carbon dioxide.

The growth of bacteria can be described in terms of the 'Specific Growth Rate', μ, where μ is the rate of increase in numbers or weight of bacteria per unit number or weight of bacteria. The rate of increase of the bacteria, dx/dt, is then equal to $\mu.x$. Integrating dx/dt gives:

$$x_1 = x_0 . \exp(\mu t) \qquad \text{or}$$

$$\ln x_1/x_0 = \mu.t \tag{1}$$

Where subscripts 0 and 1 refer to numbers of bacteria at zero time and time t.

The mean generation time (T, the time for numbers to double) is given by putting:

$$x_1 = 2. x_0$$

$$\text{when} \quad \ln 2 = \mu . T \text{ and}$$

$$T = \ln 2/\mu = 0{\cdot}693/\mu$$

The Monod theory also relates the growth rate (μ) of a bacterium to the concentration (S) of the limiting substrate by the equation:

$$\mu = \mu_m . S / (K_s + S) \tag{2}$$

where μ_m is the maximum growth rate possible on excess of the substrate (units; time unit^{-1}), K_s is the substrate concentration giving half the maximum growth rate, and S is the actual concentration of limiting substrate (weight units per unit culture volume).

The relationship between growth and utilisation of substrate is given by:

$$dx/dt = -Y\ (dS/dt) \tag{3}$$

The amount of bacterial growth is the amount of substrate used multiplied by a conversion factor for substrate to cells, Y (Wt. of bacteria formed/Wt. of substrate used). x is the concentration of bacteria in weight units per unit culture volume.

These equations can be used in construction of a model of a stirred-tank continuous-culture in which the growth of the bacteria is governed by one constituent of the culture medium.

In a continuous-culture the rate of change in limiting substrate concentration as the medium flows through the vessel is:

input minus output minus substrate used by bacteria, i.e.:

$$dS/dt = D \,.\, S_0 - D \,.\, S - \mu.x/Y \tag{4}$$

where S_0 and S are feed and effluent substrate concentrations.

The rate of change in bacterial numbers (as there are no bacteria in the feed) is:

Growth minus output, i.e.:

$$dx/dt = \mu.x - D \,.\, x \tag{5}$$

When the culture has attained a steady-state the residual substrate and the bacterial numbers remain constant with time and so dS/dt = dx/dt = 0.

From these equations it can be shown (Herbert *et al.*, 1956) that the steady-state residual concentration of substrate (S') is given by:

$$S' = D \,.\, K_s/(\mu_m - D) \tag{6}$$

and the concentration of bacteria (x') is given by:

$$x' = Y(S_0 - S) \tag{7}$$

and the growth rate of the bacteria, μ, equals D.

Herbert *et al.* (1956) showed that these equations described the growth of an aerobic bacterium on glucose over the range of growth rates they examined. The equations predict that bacterial concentration will remain constant as D increases until D = > μ_m when the concentration will fall to zero as the bacteria can no longer grow as fast as the medium is flowing through the culture and 'washout' will occur. The value of S' will increase gradually as D increases until D approaches μ_m when it will increase rapidly to infinity (or in practice the value of S_0).

The same theoretical treatment can be applied to two or more cultures in series where the effluent from one becomes the feedstock of the next. A very large number of stirred-tank cultures in series approximates to a tubular (plug-flow) continuous-culture (Herbert, 1964; Powell & Lowe, 1964). (See later).

However, although following workers found that this model was reasonably applicable to continuous-cultures, a number of deviations of actual cultures from the model were found. In particular, the model requires that Y be constant with change in μ (D). In a number of cultures, particularly with anaerobes (Hobson, 1965), the bacterial concentration, and so Y, fell rapidly as D was reduced below some value. The model also assumes that all the bacteria are viable and capable of reproduction. Postgate & Hunter (1962) and Tempest, Herbert & Phipps (1967) found that, at dilution rates lower than those used by Herbert *et al.* (1956), the proportion of viable cells in the culture decreased as D decreased.

The result of this was that the concentration of cells in the culture decreased as D decreased at low values of D (e.g. below about $0{\cdot}1h^{-1}$), as had been found for a number of cultures (e.g., with anaerobes, Hobson, 1965). This also meant that that Y was not constant but decreased with decrease in D. Powell (1965) showed mathematically that if some cells in a culture were non-viable then the viable cells grew at a rate greater than the average growth rate of the culture. It is now generally recognised that not all cells in even a rapidly-growing culture may be viable, although with cultures growing at rates approaching maximum more than 90% of the cells are viable. This concept of death of cells has been brought into models of continuous-cultures (an early example is that of Sinclair & Topiwala, 1970), and models of anaerobic digestions, by adding a term to the cell balance equation (5) of -cx, where c is a constant describing the rate of death of cells.

However, the fact that the cell yield decreased at low growth rates in many cultures also brought in the question of bacterial cell 'maintenance' and 'endogenous metabolism'. These had been thought of in connection with growth of cells in batch cultures and Herbert (1958) suggested that decrease in Y was due to endogenous metabolism; the use of some portion of the cell substance (perhaps a reserve carbohydrate) in place of external substrate for maintaining viability or other purpose. Thus the weight of cell mass produced and the yield (Y) would be lower than if all cell material produced were to remain intact. He suggested that a mathematical treatment similar to that used above for cell death would be suitable for continuous-cultures. However, the decrease in cell yield at low growth-rates could also be accounted for (Hobson, 1965) by assuming that some of the substrate was fermented to provide 'maintenance energy': that is, energy to renew cell materials that are being decomposed and rebuilt and to keep the cell intact at low growth rates where there is a long period between cell divisions. Pirt (1965) proposed a mathematical treatment for growth of bacteria to include a maintenance coefficient which was tested on data on growth of *Aerobacter* (the same type of bacterium as used by Herbert *et al.*, 1956) and of two anaerobic bacteria (Hobson, 1965). One anaerobic bacterium appeared to be anomalous. Marr, Nilson & Clark (1963) and Schulze & Lipe (1964) had also produced mathematical concepts of use of energy for maintenance, but Pirt's formulation has become the one used in most descriptions of bacterial growth. The overall rate of substrate utilisation is equated with use of substrate for growth and use of substrate for maintenance of cell structure, and other non-growth purposes, e.g. motility. The maintenance coefficient was a constant with growth rate, but other work suggested that the coefficient must vary. Pirt (1982) later proposed a modified model that included a constant maintenance term and one that decreased linearly with increase in growth rate and became zero at maximum growth rate.

The steady-state digester model

All these models dealt with the steady-state growth of a bacterium at different growth rates in continuous-cultures. To apply the models to digesters requires one more assumption than those mentioned previously, and that is that the growth of a consortium of bacteria degrading one substrate behaves as if it were one bacterium. Some of the earliest modelling of anaerobic digestions using these principles and obtaining steady-state concentrations of substrates at different RT of digesters was done by McCarty and Lawrence. McCarty (1966) showed that the equations developed to describe growth of bacteria in pure culture could describe sewage sludge digestions if it was suggested that the maximum rate of the process was determined by the rate of the slowest step; in this case methane production from volatile fatty acids. Lawrence & McCarty (1970) discussed these ideas in more detail and applied them to a stirred-tank digester, a stirred-tank with recycle of microbial solids, and a plug-flow reactor with solids recycle. McCarty (1971) modelled the steady-state breakdown of carbohydrate, protein and fat through VFA and hydrogen to methane in a simulated sewage sludge and showed that the results were similar to those obtained experimentally from a sewage digester. Kirsch & Sykes (1971) discussed the application of Monod kinetics and terms for maintenance and cell viability in the modelling of stirred-tank digesters. Amongst others who have used Monod-type kinetics in modelling steady-state digester performance is Mosey (1983, 1984)who was interested in the formation and degradation of VFA and hydrogen and the use of measurements of hydrogen concentrations as process-control parameters for digesters to show the onset of failure. Ghosh & Klass (1978) also modelled the breakdown of solids in digestion using Monod kinetics, as did one of the present authors, and some of the features of the latter's models will be mentioned to illustrate some points. These models were concerned only with the fate of the bacterial substrates not with the growth of the bacteria in the sense of the production of cell mass and the equation (6), above, was used to determine the residual substrate concentrations. Since the products from microbial substrate utilisation for growth or non-growth purposes are the same no consideration of maintenance or death of cells is required. In the first model (Hobson & McDonald, 1980) the residual VFA (mainly acetic) concentrations in a stirred-tank, continuous-flow pig-waste digester and in a laboratory continuous-culture of acetate-utilising bacteria derived from the pig-waste digestions were shown to be very similar over a range of RT from about 20 to 3 days, the concentrations increasing as the RT approached 3 days. Computor iteration showed that equation (6) could closely describe the experimental curve if μ_m were assumed to be $0{\cdot}4d^{-1}$ and K_s were 800 to 900 mg/l. The K_s value is given in terms of total acetic acid, as it is measured in digesters. The model showed that Monod kinetics could describe the growth of a simple mixed culture of methanogenic bacteria on one substrate in a laboratory culture and that this behaved in the same way as the, probably more complex, acetate-utilising methanogenic bacteria in a digester. The model also gave a minimum RT (derived from μ_m) for digestions in which the rate-limiting reaction was acetate degradation which agreed with practice and the results of others (Lawrence & McCarty [1967] gave μ_m as $0{\cdot}32d^{-1}$ for acetate degradation in a somewhat different system) and it also showed the concentrations of acid to be expected in a digester properly running at different RT. Although this minimum RT seems to be the same within experimental errors in digesters with different primary feedstocks, it might not be the same as there is some evidence that the predominant acetate-utilising bacteria differ in digesters with different substrates (Zehnder *et al.*, 1980). This might be the reason for Ghosh & Klass (1978) having found a K_s value of 4200 mg acetic acid per litre from a continuous-culture, of mixed bacteria from a sewage-sludge digester, fed with acetic acid: the pH was about the same (*ca.* 7·1) as in the pig-waste digesters and the bacterial culture used by Hobson & McDonald. On the other hand, Lawrence & McCarty (1967) quoted 154 mg/l for K_s for acetate ions in their system at pH 6·8, which is about 300–400 mg acetic/l at pH 7·1–7·2. It has been suggested (Andrews, 1968) that unionised acetic acid is the actual bacterial substrate (Acetic $\rightarrow$ acetate$^-$ + H^+). In this case the value of K_s in terms of total acetic acid will differ with the pH of the digester, assuming that the value of K_s (unionised acid) does not vary with pH, as the ratio of unionised to ionised acid will vary with pH. However, in the case of the Hobson & McDonald model and the one described next, the pH values of the digesters and the laboratory cultures were the same. It should also be noted that the residual concentration of substrate is determined without reference to the initial concentration. The Monod kinetics predicts that the final concentration of a bacterial substrate is independent of the initial concentration. This may not be true for all substrates in digesters (see later) but it would appear to hold for acid degradation as the acid is a dissolved substrate. The concept also helps in modelling of acid degradations, as although there is some acid in some digester feedstocks the actual acid being used by the methanogenic bacteria is a combination of this and acid produced by degradation of the carbohydrates and other constituents of the feed and so its initial concentration cannot be measured.

This model was followed by a more detailed one using Monod kinetics to reproduce other features of digestions of farm wastes (Hobson, 1983). It has been noted above that a plot of residual substrate at different dilution rates (or RT) in a continuous-culture of one bacterium gives a smooth curve with substrate concentration rapidly increasing as D approaches μ_m. In the more usual formulation for digesters, substrate degraded is plotted and this rapidly increases from zero at near to zero RT and approaches an asymptotic limit as RT increases to infinity (Fig. 6). It was noted that for some pig-waste digestions a graph of Volatile Solids degraded against RT showed a discontinuity at about 7 days RT and then gradually approached a maximum as RT increased. This could be modelled by assuming that the solids substrate consisted of two types of particle with different rates of degradation, and computor iteration fitted two curves, with K_s values of 782 mg/l and 500 mg/l and μ_m values of 0·139 and 0·36d^{-1}: the fractions being in the concentrations of 3·7 and 11·0g/l. The same two-component solids model, but with different values for constants (μ_m = 0·08 and 0·32d^{-1} with K_s 2200 and 1000 mg/l) could be fitted to the results of other workers for a cattle-waste digestion. Determinations of particle sizes in pig-waste did, at the time, give some grounds for dividing the solids into two groups of different degradabilities and later work by Peck *et al.* (1988) on cattle-waste particles also supported the two-component solids as a working simplification, although, obviously, farm slurries contain particles of a wide range of sizes. The values for μ_m showed that in farm-waste digestions, degradation of fibrous solids is the rate-limiting reaction, not conversion of acetate to methane. This is in accord with previous suggestions of Maki (1954), Hobson & Shaw (1971), Hobson *et al.* (1974) and with, for instance, later work by Noike *et al* (1985). The previous model of acetic acid degradation in pig-waste digesters was shown to also apply to cattle-waste digestions done by others. The same type of model was also applied to the conversion of hydrogen plus carbon dioxide to methane in digesters. In this case values for K_s and μ_m were obtained from experiments of the author and others with a rumen methanogenic bacterium (*Methanobrevibacterium ruminantium*) growing on hydrogen/carbon dioxide (μ_m was from 0·35 to 0·58d^{-1}). The kinetic equations gave the concentrations of hydrogen dissolved in the liquid, as this is the actual substrate for the bacterium, but from these figures concentrations in the gas of the digester could be calculated. The values were very low and could not be confirmed with the detection apparatus available to the author, but they seemed to be of the same order as concentrations measured in sewage digesters by Mosey & Fernandes (1984). From experiments on the effects of ammonia on methane production by a digester bacterium (*Methanobacterium formicicum*) by Hobson & Shaw (1974) and on acetate utilisation by digesting sewage sludge (McCarty & McKinney, 1961) a formulation for reduction in μ_m by increasing concentrations of ammonia was computed and applied to methanogenesis in digesters. (Hashimoto [1986] later found that ammonia inhibition of cattle-waste digestions followed a similar curve). The inhibition resulted in increases in residual acetic acid concentrations at the different RT in digesters with more than the normal concentrations of ammonia in the feedstock. Not many published results were available to test the model, but the model seemed to predict lower acid concentrations than were found in practice. However, the experimental acid concentrations contained a proportion of propionic acid and if acetic acid alone were considered the experimental results were very similar to the model ones. Similarly, the effects of ammonia on residual hydrogen concentrations were computed. At short RT hydrogen concentration was increased by inhibition of hydrogen utilisation, but the differences between inhibited and uninhibited digestions decreased as RT increased, as might be expected. Experimental evidence to directly confirm the model was lacking, but the appearance of propionic acid in digesters is associated with build-up of hydrogen because of failure of hydrogen conversion to methane and so the model could be compared with data on the appearence of propionic acid in inhibited digesters and the results seemed to support the model. Experimentally there is evidence that the degradation of pig-waste solids (Summers & Bousfield, 1980; Fischer *et al.*, 1984), and probably other wastes, is inhibited by high concentrations of solids in the feed. An inhibition of bacterial growth reflected in decrease in μ_m was postulated and an equation linking μ_m and VS in the feed was obtained from consideration of results of pig-waste digestions at different feed solids concentrations. Applied to the model predicting residual substrate concentrations the latter were increased by high feed solids concentrations. However, a somewhat similar result could also be obtained if the distribution of easily and less-easily degraded solids in the feed were altered; a situation that might have arisen in practice as the high-solids feeds were made up by adding screened-off solids to pig slurry. The effects of temperature on digestions were modelled by modifying μ_m in accordance with a formula suggested by Rathowski *et al.* (1982) from experiments on bacterial cultures growing at different temperatures. This equation appeared to apply to bacterial growth better than the 'Arrhenius Law' describing the effects of temperature on rates of chemical reactions and often used for bacterial reactions. The model predicted quite well the experimental values for solids degraded at particular RT and temperatures in pig- and cattle-waste digestions, but available data did not allow confirmation of predicted values for acid conversion to methane in digesters at different temperatures. Although the rate of digestion is increased by increase in temperature the proportion of the solids that can be degraded is not changed. So differences in solids degradation with temperature are most evident at shorter RT; at long RT all degradations approach the asymptotic maximum.

This model illustrates a number of aspects of modelling of digestions. The principles of the model would apply to any stirred-tank digester. On the other hand, results could be obtained only for systems for which there was a certain amount of data. From the data, say for a pig-waste digestion, values for the constants needed by the model could be obtained which would make the model fit the degradation of that particular waste. In Britain, the most common pig feed is barley and the excreta of the pigs used to determine the model constants came from pigs fed on barley. The model could, then, possibly be used to predict the residual solids in other pig-waste digesters in Britain. However, in the USA corn is more likely to be the basis of pig feeds than is barley and Stanogias & Pearce (1987) showed that the anaerobic digestibilities of excreta from pigs fed on different diets could differ, even though the chemical compositions of the excreta were similar. In the paper (Hobson, 1983) the model was shown to apply to a cattle-waste digestion, but the faeces from pigs and cattle differ in the feed residues and so their rate and extent of digestion varies. Thus, the constants for applying the model to a particular cattle waste differed from those for pig waste and might differ from those for other cattle wastes. This might be expected, as the Monod-Herbert model on which the present model was based describes the growth of bacteria on different limiting substrates but μ_m and K_s differ with different carbohydrate, or other, substrates. This type of limitation can be overcome to some extent by lumping together data from a large number of, say, pig-waste digestions and determining constants which will allow the model to fit, within certain limits, all these data (data too far away from the mean, perhaps because the feeds were very unusual, can be rejected). The model can then be used to predict, as accurately as is required for many purposes, the digestion of solids in the wastes from pigs fed on any diet. Hill (1983) used this approach. The Hobson model used two degradable components (undegradable material produced a constant addition to the residual solids) in the solids, the lumped-constant model uses only one component and so would not show the same detail in residual solids values.

In the Hobson (1983) model the concentration of degradable solids in the feed was defined, but the solids were not, initially, defined by analysis of components, only 'solids' were regarded as being degraded, as with the Hill and other models. For the model to predict biogas formation some generalised equation quantifying the overall breakdown of solids components to acetic acid and hydrogen ignoring intermediate reactions of conversion of other fermentation acids to acetic, or an equation directly linking solids degradation to biogas formation, is required. Although the two solid components were regarded as differing in rate of degradation they were not necessarily very different in chemical composition of the degradable components; rates could be explained by physical factors such as particle size as well as slight differences in lignification of the fibre carbohydrates. If the degradable material of the two components is regarded as similar, or if only one degradable component is assumed, then a figure for biogas (known methane and carbon dioxide composition) production per unit weight of pig-waste solids can be found from experimental data (e.g. Hobson, 1985; Hill, 1983). Such figures ignore the intermediate formation of acetic acid and hydrogen. However, experimental values for gas production for, say, pig waste solids can be of two types. One value gives a value for gas per unit of solids which includes the gas produced from the free acids usually found in pig wastes. If a particular waste contains more than the usual acid, because it has been fermenting in storage tanks for instance, the amount of gas produced per unit weight of solids could apparently differ from the norm. A better value gives the gas production from unit weight of solids corrected for gas production from acids in the waste. If the former type of value is used in a model such as that under discussion, then only the computation of residual solids, plus concentration of feed solids, is required to obtain gas production. If the latter type of value is used, then, to determine gas production, in addition to feed solids the concentration of acid in the feed as well as the residual acid concentration is required. Since hydrogen is formed only in degradation of solids, it is covered in the overall equation for conversion of solids to gas and so computations of the use of hydrogen can be ignored, except in so far as it can be assumed that the degradation of solids will produce sufficient hydrogen to be converted to methane and so the computed residual hydrogen will represent the hydrogen actually in the digester.

The production of biogas from wastes has been further generalised by equating COD with gas; x volumes of biogas or methane being produced from y weight units of COD, the composition of the material making up the COD not being defined; the usual figure adopted is 0·35 m^3 of methane per kg COD metabolised (e.g. Bryers, 1985). A more exact representation can be obtained from mass balance equations involving the chemical constituents of the waste. It is known from experimental fermentations that the conversion of anhydroglucose units in cellulose and starch to acetic acid and hydrogen is represented by the equation:

$$C_6H_{10}O_5 + H_2O \longrightarrow C_6H_{12}O_6 - (+ 2H_2O) \longrightarrow 2CH_3COOH + 2CO_2 + 4H_2$$

Similar equations can be made for pentose units in hemicellulose and for lipids and proteins. The analyses of the feed and effluent of a pig-waste digestion showed that the proportions of cellulose, hemicellulose, etc., were similar in each. The degradable solids of the pig-waste model (Hobson, 1983) could then be assumed to be of this composition and the acetic acid and hydrogen production calculated for different RT or other parameter. Account could also be taken of the acids in the pig-waste feedstock. Thus, in the model representing the digestion of this pig-waste, mass balances could be made to which the calculations of residual solids, acid and hydrogen could be applied and so steady-state gas productions could be calculated by the model for different operating parameters. The effects on gas production of ammonia inhibition, etc., could also be modelled. To calculate gas productions from other pig wastes or from cattle wastes would need data on the analysis and degradation of the particular waste or the assumption that all animal wastes are the same and that the experimental data from the one pig waste are generally applicable; rather like the 'lumped' data mentioned above.

It will thus be seen that the same basic model may differ in the number of equations that may be used to describe one overall digestion and that the data required to validate the model and apply it to other systems can also vary considerably in amount.

It was noted above that the residual substrate concentration in the model based on Monod kinetics was not related to feed substrate concentration, and, indeed, that that it was not necessary to know the feed substrate concentration. While the Monod models represented the form of the curve for residual solid-substrate concentrations against RT, the residual concentrations, especially at high feed-solids concentrations, did not exactly match experimental results where the effluent solids seemed to depend on the feed solids concentrations, although the percentage degradations of the solids were not affected by feed concentrations. This is a basic fault in the model and will be returned to later. Since the solids degradation may not be accurately represented, the gas production may be slightly incorrect, but over the range of feed solids used in farm-animal waste digestions the error introduced by the fault in calculating solids degradation may be less than the variation in analyses and performance of full-scale digesters and so not noticable when applying the model to give parameters for maximum conversion of feedstock to gas and minimum residual pollutants, etc., in the effluent.

The model discussed (Hobson, 1983) did not consider the growth of bacteria in the system, in the sense of determining the amount of bacteria produced by the various reactions. The model did, however, use the growth rates of the bacteria degrading solids and other substrates to determine the RT for optimum digestion of the wastes and the minimum RT at which the digestion would collapse through washout of the different groups of bacteria. Thus, at a certain RT there would be no degradation of the more recalcitrant solids while some degradation of the other solids fraction would take place, and at a shorter RT solids degadation would not take place but some acid conversion to gas would still occur. (Growth rates were here equated with D which may not be true; see later). However, using equation (7), above, the steady-state concentration of bacteria concerned in the various process steps could be calculated, assuming that a yield factor (Y) were known for each group. Yield factors are known for growth of some anaerobes on carbohydrates or on acetic acid or hydrogen, and such values are generally assumed to be similar to those for bacteria growing in digesters. However, although the yield value for the methanogens may be related to the defined substrates acetic acid and hydrogen, and values may be taken from laboratory experiments, the primary substrates are, as mentioned before, often defined only as 'solids'. Values for yield coefficients then have to be taken from experimental determinations of microbial mass in digesters with similar solids, which may have large experimental errors; from computer iterations to give a value that fits the other data used in the model; or from assumptions about the composition of the solids.

Because the Monod equations did not exactly fit the results for growth of bacteria in pure cultures on dissolved substrates in all circumstances, various other equations for growth of bacteria were proposed in the 1950's and 60's. Powell (1967) discusses some of the published equations such as that of Moser (1958) and his own solutions to the problems involving questions of transport of substrates across the cell membranes. Koch (1982) and Condray (1982), for instance, discuss other models describing the relations between growth of bacteria and substrate concentration. A major problem in discussions of the behaviour of bacteria in batch or continuous cultures and whether the data fit one model or another has always been the difficulties of measuring accurately cell mass and composition and the very low concentrations of residual substrates and of determining all products. There is also the problem of ensuring that all parameters that might affect the measured parameter have been controlled during the experiment (build-up of fermentation products in a batch culture might, obviously, affect rate of bacterial growth). In fact, in large-scale cultures such as anaerobic digesters the experimental errors in determinations of substrates, cells, and products, plus variations in feed composition and the mechanical variations in nominally-constant feed rate, temperature and mixing may make one model fit as well as another in spite of differences in their conceptual make-up.

Contois (1959) decided, from his experiments on growth of *Aerobacter* in batch cultures on various limiting nutrients, that the growth rate was affected not only by the concentration of nutrient but also by the density of the bacterial population. The Contois modification of the Monod equation for specific growth rate of a pure culture on a limiting substrate was:

$$R = u_m S/(BP + S) \qquad (8)$$

where R is the specific growth rate (equal to D in continuous-culture) S the substrate concentration, P the population density, and u_m and B are growth parameters that are constant under defined conditions. u_m is not the maximum growth rate of the Monod equations but is a function of the maximum growth rate. Although it has been suggested at various times that the cells in a bacterial population can affect each other's growth, Powell (1967) said that he did not believe that growth rate was affected by cell concentration except insofar as a high density of cells may bring about changes in the medium such as oxygen deficit or pH change which may affect growth, and he did not consider population density in the discussions previously referred to. Such changes might be undetected because the parameter was not being measured in a culture or might be too small to detect. Contois (1959) did say that his results and the equation might be an effect of concentration of end-products of bacterial metabolism. Contois' paper may not be entirely convincing, but the measurable bacterial concentrations in digesters are quite small compared with those usually found in laboratory cultures and they are made up of different types of bacteria which may not interact in the same way as cells in a pure culture of comparable total cell numbers (supposing such an interaction to exist). In addition, in many cases the bacteria are growing attached, or in close proximity, to solid surfaces which are the primary substrates. Such bacteria may not be affected in the same way by neighbouring bacteria as may be those suspended in liquid culture. In the digester culture, apart from the fact that the neighbouring bacteria may not be of the same type and so not have the same fermentation products, in many cases the products of one are being used by another bacterium and so do not build up to concentrations inhibitory to the first bacterium. Products may also be 'washed away' from bacteria on surfaces. Even if the Contois equation is true for the conditions under which it was evolved, it may not be applicable to mixed cultures such as digesters and to systems where fermentation products do not accumulate to inhibit growth.

Chen & Hashimoto (1980), realising the problem of the independence of residual and initial substrate concentrations in the Monod equations used the Contois equation as the basis for a different steady-state model of operation of a stirred-tank digester. However, in their derivation they seem to make two incorrect asumptions about the Contois equation. They represent the Contois equation as:

$$\mu = \mu_m S / (BM + S) \tag{9}$$

where μ is the bacterial specific growth rate and μ_m is 'the maximum specific growth rate of the microorganism'. As noted previously for equation (8), u_m in the Contois paper (written as μ_m by Chen & Hashimoto) is not equivalent to the maximum specific growth rate (μ_m) of the Monod equation (equ. (2) above) but is a function of the maximum growth rate (R_m in the Contois paper) where $(1 + b)/R_m = 1/u_m$ and b is a factor relating K_s of equation (2) (a substrate concentration) to the substrate concentration in the culture feedstock ($K_s = bS_0$). 'B' is said by Chen & Hashimoto to be 'a kinetic parameter', but B in the Contois paper is related to c, the yield factor (equivalent to Y in previous equations), and b; where $B = b/[c(1 + b)]$. The Chen & Hashimoto model is a 'steady-state' model in the same way as the Monod-based models previously described and, like these previous models, it assumes that the degradation of solids is described by a model based on dissolved substrates. It also assumes a constant growth-yield factor, Y. Chen & Hashimoto then define equations (4), (5) and (7) used in the derivation of the Monod-Herbert model of continuous-cultures above, but with slightly different notation, and then having shown that $\mu = 1/\Theta$ and M (equal to x in previous equations) = $Y(S_0 - S)$ they rewrite the Contois equation in these terms, giving:

$$\mu/\mu_m = (S/S_0)/[K + (1 - K)S/S_0] \tag{10}$$

where 'K is a dimensionless kinetic parameter equal to YB'.

Since Y is c in the Contois derivation, YB = cB. Y is a constant and B was shown above to be b/[c(1 + b)], so YB is cb/[c(1 + b)], or b/(1 + b).

b, as said before, is the factor relating K_s to the feed substrate concentration, $Ks = bS_0$ and b is a constant for different values of concentration of a particular substrate. Since b is constant then K is constant and Chen & Hashimoto (1980) said in verifying their model that K was 'independent of the influent substrate concentration', but varied with temperature. This latter might be expected as the rate of growth at a particular substrate concentration increases with temperature and so K_s and b might be expected to vary.

However, Hashimoto and Chen later said that K varied with substrate concentration and that K could be used as an index of digester performance and failure. Hashimoto (1983) used the data of Summers & Bousfield (1980) to calculate values of K at different pig-waste Volatile Solids concentrations in a digester feed. A value for μ_m was calculated from an equation previously found from consideration of data from various digestions as:

$$\mu_m = 0{\cdot}013T - 0{\cdot}129 \tag{11}$$

where T is the digestion temperature in °C. This μ_m is the reciprocal of the RT at which digestion fails and so could be related to solids degradation or to methanogenic reactions as discussed before.

The graph of the calculations shows that K is relatively constant up to an influent solids concentration of about 45 kg/m³ and then rises sharply for the last two concentrations. Hashimoto fitted an exponential curve through the points, where

$$K = 0{\cdot}50 + 0{\cdot}0043 \exp(0{\cdot}091S_0) \tag{12}$$

Chen (1983) later plotted the same data and data from a number of other digestions of pig waste in published papers and came up with a similar graph. He did say, however, that 'K remains relatively constant at a value of 0·6 for low influent VS concentrations', but 'above 55kg/m³ k increases sharply'. He fitted an exponential equation to the data, of the form:

$$K = 0{\cdot}6 + 0{\cdot}0006 \exp(0{\cdot}1185S_0) \tag{13}$$

Hashimoto (1984) plotted the same data plus some from his own experiments and found that a line of best fit through all the data was;

$$K = 0{\cdot}6 + 0{\cdot}0206 \exp (0{\cdot}051S_0) \tag{14}$$

while the data of Summers & Bousfield fitted equation (12) above.

Fischer *et al.* (1984) plotted data from their experiments and the graph shows that K was relatively constant at about 1·5 up to feed VS concentration of about 100 kg/m^3, when it suddenly shot up to about 7·8 at about 108 kg VS/m^3. The authors fitted an exponential curve to the points, although the fit was not good. The equation for the curve was:

$$K = 0{\cdot}5 + 0{\cdot}03 \exp (0{\cdot}05S_0) \tag{15}$$

Experiments thus seem to show that K varies with substrate concentration, although theoretically it should be constant. For the various experimental reasons discussed before, there is always some scatter in data from digesters and, although all the workers were using pig waste the data from the experiments may refer to slightly different wastes (the efffects of pig feed, storage of wastes, etc., have been mentioned before). Considering the sets of points on the graphs separately one might conclude that K is slightly different for different sets , which would be expected if the substrates differ to some extent, and that K is, as Chen pointed out, constant (within experimental error) over a range of VS concentrations. The increase in K at high substrate concentrations may then be, as Hashimoto and Chen do indicate, due to an increase in some inhibitory substance in the feed; perhaps copper in the Summers & Bousfield wastes or ammonia or VFA in these and other feedstocks. In effect, the dependence of K on substrate concentration may be an artifact because the digestions being compared are not all exactly the same and so parameters for degradation of the waste components will not be the same in each case. For instance, if the limiting substrate changes in some way, or inhibitors are present, then μ_m will not be constant.

Some time has been spent on the discussion not to denigrate the model, but to show some of the problems of modelling a microbial system with mixed bacteria, a mixed and poorly defined substrate or, rather, substrates, and a mechanical system that is subject to problems in keeping constant feed rates, constant feed composition, and so on. The variations in data are such that curves may be mathematically fitted to sets of data that microbiologically differ; factors may be put into equations which have no basis, or no known basis, biologically, and a factor may lump together the effects of a number of biological parameters.

The more accurately a model represents the biological events in a mixed fermentation the more data are needed to make the model work and the more limited is the application of the model to different systems. The simpler model may be effective for many applications; whatever K in the models above actually represents, it increases when some factor inhibits digestion and calculation of a high value of K for a proposed digestion could suggest that the digestion might not be efficient. Chen (1983) showed how the model might be used in design of the optimised digestion plants for pig wastes.

Lawrence (1971) had earlier used a steady-state model based on methane production from acids as the rate-limiting step in digestion for suggesting the optimum parameters for running anaerobic digesters and McCarty (1966) used his model, based on Monod kinetics in batch digestions, to suggest optimum retention times for digesters. This model considered hydrolysis, fermentation and methane production as possible rate-limiting steps. He also pointed out that his model, like others, supposed that the organic matter was the limiting substrate for the bacteria, but that growth might be limited by N or P or trace-metals, etc., and so the digestion might fail before the predicted failure RT of the model. Gaudy *et al.* (1967) modelled aerobic sewage treatment systems by Monod kinetics and compared the results with experiments on glucose degradation in a simulated activated-sludge system inoculated from a full-scale plant. They found that μ_m, K_s and Y varied with dilution rate and could not be considered as constants. They also found, as Hobson (1965) had also pointed out, that Y values found in batch experiments were lower than those from continuous-culture experiments. Obviously, this was a simpler system than anaerobic digestion, but it does show the problems involved in modelling in that although the Monod equations described the growth of bacteria in the system there were difficulties in obtaining suitable values for the various factors used in model equations.

Dynamic models

The models described so far have predicted the steady-state behaviour of digesters and have been used to predict the residual substrate levels (and so residual pollutants in the effluents), gas productions and optimum or minimum RT for degradation of different wastes. However, although these aspects are important, one of the main times when failure of digestion may occur is in the start-up of digesters and the attainment of steady-state working. Emperical experiments have suggested that start-up should be at low-loading rate and long RT and that loading-rate should be gradually increased and RT decreased until the desired running conditions are attained. Such a start-up may take some months with a sludge digester. The steady-state model cannot predict the transient behaviour of a digester during this time, neither can it predict the behaviour of a digester during the time after RT, loading rate, or other parameter is changed and before a new steady-state is attained. Models which could simulate such states would have a use in suggesting start-up conditions for digesters and predicting possible ill-effects of changes in loading rates, etc.

The equations used in steady-state modelling can be solved by ‘paper’ calculations, although computers may speed up the processes of curve-fitting and computation of suitable constants for the equations. Although similar kinetic considerations may be used in dynamic models, the equations used in such models may not be solvable mathematically but only by iterative processes that can be done most easily by computer. The development of dynamic models of microbial systems has, thus, followed somewhat the increasing power and availability of computers in laboratories. A further advantage of the computer in modelling is that it can compress the months mentioned above as required to start up a sludge digester into seconds of computing time, thus a proper model could take the place of months of experimentation in determining the optimum start-up conditions for a digestion.

Andrews pointed out these uses of modelling and of computers in modelling in papers (Andrews, 1969; Andrews & Graef, 1971) on development of a dynamic model of digestion in a stirred-tank digester. This was similar to steady-state models in that it considered only conversion of acids to gas as the controlling step in digestion. However, instead of the Monod equation (equation (2) above) for growth rate of the bacteria a similar equation proposed by Haldane (1930) for the inhibition of rates of enzyme action by high substrate concentrations was used. (This reduces to the Monod equation for low substrate concentrations or for substrates that are less inhibitory). The Andrews model, then, used unionised volatile acids as the growth-limiting substrate and the inhibitor of methane production. Similar equations to those used above for the balance of influents and effluents in a continuous-flow digester were used, but equations linking the concentrations of gaseous carbon dioxide, dissolved gas and bicarbonate, and so pH, were also incorporated in the model. Computations showed that the model simulated the failure of some digestions with overloading and similar changes in running conditions and that start-up failure could be prevented by slowly increasing loading rate, and so on. The model thus simulated digesters and could be used to suggest parameters that could be recorded in process control of digestion and changes in these which could give warning of failure, and in setting conditions for changes in loading rates or for starting up digestions; provided the necessary values for model constants were available. As Andrews pointed out, modelling is a dynamic process in itself in that a model is proposed and tested against practice and modified as new values for factors are found or new aspects of the process are recognised.

It might be pointed out here that while high substrate concentrations can inhibit enzymes, and so bacterial activity, this applies only to substrates in solution. A bacterium on the surface of a particulate substrate and digesting the surface surrounding the bacterium (see later) is not affected by the number of particles in the digester liquid. Thus, a high concentration of a particulate substrate *per se* would not cause any substrate inhibition of particle degradation. Since the rate of particle degradation is not affected by other particles, the rate of production of metabolites (such as VFA) and their metabolism by nearby, secondary, methanogenic or other bacteria should also not be affected. Only if the rate of primary colonisation of a suddenly increased number of particles in a feedstock is not equalled by the rate of colonisation by the secondary flora should acids or gaseous products of the primary flora begin to diffuse away and build up in the liquid and possibly cause inhibitions.

The Andrews model was limited to one microbial reaction and considered only microbial growth and not growth and death. Hill & Nordstedt (1980) extended the Andrews model in that they introduced another bacterial reaction and another inhibition factor and a factor for death of bacteria which decreased their growth rate. The digestion process was considered to be first a solubilisation of solid feed, which was taken to be at a rate proportional to the mass of acid-forming bacteria. The soluble products were then converted to acids and the acids were used for production of methane. Balance equations for substrates similar to those given above for continuous-cultures, of the form:

'Change in concentration equals input minus output minus amount used by bacteria'

were constructed for the primary substrate, the acids were produced from this substrate, and the acids used to produce methane. Balance equations for bacterial production were also constructed. The growth rates of fermentative and methanogenic bacteria were both described by the Haldane equation used by Andrews but with an extra term for inhibition by ammonia:

$$\mu = \mu_m/(1 + Ks/C + C/KI + N/KIN) \tag{16}$$

where μ = actual growth rate; μ_m = maximum growth rate; Ks = the concentration of substrate to give half maximum growth rate; C = concentration of unionised acid; N = concentration of unionised ammonia; KI = an inhibition constant for the unionised acids; KIN = an inhibition constant for unionised ammonia.

Ammonia concentration was determined by a balance equation of the form:

'Change in concentration equals input minus output plus ammonia produced from the waste plus ammonia transferred to or from the gaseous phase'.

A carbon dioxide balance included terms for CO_2 produced with methane from acids and from fermentation, and CO_2 and bicarbonate and pH equilibria were similar to Andrews' considerations. The values for the constants needed by the model were obtained from data on actual digesters and by computer iteration.

The model simulated, for example, the increase and then decrease in VFA as a digestion of pig waste was started up at a relatively low loading rate and as the digestion developed towards a steady-state working. It also simulated acidification and breakdown of digestion during start-up of a heavily-loaded digester.

This model, like others, simplified the microbial reactions. Instead of the complex population of different species and strains degrading different constituents of the solids to different endproducts, one 'group' of bacteria degrading an undifferentiated 'solid' to one kind of acid endproduct was assumed. Similarly, one 'group' of methanogenic bacteria was used. These 'organisms' had generalised growth rates, etc., based on analyses of working digesters and the behaviour of the model. This kind of simplification and fitting of constants is done with all models, as said before, but even so the model needed some 20 values of different constants to be determined before the model ccould run. As previously pointed out, the better a model describes the actual reactions of a complex microbial system, the more difficult it is to obtain the data needed by the model and the more specialised are the applications of the model. Hill (1983*a*) tried to overcome some of these problems by proposing a model using Monod kinetics with the Haldane inhibition function, but with 'lumped' parameters. Thus, the primary substrate was defined as the biodegradable VS in, say, 'pig waste'. This was a factor determined by computer iteration on all available data on the maximum degradation of solids in working pig-waste digesters. Similarly, a generalised factor for the acid contents of 'pig waste', 'dairy-cattle waste', etc., was obtained. The Monod half maximum growth-rate constants and yield coefficients for acid- and methane-forming bacteria were determined for the waste types (pig, beef cattle, dairy cattle, etc.) by computer iteration as before and values of μ_m were computed from a function derived by Hashimoto *et al.* (1979) and the same value was used for acid-forming and methanogenic bacteria. Death rates of bacteria were assumed to be 10% of the μ_m values for the different wastes. The values for the constants were refined by computer iteration fitting model steady-states to analyses in the literature of actual digesters running on the different feedstocks. The inputs needed by the final model were influent total Volatile Solids (VS) concentration, temperature, retention time and waste type. The model was dynamic and could predict transient-and steady-states. Minimum RT, VFA and VS concentrations in digester effluents and methane production rates could be predicted, as would be expected. However, Hill *et al.* (1983) later found that the model did not adequately describe death of bacteria. So, on the basis of experiments with 'Continuously-Expanding' (Fed-Batch, see Chapter 6) digesters they decided that death of bacteria was caused by the acids which inhibited the bacteria and that the death rate, with a maximum specific death rate, was related to the VFA concentration in the same way as the growth rate is related to substrate concentration; there was not a constant death rate as used in the previous models. Hill (1983*b,c,d*) used models similar to the ones described to predict optimum operating conditions for digestions of pig-, cattle- and poultry-wastes in terms of RT, temperature, and loading rate to give certain methane productions and VS reductions, etc.

The process of model making and comparison with actual digestions can lead to modifications to a model to more adequately describe some new data. The modifications to the model may be based on some 'biological' aspect; e.g, bringing in another group of bacteria or bacterial reactions which are known to play a part in the reactions but which have previously been left out for reasons of model simplification or because no data on the reactions of the bacteria were available. Other modifications may be based not on defined bacterial reactions but on mathematical modifications to an equation to make it fit better the known data although the actual mechanism of the microbial reaction or behaviour is not known. That models may be quite emperical in many ways is emphasised by the remarks of Hill (1983*b*). Comparing the steady-state output of his model of pig- and poultry-waste digestions with that of the Chen and Hashimoto model he remarks on the good agreement and says that although the models are 'based on entirely different mathematical representations' the agreement reflects the 'advancement made in recent years in modelling and understanding the anaerobic fermentation'. It could also be said to be a reflection of the experimental errors involved in analysing digestions which allow various curves to be fitted more or less to data and the limitations of the predictions made by the models compared with the data used to determine the various factors used in the models. Both models use derivatives of the Monod equation as the basis for their descriptions of bacterial growth.

Bryers (1985) proposed a model in which hydrolysis of particulate feedstock, fermentation of amino acids and sugars, degradation of long-chain fatty acids from lipids, conversion of propionic acid to acetic and hydrogen and conversion of these latter compounds to methane, were considered. All the components were reduced to COD equivalents. The hydrolysis rate of the particulate feed was assumed to be first-order on the remaining particle concentration and the reaction rates for the other reactions were assumed to be of the form:

$$r = uX/Y = umX/(Y(K_s + S)) \qquad (17)$$

where X is the bacterial concentration and Y, Ks and S are as before. Material balances were of the form 'input minus output plus production minus use', as before.

A further equation for calculating bicarbonate alkalinity was also included. The values for the constants used in the models were taken from the literature ('after appropriate temperature adjustments'), except for certain values such as the particle-hydrolysis rates which were calculated from the data of the experiments used. The model reproduced the fall in gas production and pH during startup of digesters which were allowed to become acid and so only solids were degraded (Eastman & Ferguson, 1981), and the acetic and propionic acid concentrations at steady-state operation of these digesters at different RT. The model overestimated the acid concentrations and did not describe the compositions of the acid mixtures correctly, probably due to the assumption that propionic was the only acid, other than acetic, formed. The model predicted quite reasonably the behaviour of batch- and continuous-thermophilic digestions of activated sludge (Hamer, Bryers & Berger, 1985). The author points out that while the model predicted acid concentrations and gas productions quite well, it did not predict the apparently random fluctuations in these parameters that were found in the actual systems. This is, as pointed out before, a feature of all models; they tend to give 'smoothed' curves of the gas and effluent productions of digesters. This might be expected, though, as all models are simplified descriptions of systems which are very complex in microbial populations and microbial substrates and do not consider mechanical fluctuations in digester running.

Some other models

Other models have been proposed, although most are basically similar to the ones mentioned here, and, for instance, Rozzi & Passino (1985) and Hanaki,

Noike & Matsumoto (1985) have produced reviews on models of anaerobic processes.

Billington (1988) reviewed various models of stirred-tank digesters based on Monod kinetics, modified Monod kinetics allowing for end-product inhibition, Contois kinetics and first-order kinetics. He concluded that none was adequate for dynamic modelling and he also pointed out that the models really described degradation of dissolved substrates whereas farm wastes contained large amounts of particulate substrates. Some of the models considered in the present chapter were discussed by Billington and he concluded that more work was needed to produce a completely acceptable model.

Roques *et al.* (1982) had previously concluded that Monod kinetics were inadequate and they proposed a model derived from Contois kinetics, which included end-product inhibitions, but, again, they considered substrate only as BOD or COD and not as defined entities.

Cecchi *et al.* (1990, 1991) modelled the digestion of sorted Municipal Solid Waste (MSW) by a first-order reaction for the whole waste and a 'Step-Diffusional' model which considered the breakdown of the acids, the easily-degradable and the more recalcitrant organic matter separately (there was also undegradable material). The stirred-tank experimental digester was fed twice a day and gas production was not constant, gas being produced first from easily-degraded and then from more recalcitrant components over the period between feeds. With suitable factors this model reproduced the gas productions when the digester was fed with various MSW fractions and sewage sludge better than the first-order model.

The limitation to modelling of stirred-tank anaerobic digesters seems now to be not in computational power but in lack of detailed data of the rates of the processes involved and of the contributions of different bacterial species and groups to the overall reactions, combined with lack of detailed analyses of the feedstocks of the digesters. Mathematical modelling of the rumen, a similar microbial system, has proceeded to much greater detail than has modelling of digesters as the knowledge of the intermediate reactions and of the microorganisms involved, as well of the composition of ruminant feeds, has become much greater than knowledge of digesters. Models of rumen function have recently been reviewed (Hobson & Jouany, 1988), but two models will be mentioned here. Baldwin, Thornley & Beever (1987) extended a previous model of France, Thornley & Beever (1982) and incorporated ideas from previous models of their own and others to model digestion in the rumen of a lactating cow. The kinetics were similar to those used in digester models, but the number of reactions and the mass balances could be much more detailed than in the digester models. The degradations of starch, cellulose, hemicellulose, pectins, lipids, proteins and other dietary components (in soluble form and small- and large- particle forms; although the 'particles' were not considered as actual particles [see later]) and fermentations of the degradation products, were considered. Formation of different fermentation products, with ATP and microbial cell yields, inter-conversions of products, absorption of products across the rumen wall and passage of residues to the lower gut were amongst the many reactions of the models. The models were run by computer integration of the various equations over short-time step lengths and were 'dynamic' in that they predicted the changes in feed constituents and digestion products in the rumen over time after a change in the 'ruminant's' diet to certain mixtures of forages and concentrates, and the steady-state rumen contents which were attained after some days of feeding the diets. Feeding was either continuous or intermittent. Data from published feeding experiments on some animals and from laboratory microbiological experiments were used to obtain values for the various constants used in the models and the model predictions were then compared with feeding experiments on other animals. However, even with the data available on the rumen functions France *et al.* (1982) concluded that although the model showed 'a measure' of agreement with experimental results there were some large discrepancies, particularly with intermittent feeding, and that (as mentioned before) much of the experimental data available did not contain enough detail to suggest how the model might be modified. The model of Baldwin *et al.* (1987) behaved reasonably on intermittent feeding, but discrepancies between model and experiments were greater on concentrate diets than on forage diets. The authors suggested that while the constants used were derived from a large number of experiments and might be expected to apply to a range of diets they evidently did not apply to high-concentrate diets. In deriving parameters for rumen models both particular and lumped parameter (Hill, previously mentioned) methods have been used. In all modelling of microbial systems there is the problem that to reproduce one system perfectly the model needs to be supplied with a large amount of data pertaining to that particular system. The agreement of the model with the experimental results then shows that the concepts of the microbial reactions taking place in the system are correct, but the model can predict the results of only very little deviation from the operating parameters of the initial system. If more generalised values are used for model parameters then it will not produce such a good representation of a particular system but it may produce reasonable, but rather generalised, representations of a number of similar systems. However, given the initial data it should be rather easier to make a model of digesters than of rumens, as the latter have 'animal factors' as well as complex microbial systems and feedstocks to contend with.

Retained-biomass digesters

Most modelling has been done on stirred-tank digesters as these were the original type and are still most common. However, some modelling of other types has been attempted.

The modelling of retained-biomass digesters is based on considerations similar to those used in modelling stirred-tank digesters in that the system may be considered to have one rate -limiting step, e.g. methanogenesis, or a series of bacterial reactions (fermentation, methanogenesis, etc.) may be modelled. The application of 'standard' (i.e., Monod, etc.) equations for bacterial growth to retained-biomass digesters where the bacteria are growing in suspension in the liquid phase is more exact than is their application to stirred-tank digesters as the substrates for retained-biomass digesters are almost entirely in solution; the same situation as in the pure-culture systems from which the equations were first derived. As a first approximation the bacteria are regarded as having free access to substrates, the complications of floc formation being ignored. The exact modelling of digesters in which the biomass is retained on some support is more difficult. Although the bacteria are in a surface film exposed to substrate in solution and the kinetics of uptake of substrate should be the same in both cases, in the case of the filter digester there will be a decrease in substrate concentration in the direction of the substrate flow in the filter, while in fluidised-beds mixing will suspend the bacteria and carrier in a more uniform concentration of primary substrate. In both cases only the biomass in the outer layers of the surface films may be actively taking up and metabolising the substrates. Similarly, only the cells in the outer layers of bacterial flocs may be active, and in both cases there will be a decrease in available substrate with distance from the surface of the biofilm or floc. Some of the equations derived in the modelling of the digesters considered in the next sections are rather complicated, so, for reasons of brevity only the basic premisses on which the models are based will be presented.

Lawrence & McCarty (1970) considered the stirred-tank with solids recycle (thc Contact digester) along with their previously-mentioned steady-state modelling of a continuous-flow stirred-tank. It was assumed that all the biological reactions occurred in the digester tank (the separator is small and recycle continuous). The kinetics were the same as for the stirred-tank, but the solids (bacterial) retention time was greater than the liquid retention time by a factor governed by the solids recycle rate. The recycled solids were assumed to be bacteria. Pirt & Kurowski (1970) also modelled fermentors with bacterial recycle and this model could be applied to digesters, although it was meant for laboratory and industrial fermentations of a defined substrate by one microorganism. The experiments on which the model was based were conducted with a fermentor with a filter outlet as well as a direct outlet so that various proportions of the biomass could be retained at different liquid RT. This is slightly different in practice from the usual Contact digester, although the principle is the same: digesters with some form of 'filter' to retain some effluent solids have been advocated. (However, there is a distinction between retention or recycle of bacteria to increase the bacterial concentration and RT in the digester while liquid [and so substrate] RT remains low, and retaining or returning substrate solids with attached bacteria so as to increase the RT for which the solids are available for degradation). The Pirt & Kurowski models also considered a 'sedimentation zone' fermentor, which is the same as thc sludge-blanket digester. In all the cases the modelling is similar and the authors used Monod kinetics and balance equations similar to those described earlier for the Monod-Herbert continuous-culture models. Some substrate, as well as bacteria, was assumed to be fed-back, and steady-state fermentor concentrations of biomass and substrate (dissolved) were determined as before. The models could be used to described the Contact digester as the bacteria therein are generally regarded as non-flocculating, but they would be a simplification of the sludge-blanket digester, as pointed out above.

Gates *et al.* (1967) also used Monod kinetics (with μ reduced by a death rate of bacteria) as the basis for a steady-state model of a contact digester and showed theoretical curves very similar to those of Pirt & Kurowski (1970) for residual growth-limiting substrate and biomass in the digester tank with different proportions of the effluent fed back. The model was similar to that of Pirt & Kurowski in that it considered there was only one microbial species and one reaction in the digester. The model was tested against digestion of a synthetic waste of skimmed milk plus P and K in a stirred, bottom-fed, tank with baffles at the top for return of solids from the outflowing waste stream. The digester seemed to be a possible form of contact digester and data from it fitted quite reasonably to the model. COD or BOD and VSS were used for substrate and microorganism concentrations. Values for K_s, μ_m, Y and a death rate were determined from this data and also from published data referring to contact-digester treatment of a meat-processing waste. The authors did remark at the end of their paper that 'The values of the model's constants as determined from overall process evaluation should be used only in the context of process kinetics. They imply nothing about the specific biological components of the process.'

Pirt (1973) also derived a model for the metabolism of bacteria on a column packed with inert material. This was related mainly to aerobic trickling filters, but anaerobic metabolism on the column was also considered and the model could describe an anaerobic filter. One substrate metabolised by one bacterial species or one group of similar bacteria was used, but, as before, the treatment could be extended to a number of substrates and bacteria. The column was supposed to consist of a number of theoretical 'compartments' each containing a film of bacteria over the cross-sectional area of the column which was covered with a layer of liquid medium and a gas layer. The liquid percolating down the column was supposed to spend an equal time in each compartment and substrates in the medium penetrated the bacterial film only by diffusion. The thicknes of the bacterial film actively metabolising substrate was calculated assuming that substrate consumption rate followed zero-order kinetics and so a constant rate could be used for substrate metabolism at any depth in the film. From Monod's (1942) work on the relation of substrate concentration to bacterial growth this assumption was said to be true for carbohydrate substrates at concentrations above about 20 mg/litre. The equation for the film thickness (y) to reduce the substrate concentration to zero at the support surface was determined as:

$$y = (2DS/qp)^{1/2}$$

where D is the diffusion coefficient of the substrate in the biofilm; S is the concentration of the substrate in the feedstock; q is the specific rate of substrate consumption by the bacteria; p is the concentration of bacteria in the film (grams dry weight per cm^3).

If a volume V/n of medium passes through the column then the amount of substrate consumed on passage through the first theoretical bacterial film (a) will be given by:

$$a = qxt/n$$

where V is the volume of medium filling the column; n is the number of theoretical films; x is the weight of active bacteria in the film; t is the retention time of the medium in the column.

The substrate left is (S – a/(V/n)) and this passes through the second film where the substrate removal is as in the first film and the substrate left is (S – (a_1 + a_2)/(V/n)) and so on for n films. The substrate is removed stepwise as the feedstock percolates down the column (theoretical flow with no mixing is, of course, assumed). From this consideration an equation for the number of theoretical films needed to bring the feed substrate concentration down to a particular value was obtained. Equations for the film area and the total active and inactive biomass at steady-state were also derived.

This theoretical treatment of a filter column is analagous in some ways to the modelling of a tubular (plug-flow) fermentor as a large number of linked stirred-tank fermentors in which the effluent from one provides the feed to the second, and so on. Substrate is removed stepwise as the feedstock flows along the tube, but, in contrast to the filter, bacteria flow with the residual substrate from tank to tank.

Canovas-Diaz & Howell (1988) produced a steady-state model of an anaerobic filter with a two-part biofilm. The outer film produced acids by sugar metabolism while the inner film produced methane from the acids. Hydrogen metabolism was not considered. Zero and first-order kinetics were used and all substrates were measured in COD units not different sugars or acids. Denae *et al.* (1988) considered two models of an anaerobic filter using Monod kinetics. Unlike the previous authors they assumed a homogeneous distribution of all the organisms in the biofilm. One model incorporated biofilm diffusion while the other was a homogeneous system with no diffusion. They found that diffusion effects were negligible and that the homogeneous model was satisfactory in describing the digestion of molasses wastewater.

Bolte *et al.* (1984) developed a model of an upflow anaerobic filter partly based on the model of Rittman & McCarty (1980) of steady-state biolfilm kinetics. The model of Bolte *et al.* had a bulk liquid flow through the filter and in this liquid there was no bacterial action. Between the bulk liquid and the biolfilm there was a zone, also with no bacteria, from which substrates diffused into the biofilm. The biofilm was assumed to be homogeneous with regard to bacterial concentrations and of uniform thickness. The substrates diffused into the biofilm and were metabolised. Complex organic molecules were hydrolysed to simple molecules by the fermentative bacteria and these molecules were fermented to acetic acid and then to methane. The complex organic pool was added to by death of bacteria, which was assumed to be proportional to the concentration of the bacteria. The growth rates of the bacteria were modelled by Monod kinetics with an acid inhibition term and the rates of reactions were affected by digester temperature. The filter was divided into compartments vertically and horizontally. Substrate concentrations at the outside of the biofilm were those of the bulk liquid, while substrate concentrations at the filter matrix surface were zero. In all, 15 equations were constructed which were not solvable analytically, so a computer program was made to numerically integrate the equations over simulated space and time intervals. Constants for the model, other than filter dimensions, were taken from papers by various authors and the model was tested against the practical results of Young & McCarty (1969) for filters treating synthetic wastes containing protein plus carbohydrate or a mixture of acetic and propionic acids. The model predicted quite well the trends in increased COD removal with increased RT at different loading rates and the removal of COD as the liquid passed up the column.

Models like the last one are complex, and so the number of terms for which values have to be obtained before the model can be run is large. So Bolte & Hill (1985, 1990) produced simpler steady-state models. The later one assumed that there was one degradable substrate and one methanogenic culture in the filter (degradable VS is converted straight to methane) and Monod kinetics were taken for bacterial growth. The bacteria were assumed to be homogeneously distributed in a completely-mixed reactor, but the bacteria were attached and so the bacterial concentration was independent of feed flow and based only on the feed substrate concentration and the maximum bacterial concentration supported by the filter matrix. The model gave reasonable predictions of methane production for various filter systems described in the literature.

Yang & Guo (1990) studied the formation of methane from the lactose of whey and the acids from the primary fermentation in batch cultures. From this they produced a three-step reaction mechanism for modelling whey digestion in a filter type of reactor.

A further digester system using attached growth is the Anaerobic Rotating Biological Contactor. This is a totally-enclosed version of the earlier aerobic apparatus in which the bacteria are attached to discs which rotate in the feedstock liquid. It can be used for low-solids wastewaters (see Chapter 7). Models of the system have generally been based on ealier models of aerobic rotating-disc systems, such as the steady-state models of Hansford *et al.* (1978), Paolini *et al.* (1979) and the simplified model of Andreadakis & Cailas (1984). This model was applied to an anaerobic rotating-disc digestion of domestic-sewage waters by Deshpande *et al.* (1991). Experiments provided the data to construct and test the model and the authors concluded that the model, with a simple Monod-type rate expression, could successfully describe the steady-state running of their digester.

Trulear *et al.* (1982) discussed the growth of biofilms in general (i.e. contaminating microorganisms on pipes and heat-exchangers, etc., as well as bacteria in biological reactors) and derived equations for substrate removal and biofilm accumulation and detachment ('sloughing') and tested these against experiments with a reactor consisting of a stationary outer cylinder and a close-fitting, rotating, inner cylinder and supplied with a glucose medium inoculated from a domestic (presumably aerobic) wastewater treatment plant. The thickness of the microbial film and shear stresses could be determined, as well as the effects of glucose concentration while the medium was completely mixed around the films. Amongst other results which could be related to biofilm development in anaerobic filters they found that glucose removal was proportional to active biofilm thickness and that the depth of the active part of the biofilm increased with increased glucose concentration. Biofilm density increased with glucose loading rate, and low-density film had a filamentous structure while high-density film was more compact. Biofilm accumulation was also affected by medium velocity.

Nilsson & Karlsson (1989) used a specially-designed apparatus to measure the diffusion rates of substrates into a matrix of methane-producing bacteria and found that the diffusion rate in the matrix was only about 22% of that in water. Presumably this was due to the polysaccharides and other bacterial polymers between the cells.

Two-phase and two-stage digesters

Two-phase digesters will be described later, but, essentially, they consist of two linked digesters. The running of the digesters is arranged so that the initial hydrolysis and fermentative steps in digestion can take place in the first digester and the conversion of the products of these reactions (principally volatile fatty acids) to methane can take place in the second tank. In this way the growth conditions for the two types of bacterial populations may be more optimised than they can be in a system in which all the reactions take place in one tank. However, it is not easy to separate the two reactions by the running conditions of the digesters. Both populations grow in the same temperature range. Although separation of the reactions was done on a laboratory scale by membrane fermentors, the only really practical means of separating the reactions to some extent has been by retention time. This has limited the most effective two-phase systems to digestion of sugar-containing wastewaters. In this case, fermentation of sugars can take place in a few hours at a pH between 5 and 6. The retention time of such a fermentor is too short for the growth of acid-using methanogens and so they are excluded from the tank. The pH is also too low for active methanogenesis. Methanogenesis from the fermentation products can then take place in a tank digester of RT 3 days or more or in some form of retained-biomass digester at a lower HRT. Details of such two-phase digesters will be described later, but for the moment it is sufficient to say that the systems can be modelled by linking two models of the individual digesters constructed as previously described with the effluent from the first tank being the feedstock for the second (e.g. Ghosh, 1978; Zoetemeyer, 1982; Cohen, 1981, applied Monod-type kinetics to one or both of the reactions in stirred-tanks). If fibre hydrolysis is to take place the pH optima for hydrolysis and methanogenesis are similar and long retention times are required for fibre hydrolysis. Such RT allow the growth of methanogenic bacteria, so that methanogenesis takes place in the first tank and little of the fermentation products may be left for methanogenesis in the second tank. A batch process for solid feedstocks in which liquid from a hydrolysis tank is circulated through a retained-biomass methanising digester and then returned to the hydrolysis tank is described later. Mata-Alvarez (1987) modelled such a system with batch loading of solids and the leachate recirculated through a methaniser. He used first-order kinetics (see later) for the hydrolysis and Monod kinetics for methane production from acids in the methaniser and in the hydrolyser.

A different concept is the 'multi-stage' digester. Two-stage stirred-tank fermentors have been suggested and used where a product is made by non-growing cells. In this case growth of the cells could take place in the first fermentor where, say, nitrogen could be limiting growth in a liquid medium. The cells could then pass on to a second fermentor where excess sugar in the medium could be used to form some product by cells which would not grow because of lack of nitrogen. The size of the second fermentor could be larger than that of the first so that the retention time of the cells could be long for product formation but short for rapid growth in the first fermentor. On the other hand the same reaction can take place in both fermentors and this was the principle on which two-stage digesters were run by one of the present authors. In the ideal two-stage fermentor one type of bacterium is growing on one limiting substrate and carrying out the same reactions in both tanks. The construction of the digesters will be discussed in more detail later but as in other cases in applying the principle to digester modelling a consortium of bacteria is considered as a pure culture.

Following Herbert (1964) Hobson, Bousfield & Summers (1980) and Hobson, Summers & Harries (1984) modelled the degradation of a dissolved substrate in two-stage stirred-tank digesters where the tanks were either equal or dissimilar in size. The model was a steady-state one based on Monod kinetics and the continuous culture model developed by Herbert *et al.* (1954). The equation for residual substrate in the effluent from the first tank is the same as that for a single continuous-culture (Equation 6), but the equation for the residual substrate from the second tank is more complicated and involves the roots of a quadratic equation. It is found that the growth rate of the bacterium is not the same as the dilution rate in tank two and that as long as bacteria are growing in the first tank no washout of the bacteria in the second tank can take place whatever the second dilution rate. The residual substrate in the first tank is independent of the feed substrate concentration but in the second tank depends on the feed substrate, and increasing the feed substrate concentration tends to lower the residual substrate concentration. With a polluted wastewater as feed it should be possible to reduce residual pollutants to the low level obtained by a single tank running at very long RT by running a two-stage system at a lower RT. Various systems were modelled and some were tested. One system could be the normal one used in sewage and farm digesters where an effluent from a heated digester flows to a lagoon or tank where it is stored before disposal. The lagoon acts as a second digester running at lower temperature and longer RT than the first stage. A farm-waste digester and storage lagoon were shown to behave, for dissolved acids and BOD, like the model so far as the accuracy of measurements on the lagooned effluent allowed. More detailed tests were carried out with two heated digester tanks with the second stage of volume such that the RT was half that of the first stage. Tests were made with a settled pig slurry of TS about 1·5% but with up to 8000 mg VFA per litre. In two experiments with steady-state RT of around 15 days in stage one, the residual VFA in stage one effluent were calculated by the model for acid degradation previously mentioned (Hobson & McDonald, 1980) as 228 and 153 mg/l with values of 336 and 144 determined. The second-stage acids were calculated as 15 and 17 mg/l and actual values were below the level of determination. The effects of high ammonia concentrations in one batch of slurry were reasonably modelled by decreasing the bacterial growth rates as in the earlier single-stage model (Hobson, 1983). With this settled slurry the model conditions of one substrate fed into the first stage were obtained as there was little, if any, substrate for generation of acids in the feed to the digesters. Further tests were done with a 4% TS slurry, but the same RT of 15 and 16 days for the first tank were chosen because over 90% of the degradable solids would be degraded in the first tank. In the first tank the acids substrate would be acids in the feed plus acids generated by degradation of the solids. In the Monod model the residual substrate is independent of the feed for the first tank, but the total of feed and generated acid could be calculated for modelling the second stage and virtually no further acid would be added to the effluent from the first stage by solids degradation in the second stage. Again model and actual values for acid concentrations agreed within the limits of determination. Later experiments (unpublished) showed that the fibre degradation in two stages when the first stage was not of long enough RT for complete fibre degradation did not behave as the model. This is most likely because a fibre is colonised in the first stage and degradation starts. The fibre, with colonising bacteria, is washed from the first to the second stage but this makes no material difference to the bacteria, which are growing independently of the dilution rates in the system (see later). Thus, the fibre degradation to be expected in two stages is that expected for a single stage of RT that of stages one and two added. The acid degradation in a two-stage digester where digestion to acids of some primary substrate (preferably dissolved) is proceeding in the second stage can be modelled by assuming that it is a two-stage fermentor with substrate (acids) being fed into both stages (Herbert, 1964).

Tubular digesters

The ideal tubular fermentor or digester is a 'plug-flow' system where a volume of medium added to one end of the tube (Fig. 2b) is inoculated and passes along the tube without mixing with medium before and behind it while substrates are fermented. In practice this is never attained: friction on tube walls can induce turbulence, gaseous fermentations mix the liquids, suspended solids may settle or float, fermentor heating may cause thermal movements, and so on. Probably most practical 'tubular' digesters can be more reasonably regarded as elongated stirred tanks with some solids settlement so that solids RT is not equal to liquid RT.

A tubular digester acting as second stage in a two-stage digester system would fulfil the conditions for an inoculated feedstock, as would a tubular digester with feedback, from the effluent to the feed, of separated suspended solids. Of course, the previous comments about mixing in the fermentor would apply to both digesters.

The theoretical tubular fermentor can be considered as a batch culture in space as well as in time, the bacteria growing and substrate being utilised as the medium passes along the tube. Thus, for any feed input the increase in bacterial numbers (B) by growth over the length of the tube will be:

$$(x_1 \exp(\mu t) - x_1)$$

where x_1 is the inoculum concentration of bacteria and t is the RT of the fermentor, while $x_t = x_1 \exp(\mu t)$

Substrate utilisation will be:

$$B/Y$$

where Y is the yield coefficient. This theoretical treatment holds only if μ is constant; i.e., that substrate concentration is initially much above K_s or that t is short so that there is little change in substrate concentration. Y will then be constant.

The equations could hold, approximately, if substrate concentration were very small. This could happen, for instance, if the tubular fermentor were the second stage in a two-stage digester and was removing the residual acids from a first stage running at long RT for degradation of pig slurry, as in the previous example. Here the bacterial growth rate on the residual acids would be very low, and while it would decrease as acid concentration fell, overall it could probably be approximated to by some mean growth rate.

Lawrence & McCarty (1970) treated a tubular reactor with solids recycle (Fig. 2b) in a similar manner to a stirred-tank with solids recyle and applied Monod-type kinetics and balance equations similar to those earlier outlined. They made the assumption that microbial mass did not change over the length of the tube and that the influent substrate concentration was the substrate from the feed plus the residual substrate from the recycle of the effluent. They obtained equations linking the RT and influent and effluent substrate after making some further simplifying assumptions in the mathematics. Powell & Lowe (1964) extended the theoretical treatment of two-stage fermentors given above to multi-stage fermentors, the limit of which is a tubular fermentor (i.e. an infinite number of fermentors each of volume approaching zero). They also considered feed-back, with large and small proportions of the bacteria being fed back, as either effluent itself or concentrated effluent where the volume fed back can be negligible compared to the volume of feed. These considerations showed that the effluent from a tubular fermentor could contain, effectively, no residual substrate. This is because the bacteria can continue to grow at lower and lower rates as the substrate is decreased, as there is no relation between growth rate and RT. The tubular fermentor is then a theoretically ideal system for secondary treatment or 'polishing' of effluent from a primary treatment of a waste.

CONCLUSION

The preceding sections will have shown some of the scope of digester models: they range from simple to complex but all present a simplified representation of the bacterial reactions occurring in digesters, particularly in those with a complex, particulate feedstock. The limitations of computation which required earlier models of microbial growth and reactions to be mathematically relatively simple no longer apply with the production of more and more powerful computors, and, indeed, the models do not need to be mathematically solvable. Models of the similar rumen system have progressed to a much more detailed and accurate description of the microbiology in which not only all the reactions of the chain of fermentations are considered, but they are considered to take place by the agency not only of groups of bacteria but of individual species and genera within these groups each with their own growth characteristics. Such modelling has been possible because of the large amount of data on the microbiology of the rumen and the composition of ruminant feeds that is available. Even here, however, modelling has shown gaps in the data that preclude the models from extreme accuracy, and the data available on digesters, although increasing, is much less than that on the rumen. However, although the digester models may not be as microbiologically detailed as those of the rumen, because the digester is physically a mechanical fermentor with defined operational parameters its overall behaviour can be predicted with a relatively simple model much better than can the behaviour of the complex, natural, rumen fermentation system. Thus, the objective of the earliest digester models, and of most subsequent ones, to define conditions for optimising digester running, has been met by many of the models. Such models may not be directly applicable to a particular feedstock, but illustrate a general principle; a simple one being the increase in waste purification with increased RT of a stirred-tank digester. If sufficient data are available then the optimum RT, etc. for digestion of a particular waste may be determined.

In the case of the simplified model, with feedstock defined only vaguely as VS, COD, etc., the experimental determination of the required constants may make the model almost superfluous so far as designing an optimised digester is concerned, but the model may predict parameters of failure of the system. The more complex model, such as those mentioned for the rumen, where the degradation of every chemically-defined component of the fedstock is carried out by defined bacteria should, in theory, be able to model digestion of any waste provided its composition is known in detail.

In a wastewater the substrate molecules are in solution and their degradation depends only on their chemical composition. The total degradation of different mixtures of these molecular species could, theoretically, be quite accurately forecast from the compositions of the mixtures. However, the constituents of a particulate feed are not separate molecules and the degradation of a particular molecular species depends not only on its basic chemical composition but on its physical relationship with other degradable and non-degradable constituents of a particle. This complicates the forecasting of the degradation of a particulate feed from its composition in terms of molecular species. Particles may be homogeneous in composition and of relatively uniform size, as in the case of straw or wood ground for digestion, or paper-making fibres. On the other hand, the particles in, say, a farm slurry fed to a digester will be much more varied in size and shape, as they are the residues of any initial treatment of the animal feed with the added partial degradation in the animal alimentary tract and in waste tanks and pipes. The excreta particles may also vary in composition and have different degradabilities because of different degrees of digestion of the initial particles in the animal (Peck, Hawkes & Hawkes, 1988). The modelling of the microbial degradation of particulate substrates has not so far been fully considered.

In a number of the models considered here the substrates are described as 'particles' but models of their degradation do not really consider the substrates as particles. In most models the degradation is based on kinetics, which, as previously pointed out, were developed for dissolved substrates. Noike *et al.* (1985) when modelling a stirred-tank continuous-flow digester fed with cellulose, dissolved starch or glucose concluded that the rate of degradation of cellulose was best described by an equation like that of Contois for microbial growth rate, while the degradation of the dissolved substrates was best described by Monod-type equations. Bryers (1985) said that 'Very little is known about the governing mechanisms, much less the kinetics, of biological particulate hydrolysis in microbial systems. Traditionally, biodegradable particle utilisation has been emperically modelled using simple kinetics that are first order on the remaining particle concentration.' He used first order kinetics in his digestion model (described earlier). Gujer & Zehnder (1983) followed Imhoff & Fair (1956) and Eastman & Ferguson (1981) in using first-order kinetics for describing particle degradation. Torre & Stafanopoulas (1986) followed Eastman & Ferguson in using first-order kinetics for hydrolysis of sludge solids and Monod kinetics for growth of fermentative bacteria. A number of the functions and parameters were taken from papers discussed here. Pavlostathis & Gossett (1986) used first-order kinetics for degradation of particles in sludge, with Monod kinetics for growth of bacteria and substrate utilisations. The particles in the activated-sludge feed to the digester were considered to be microorganisms which died and lysed and were also degraded by digester bacterial enzymes. The soluble products were then converted to acids and gases. All concentrations were in COD units.

Early light microscopic observations of rumen contents (e.g. Baker, 1942) showed that bacteria were attached to particles of plant fibres and starches and were producing pits around the bacterial cells, presumably by hydrolysis of the solid polysaccharides. These observations have been extended using light and electron microscopy and the observations show that particle breakdown is by bacteria attached to the particles and that factors possibly controlling growth are complex. A bacterium on a cellulosic fibre in a digester is growing in a medium where carbohydrate is supplied by hydrolysis of the fibre and other nutrients by diffusion from the digester liquid. Since in animal-waste digestions ammonia (the N-source) is in excess (Hobson, 1983), as are probably other minor nutrients, the bacterium is probably growing at a rate which may be determined by the rate of hydrolysis of the fibre polysaccharides. The fibre polysaccharides may be structurally difficult to hydrolyse and the rate of growth of the bacterium dependent on the low rate at which sugars become available from the hydrolysis. Or, the polysaccharide may be easily degraded, in which case sugar supply will be in excess and the bacterium growing at its maximum rate. In either case the bacterium will be degrading the particle at a constant rate in terms of depth of pit produced per unit time and so, as the area of the bacterial pits is fairly constant, in terms of weight of particle per unit time. This will be independent of the digester dilution rate. However, the rate may vary as the bacterium 'digs' deeper into the particle and encounters layers of different ease of degradation. For an individual bacterium the rate of degradation of the particle will be independent of the weight of particle remaining. For a number of bacteria entirely covering a particle of plate form and attacking the surface from one or both sides the overall rate of hydrolysis will again be independent of the weight of particle left, as the surface area, and so the polysaccharide available to the bacteria, will not change as the particle is degraded. In the degradation of particles of other shapes (cylindrical, etc.) by bacteria covering the surface the overall rate of hydrolysis of the particle may change as degradation proceeds, as in some cases the surface area increases or decreases as the particle is degraded (Hobson, 1987). However, the form of a plant fibre particle changes little during degradation by surface bacteria (the fragmentation of fibres in the rumen is brought about by chewing rather than microbial action) and for substrates such as animal excreta the fibre is not completely covered with bacteria and the total degradation of the fibre particles is not great. So, to a reasonable approximation once the fibres are colonised as much as possible the rate of degradation will be constant and independent of the weight of fibre remaining. Hobson (1985) modelled the digestion of pig slurry in a batch digester on these lines. The experimental gas production was not constant with time and this suggested that the substrates were, besides the VFA initially present in the slurry, a number of groups of fibre particles of different susceptibilities to degradation. The experimental digesters were inoculated with bacteria from a working digester and the model was given an inoculum of acid- and fibre-degrading bacteria. The bacteria grew according to Monod kinetics on the acid. On the solids the bacteria were generally growing at a constant rate and grew from the inoculum to colonise the fibres. Thus, although the degradation of each micro layer of the particle was at a constant rate (above) by each bacterium, the rate decreased as more lignified layers were encountered by the bacteria moving into the fibre surface and the overall rate of degradation varied as the number of bacteria increased. For some particles the growing bacteria could be accommodated on the surface, for one group of poorly-degradable particles the bacteria colonised the available surface before the particle was degraded, so daughter cells then dropped off the particle. The experimental digestion lasted 32 days and a computer program calculated degradations and gas productions and microbial growth at 0·25 day intervals and altered parameters as required to make a model which reproduced the experimental gas productions. It is impossible to determine experimentally the numbers of bacteria on particles in digesters, so the model inoculum was somewhat arbitrary but based on counts of bacteria in digester liquids, and the relative increases in bacterial numbers were calculated.

Although the rate of degradation of fibres by a bacterium or group of bacteria is not related to the weight of particle remaining, in a continuous-flow digester when particles have been colonised the overall rate of gas production per digester volume will depend on the weight of particles in the digester.

The degradation rate of uniform particles in a stirred-tank digester feedstock can be set in terms of weight per unit weight of particles per unit time. A model can them be made based on digester feeding at uniform intervals where at each feed a certain volume of particle suspension is mixed into the digester and the same volume containing partially-degraded particles from previous feeds and a proportion of the undegraded particles from the new feed is overflowed. The weight of particles in the digester can be then calculated after every feed input and equations can be formulated which give the residual solids in the digester after n feeds or after an infinite number (the steady-state) starting from a digester filled with water plus a bacterial inoculum of negligible mass or a solids-containing slurry (Hobson, unpublished). The model is dynamic and shows residual solids approaching the steady-state or if loading rates or RT are changed. It reproduces the experimental curves for steady-state residual solids at different RT and the residual solids also depend on the feed solids concentration. A feedstock composed of particles of different digestibilities and rates of degradation can be modelled, as can digestion of a particle composed of layers degraded at different rates (based on the fact that the second layer cannot be degraded unless the RT is such that the first layer can be removed). From the preceding arguments it can be seen that describing solids degradation in a digester as overall of first order can be reconciled with the actual manner of degradation of fibre surfaces by bacteria. Some attempt can also be made to calculate the bacteria involved in the degradations in the last model, assuming that the bacterial growth is at a constant rate independent of the digester RT, as previously described. Models such as the batch culture described previously become extremely complex when applied to continuous-cultures and generally must be simplified. It is really almost impossible to apply them to particular feedstocks as the full implementation requires a knowledge of the degradation rates of all the major groups of particles in, say, a pig slurry, and the variation in composition of the particles and other data.

More recently, Font & Lopez (1990) have published a model of degradations of particulate substrates in continuous stirred-tank reactors which can be applied to anaerobic digesters and which takes into consideration some of the aspects of particle degradation mentioned above. The model applies to defined HRT. However, it is rather complicated and only an outline of it can be given here. The distribution of residence times of particles about the mean (hydraulic retention time) is considered and the extent of degradation of a particle depends on its residence time. Among the assumptions made is that the rate of degradation of the particles is proportional to the concentration of bacteria in the digester. As they point out this may not be true for some of the reasons mentioned above; e.g. the surfaces of the particles are saturated with bacteria. The authors consider two cases. In the first, the degradation rate is proportional to the mass of non-reacted substrate, all particles have the same mass and the particles are of uniform composition. As the residence time of the particle determines its degradation there is a range of conversions of substrate. The maximum growth rate of the bacteria is taken as the rate found in a batch culture of the particulate substrate. The maximum growth rate is, of course related to the minimum residence time of the feed particles. The equation for the mean conversion of the particles is the same as that for a dissolved substrate. In the second case, the degradation of the particle does not depend on the mass of non-reacted substrate in the particle, and all the particles are identical. This takes flat-plate particles as substrate and, again, degradation is related to residence time. The microbial growth rate is assumed to be constant. In the two cases the particles are assumed to be all of the same size and shape and, again, the model produces equations for the extent of degradation of the substrate. If the substrate is particles of different size (but not composition) the first case equations still apply, but the second case is modified. The authors point out that other situations could be considered where the particles are cylindrical or spherical, but these would be complicated, and if the substrate particle size and shape is not known then their mathematical procedures cannot be applied. A further complication pointed out earlier is that, in practice, particles in most waste slurries are not homogeneous and the bacterial growth rate and the rate of particle degradation can vary with the initial size of the particle and with the extent of degradation of the particle as it resides in the digester.

The authors also consider their model in relation to the model of Chen & Hashimoto (1980) previously described. Without defining the 'kinetic constant', K, they say that if the value of K is between 0·3 and 0·4 then the data used in the model will fit the second-case equations mentioned above; i.e. a constant growth rate of the bacteria. Values of K less than 0·3 cannot be explained. When $K = 1$ this corresponds to the case above where the reaction rate is proportional to the substrate mass. How well the model would, or could, apply to digestions of different animal and other wastes remains to be seen. In the paper some data of O'Rourke (1968) on the volume of methane produced per unit weight of COD added to a sewage digester at different retention times were fitted to the second-case model.

Exact modelling of retained-biomass digesters may be complicated by diffusion of substrates into layers of bacteria in suspended particles, as has been mentioned, but a further complication may arise in the case of fixed-film digesters. Fiestas *et al.* (1990) found that the kinetics and the yield coefficients of bacteria degrading olive-mill wastewaters were influenced by the support materials. This would seem to pose a further complication in using a model of a filter with one type of packing to predict the behaviour of a filter with another packing.

A further complication with the UASB type of digester is that the bacterial granules may be of different sizes and models have tended to consider granules of one size. Beeftink & Van Den Heuvel (1990) have produced a model with aggregates of different sizes and densities and investigated the results of these variations on biomass growth and washout and other factors.

The more complex models attempt to reproduce the microbial community and its actions; the simpler models are generally adequate for predicting overall digester behaviour and many aspects of design of digesters and running parameters. But even here there may be complications.

REFERENCES

Andreadakis, A.D. & Cailas, M. (1984). In *Fixed-Bed Biological Processes*, p. 664.
Andrews, J.F. (1968). *Proc. 23rd Indust. Waste Conf.*, Purdue Univ. p.285.
Andrews, J.F. (1969). *J. San. Eng. Div. Proc. Am. Soc. Civil Eng.* **95**, SA1, 95.
Andrews, J. F. & Graef, S.P. (1971). *Ad. Chem. Series*, **105**, 126.
Audsley, E. & Wheeler, J. (1978). *J. Agric. Eng. Res.* **23**, 189.
Baker, F. (1942). *Nature*, **150**, 479.
Baldwin, R.L., Thornley, J.H.M. & Beever, D.E. (1987). *J. Dairy Res.* **54**, 107.
Bala, B.K. & Satter, M.A. (1990). *Biol. Wastes* **34**, 21.
Beeftink, H.H. & Van Den Heuval, J.C. (1990). *Chem. Eng. J.* **44**, B1.
Billington, R.S. (1980). *J. Agric. Eng. Res.* **39**, 71.
Bolte, J.P. & Hill, D.T. (1985). In *Agricultural Waste Utilisation and Management* (ASAE Pub. no. 13–85), ASAE, St. Joseph, MI, USA. p. 104.
Bolte, J.P. & Hill, D.T. (1990). *Biol. Wastes*, **31**, 275.
Bolte, J.P., Norstedt, R.A. & Thomas, M.V. (1984). *Trans. ASAE*, **27**, 1483.
Bryers, J.D. (1985). *Biotech. Bioeng.* **27**, 638.
Canovas-Diaz, M. & Howell, J.A. (1988). *Biotech. Bioeng.* **32**, 348.
Cecchi, F., Mata-Alvarez, J., Marcomini, A. & Pavan, P. (1991). *Biores. Technol.* **36**, 261.
Cecchi, F., Traverso, P.G., Mata-Alvarez, J., Medici, F. & Fazzini, G. (1990). *Biomass*, **23**, 79.
Chen, Y.R. (1983). *Agric. Wastes.* **8**, 65.
Chen, Y.R. & Hashimoto, A.G. (1980). *Biotech. Bioeng.* **22**, 2081.
Cohen, T. (1981). Thesis, *Optimization of anaerobic digestion of soluble carbohydrate containing wastewaters by phase separation*, University of Amsterdam, The Netherlands.
Condray, R.E. (1982). *Biotech. Bioeng.* **24**, 1705.
Contois, D.E. (1959). *J. Gen. Microbiol.* **21**, 40.
Denae, M., Miguel, A. & Dunn, I.J. (1988). *Biotech. Bioeng.* **31**, 1.
Deshpande, V.P., Kaul, S.N. & Deshpande, C.V. (1991). *Biores. Technol.* **38**, 31.
Eastman, J.A. & Ferguson, J.F. (1981). *J. Water Pollut. Cont. Fed.* **53**, 352.

Fazzini, G. (1990). *Biomass*, **23**, 79.
Fiestas, J.A., Martin, A. & Borja, R. (1990). *Biol. Wastes*, **33**, 131.
Fischer, J.R., Iannotti, E.L. & Porter, J.H. (1984). *Agric. Wastes*, **11**, 157.
Font, R. & Lopez, J.M. (1990). *Chem. Eng. Sci.* **45**, 3097.
France, J., Thornley, J.H.M. & Beever, D.E. (1982). *J. Agric. Sci.* (Camb.), **99**, 343.
Gates, W.E., Smith, J.H., Lin, S-D. & Ris, C.H. (1967). *J. Water Pollut. Cont. Fed.* **39**, 1951.
Gaudy, A.F., Ramanathan, M. & Rao, B.S. (1967). *Biotech. Bioeng.* **9**, 387.
Ghosh, S. & Klass, D.L. (1978). *Proc. Biochem.* **13**, 15.
Gujer, W. & Zehnder, A.J.B. (1983). *Water Sci. Technol.***15**, 127.
Haldane, J.B.S. (1930). *Enzymes*, Longmans, London.
Hamad, M.A., El-Gammal, M.A., Mitry, N.R. & El-Halwagi, M.M. (1989). *Biol. Wastes* **28**, 53.
Hamer, G., Bryers, J.D. & Berger, J. (1985). *Acta. Biotechnol.* **5**, 225.
Hanaki, K., Noike, T. & Matsumoto, J. (1985). *Developments in Environmental Modelling*, **7**, 583.
Hansford, G.S., Andrews, J.F., Grieves, C.G. & Carr, A.D. (1978). *Water Res.* **12**, 855.
Hashimoto, A.G. (1983). *Agric. Wastes*, **6**, 175.
Hashimoto, A.G. (1984). *Agric. Wastes*, **9**, 299.
Hashimoto, A.G. (1986). *Agric. Wastes*, **17**, 241.
Hashimoto, A.G. & Chen, Y.R. (1981). In *Proceedings of the 4th. Int. Symp. on Livestock Wastes*, A.S.A.E., St. Joseph, Michigan, USA. p. 129.
Hashimoto, A.G., Chen, Y.R., Varel, V.H. & Prior, R.L. (1979). ASAE Papers no. 79–4066, ASAE, St. Joseph, MI, USA.
Hawkes, D.L. (1980). In *Anaerobic Digestion*, ed. D.A. Stafford, B.I. Wheatley & D.E. Hughes, Applied Science Publishers, London. p. 131.
Hawkes, D.L. & Rosser, B.L. (1983). In *Proc. 3rd. Int. Symp. An. Digestion.* 3rd Int. Symp. An. Dig., 99 Erie St., Cambridge, Mass., USA. p. 465.
Herbert, D. (1958). In *Recent Progress in Microbiology, VII Int. Cong. for Microbiol.*, Stockholm. p. 381.
Herbert, D. (1964). In *Continuous Cultivation of Microorganisms*, ed. I. Malek *et al.*, Academia, Prague. p. 23.
Herbert, D., Elsworth, R. & Telling, R.C. (1956). *J. Gen. Microbiol.* **14**, 601.
Hill, D.T. (1983*a*). *Agric. Wastes*, **5**, 1.
Hill, D.T. (1983*b*). *Agric. Wastes*, **5**, 157.
Hill, D.T. (1983*c*). *Agric. Wastes*, **5**, 205.
Hill, D.T. (1983*d*). *Agric. Wastes*, **5**, 219.
Hill, D.T. & Norstedt, R.A. (1980). *Agric. Wastes*, **2**, 135.
Hill, D.T., Tollner, E.W. & Holmberg, R.D. (1983). *Agric. Wastes*, **5**, 105.
Hobson, P.N. (1965). *J. Gen. Microbiol.* **38**, 167.
Hobson, P.N. (1983). *J. Chem. Technol. Biotechnol.* **33B**, 1.
Hobson, P.N. (1985). *Agric. Wastes*, **14**, 255.
Hobson, P.N. (1987). *J. Ferment. Technol.* **65**, 431.
Hobson, P.N., Bousfield, S. & Summers, R. (1974). *Crit. Rev. Environ. Cont.* **4**, 131.
Hobson, P.N., Bousfield, S. & Summers, R. (1980). *Biogas Production from Sewage and Agricultural Wastes*, Applied Science Publishers, London.
Hobson, P.N. & Jouany, J-P. (1988). In *The Rumen Microbial Ecosystem*, ed. P.N. Hobson, Elsevier Applied Science, London. p. 461.
Hobson, P.N. & McDonald, I. (1980). *J. Chem. Technol. Biotechnol.* **30**, 405.
Hobson, P.N. & Shaw, B.G. (1971). In *Microbial Aspects of Pollution*, ed. G. Sykes & F.A. Skinner, Academic Press, London. p. 103.
Hobson, P.N. & Shaw, B.G. (1974). *Water Res.* **10**, 849.
Hobson, P.N. & Summers, R. (1967). *J. Gen. Microbiol.* **47**, 53.
Hobson, P.N., Summers, R. & Harries, C. (1984). In *Microbial Methods for Environmental Biotechnology*, ed. J.M. Grainger & J.M. Lynch, Academic Press, London. p. 119.

Horton, R. (1980). In *Anaerobic Digestion*, ed. D.A. Stafford *et al.* Applied Science Publishers, London. p. 321
Imhoff, K. & Fair, G.M. (1956). *Sewage Treatment*, 2nd. ed., J. Wiley & Sons, New York.
Kirsch, E.J. & Sykes, R.M. (1971). *Prog. in Ind. Mirobiol.* **9**, 155.
Koch, A.L. (1982). *J. Ther. Biol.* **98**, 401.
Lawrence, A.W. (1971). Ad. *Chem. Ser.* **105**, 163.
Lawrence, A.W. & McCarty, P.L. (1967). Tech. Rep. 75, Stanford University, Stanford, USA.
Lawrence, A.W. & McCarty, P.L. (1970). *J. San. Eng. Div.* **96**(SA3), 757.
Lusk, P.D. (1991). *Biores. Technol.* **36**, 223.
McCarty, P.L. (1966). *Developments in Industrial Biology*, **7**, 144.
McCarty, P.L. (1971). *Adv. Chem. Ser.* **105**, 91.
McCarty, P.L. & McKinney, R.E. (1961). *J. Water Pollut. Cont. Fed.* **33**, 223.
Maki, L.R. (1954). *Antonie van Leeuwenhoek*, **20**, 13.
Marchaim, U. (1983). In *Proc. 3rd Int. Symp. An. Digestion.* 3rd Int. Symp. An. Dig., 99 Erie St., Cambridge, Mass., USA. p. 343.
Marr, A.G., Nilson, E.H. & Clark, D.J. (1963). *Ann. New York Acad. Sci.* **102**, Art. 3, 536.
Mata-Alvarez, J. (1987). *Biotech. Bioeng.* **30**, 844.
Monod, J. (1942). *Rescherches sur la Croissance des Cultures Bacteriennes.* Herman et Cie, Paris.
Moser, H. (1958). *The Dynamics of Bacterial Populations Maintained in the Chemostat*, The Carnegie Institute, Washington D.C., USA.
Mosey, F.E. (1983). *Water Sci. & Technol.* **15**, 209.
Mosey, F. & Fernandes, X.A. (1984). In *Microbial Methods for Environmental Biotechnology*, ed. J.M. Grainger & J.M. Lynch, Academic Press, London.
Nilsson, B.K. & Karlson, H.T. (1989). *J. Chem. Technol. Biotechnol.* **44**, 255.
Noike, T., Endo, G., Chang, J-E., Yaguchi, J-I. & Matsumoto, J-I. (1985). *Biotech. Bioeng.* **27**, 1482.
Oliver, B., Pain, B.F. & Phillips, V.R. (1986). *J. Agric. Eng. Res.* **34**, 229.
O'Rourke, J.T. (1968). Ph.D. Thesis, Stamford University, USA.
Paolini, A.E., Sebastini, E. & Vaviali, G. (1979). *Water Res.* **13**, 751.
Parsons, D.J. (1986). *J. Agric. Eng. Res.* **35**, 259.
Pavlostathis, S.G. & Gossett, J.M. (1986). *Biotech. Bioeng.* **28**, 1519.
Peck, M.W., Hawkes, F.R. & Hawkes, D.L. (1988). In *Fifth Int. Symp. An. Digestion: Poster Papers*, ed. A. Tilche & A. Rozzi, Monduzzi Editore, Bologna. p. 411.
Pirt, S.J. (1965). *Proc. Royal Soc. B*, **163**, 224.
Pirt, S.J. (1982). *Arch. Microbiol.* **133**, 300.
Pirt, S.J. (1973). *J. Appl. Chem. Biotechnol.* **23**, 389.
Pirt, S.J. & Kurowski, W.M. (1970). *J. Gen. Microbiol.* **63**, 357.
Postgate, J.R. & Hunter, J.R. (1962). *J. Gen. Microbiol.* **29**, 233.
Powell, E.O. (1965). *Lab. Practice.* **14**, 1145.
Powell, E.O. (1967). In *Microbial Physiology and Continuous Culture*, ed. E.O. Powell, C.G.T. Evans, R.E. Strange & D.W. Tempest, HMSO, London. p. 34.
Powell, E.O. & Lowe, J.R. (1964). In *Continuous Cultivation of Microorganisms*, ed. I. Malek *et al.*, Academia, Prague. p. 45.
Rathowski, D.A., Olley, J., McMeakin, T.A. & Ball, A. (1982). *J. Bact.* **149**, 1.
Rittmann, B.E. & McCarty, P.L. (1980). *Biotech. Bioeng.* **22**, 2343.
Roques, H., Yue, S., Saipanich, S. & Capdeville, B. (1982). *Water Res.* **16**, 839.
Rozzi, A. & Passino, R. (1985). *Developments in Environmental Modelling* **7**, 637.
Safley, L.M. & Westerman, P.W. (1990). *Biol. Wastes* **34**, 113.
Schulze, K.L. & Lipe, R.S. (1964). *Arch. Microbiol.* **48**, 1.
Sinclair, C.G. & Topiwala, H.H. (1970). *Biotech. Bioeng.* **12**, 1069.

Stanogias, G.B. & Pearce, G.R. (1987). *Biol. Wastes*, **21**, 101.
Summers, R. & Bousfield, S. (1980). *Agric. Wastes*, **2**, 61.
Tasdemiroglu, E. (1991). *Biores. Technol.* **36**, 147.
Tempest, D.W., Herbert, D. & Phipps, P.J. (1967). In *Microbial Physiology and Continuous Culture*, ed. E.O. Powell, C.G.T. Evans, R.E. Strange & D.W. Tempest, HMSO, London. p. 240.
Torre, A.D. & Stephanopoulos, G. (1986). *Biotech. Bioeng.* **28**, 1106.
Trulear, M.G. & Characklis, W.G. (1982). *J. Wat. Pollut. Cont. Fed.* **54**, 1288.
Videla, S., Chamy, R., Navarette, E. & Lema, J.M. (1990). J. *Chem. Technol. Biotechnol.* **49**, 223.
Yang, S-T. & Guo, M. (1990). *Biotech. Bioeng.* **36**, 427.
Young, J.C. & McCarty, P.L. (1969). *J. Water Pollut. Cont. Fed.* **41**, R149.
Zehnder, A.J.B., Huser, B.A., Brock, T.D. & Wuhrmann, K. (1980). *Arch. Microbiol.* **124**, 1.
Zoetmeyer, R.J. (1982). *Acidogenesis of Soluble Carbohydrate Containing Wastewaters*, Thesis, University of Amsterdam.

Chapter 4

TYPES OF DIGESTERS AND THEIR CONSTRUCTION

DESIGN AND OPERATION

Introduction

There are several basic types of anaerobic digester (Fig. 2) and the choice between these depends on the type of waste to be treated. Therefore, in the following chapters where examples of the treatment of high-solids and low-solids wastes are given a brief description of the types of digesters has been repeated. In the present Chapter more details of the designs, materials, methods of construction and performances of full-scale plants are given.

The basic designs are considered here: in the following chapters digesters using combinations of these basic designs are described. In such cases, where construction has progressed to full-scale plants the materials and methods of construction are much the same as those described in this Chapter.

Some details of costs and running parameters of digesters are also given here. However, available data are somewhat limited. Some farm digesters have been constructed, or partly constructed, by the farmers and no adequate measure of true costs exists. For many plants it is often difficult to separate given costs into the costs of the digester, of ancillary equipment and of civil engineering work involved. For more details of costs, power requirements, etc., reviews such as that of Noone (1990) and the papers quoted in the present text should be consulted, but costs continually change with inflation and currency values of different countries. Some digesters have been run with little measurement except for gas production. Firms building and running digesters are sometimes, for proper commercial reasons, not willing to give details of design and running or problems associated with start-up or other aspects of the digester. In general, most details of digesters are obtained from plants which have been built with the aid of government or other public funds. In these cases the digesters have been built entirely as experimental plants or on the basis that their running can be monitored by some university or other research unit.

The majority of wastes treated by anaerobic digestion are liquids, although digestion of solid wastes is attracting increasing attention. Digesters for the treatment of liquid wastes form, therefore, the majority of plants. As shown later, the design of digester then depends on whether the liquid feedstock contains particles and colloidal materials (a 'sludge' for sewage, a 'slurry' for farm wastes) or whether it is an almost particle-free wastewater. In practice, the former feedstocks are, in the main, sludges from sewage-treatment plants, untreated human excreta, animal excreta from intensive farms and vegetable matter macerated with water. The latter feedstocks are wastewaters from various industrial processes. However, industrial processes may produce sludges and in both cases mentioned above the primary waste may be pretreated to alter its physical form, or waste streams may be combined with the same objective, as is described in Chapter 5.

The simplest form of anaerobic digester is the deep 'anaerobic lagoon'. These are still used on farms in some parts of the world, but they are not very efficient and pose problems. Some problems and performance data on anaerobic lagoons are given in Chapter 6. Lagoons are essentially holes, in the ground or earth banked if excavation is not possible. Depth may be some three metres and they may be some tens or hundreds of metres in length and breadth. They should have bottom and sides impervious to liquids, and this can be done by puddled clay, or, more usually now, rubber or plastic sheeting. Influent and effluent control is usually by gravity and weirs, and mixing is by gas production and convection currents caused by climatic heating: retention times are often indefinite, but long. This type of digester will not be further considered here.

In countries with small peasant farms with few animals and low incomes digesters have to be as simple and cheap as possible to construct and run. Heating is climatic and loading and unloading are manual. The largest numbers of these types of digesters are found in China and India, and some description of the construction and use of these digesters is given in Chapter 6.

In the present chapter digesters required by large farms and large industrial and municipal concerns, in any part of the world, are described.

With wastes containing particles, the designed retention time of the feedstock in the digester is governed by the rate of breakdown of the solids. Typical retention times in sludge and slurry digestions are 10 to 25 days. This allows time for growth and division of the slowest-growing bacteria (see Chapter 2) and so a stable digestion, without bacterial washout, can be obtained. The flow rates of many industial wastewaters are such that a retention time of 5–10 days would result in an impossibly large digester and so reactor designs have been developed for wastewaters in which the bacteria are held back within the digester or separated from the effluent stream and returned to the digester.

Sludge- and slurry-digesters

Sewage-sludge digesters were the first plants to be erected and run (a short history of sewage digestion is given in Chapter 6) and the sewage digester was the basis for the earlier farm-waste digesters, although there has lately been feedback from farm to sewage digester. Thus sewage-sludge digesters will be considered first, but many aspects of design and construction apply to sludge and slurry digesters on any site.

Sludge digesters are large, completely-mixed, heated tanks. The most important features of design are the efficiency of mixing, the effectiveness of heating, and, because most of the substrates are in the solids of the feedstock, the solids retention time.

Mixing

Mixing is necessary to ensure dispersion of the feed sludge throughout the tank; to promote good contact between the waste and the bacteria; to avoid dead zones, thermal stratification and grit and scum accumulation. It should also help to keep heating systems free from a build-up of sludge covering heating surfaces. Gas production within the digester causes some stirring, but this is usually insufficient to mix sludges, although industrial digesters operating at high rates and which produce more than 1·5 digester volumes of gas per day can be completely mixed by gas evolution (Tilche & Vieira, 1991). Surveys of traditional concrete sewage digesters have many times found poor performance to be linked to poor mixing (Swanwick *et al.*, 1969; Brade & Noone, 1981; Sidwick, 1986). Most problems were traced to blockage, fouling and corrosion of mechanical mixers. Rectification of these problems was a first step to improving the performance of the digesters.

There are three methods of mixing sludge digesters: recirculation of sludge; recirculation of gas; mechanical mixing (Fig. 7). Systems for external recirculation of the sludge, although easy to maintain, do not produce thorough mixing. The digester contents are turned over 'en bloc' rather than completely mixed and short circuiting and dead zones occur. Sludge recirculation is not used as the sole source of mixing and traditionally circulation is assisted by internal mechanical mixing. In the most common system sludge is sucked up a fixed tube by an impeller (usually of archimedean-screw type) and sprayed out over the digester surface. This action avoids scum formation. Other systems use large paddles rotating on a vertical shaft in the centre of the digester. Blades near the bottom and the liquid surface help to reduce scum and sediment formation. There are obvious mechanical limits to the size of this type of stirrer, but as the stirring does not need to provide gas as a bacterial substrate, as in aerobic systems, the mixer speed can be low: a few rpm. Mechanical mixers are the most efficient type of mixer when new (Brade & Noone, 1981), but rapidly become choked with rags in sewage sludge and are eventually damaged by corrosion. Cleaning impellers is expensive and unpleasant. Current preferred practice is to use gas-lift mixing. Gas is collected from the digester head-space, cleaned (although this is not usually done for farm digesters, for example), dewatered (dewatering is by condensate traps on low-cost digesters) and compressed at low pressure and reintroduced at the bottom of the digester tank. The gas may be injected into one or more vertical draft tubes which act as gas-lift pumps and may also be double-skinned to provide hot-water heating of the sludge (confined gas-lift mixing). Or, the gas may be released from jets placed at intervals around the digester bottom (unconfined mixing). Experiments have shown that unconfined gas mixing from jets evenly spaced over the base of a tall narrow tank gives the best mixing and least scum (Rundle & Whyley, 1981; Baumann & Huibregtse, 1981). Unconfined gas-lift mixing is also significantly cheaper than other types of mixing. Gas mixing systems require simple non-return valves on the gas jets, or some other system to prevent back-flow of digester contents up the gas pipes, has to be fitted, particularly as gas mixing is usually intermittent as is the operation of mechanical mixers.

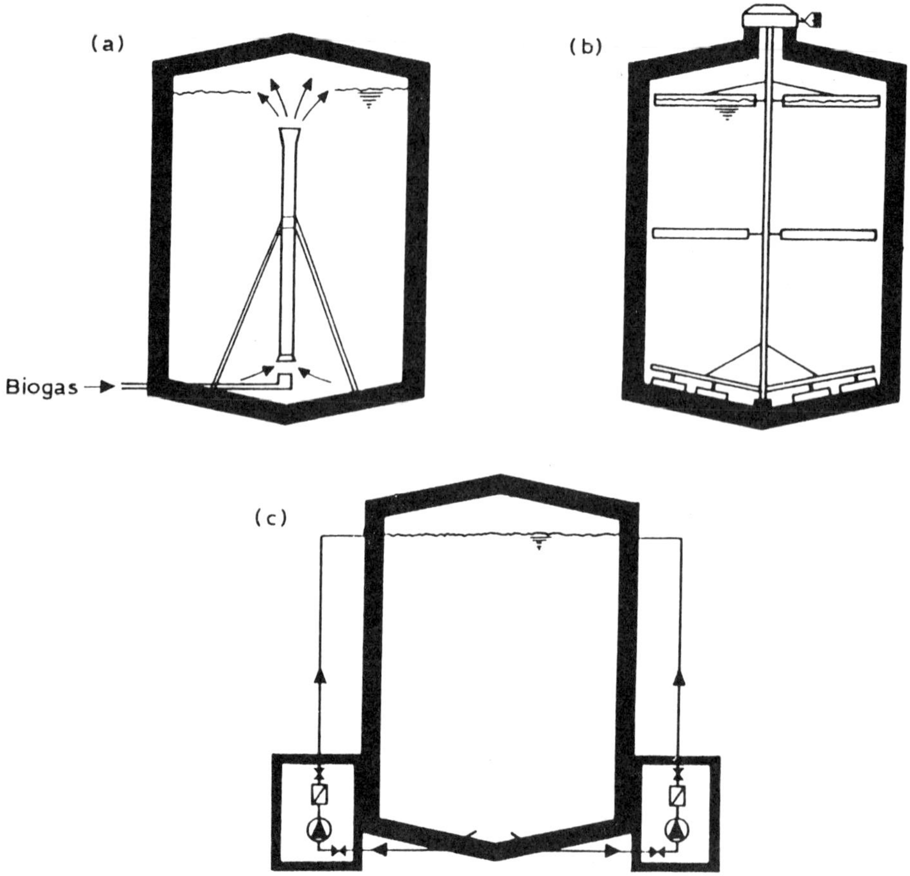

Fig. 7. Mixing systems used in stirred-tank digesters. (a) confined gas-lift mixing: (b) impeller mechanical mixing: (c) sludge recirculation. Sludge lift up the central tube in (a) can also be produced by a mechanical, screw-type impeller.

Many older digesters, over 1000 m^3 volume, were too large for easy mixing. This was partly because the shape of concrete tanks was dictated by providing the largest volume at the lowest cost and so digesters were large-diameter shallow tanks. This was based on storage-tank design. Some tall and narrow concrete digesters have been, or are being, built, and this design is better suited to gas mixing. The glassed-steel prefabricated digesters now preferred have this better aspect ratio. Large volumes can be provided for by using several of these digesters. This modular approach has maintenance advantages as well, each digester can be serviced in turn without disrupting treatment.

Heating

As explained in other chapters, most digesters are run at higher mesophilic temperatures around 35°C. The climatically-heated digesters of peasant farms in warm countries attempt to obtain a temperature near to this, but in many areas digester temperature can fall drastically at night or during winter seasons. Unheated digesters under cattle sheds may run at about 20°C (Chapter 6) and anaerobic lagoons in cool climates may never exceed about 12°C and so the rates of reactions in these systems are slow. Sewage digesters and most farm and factory digesters in temperate climates are, then, heated. With a well-lagged digester the largest heat requirement is for raising the feed sludge from ambient to digester temperature.

Heating can be internal to the digester or external. External heating has been common for sewage-sludge digesters and some farm digesters; an example of the latter is the digester run for many years by one of the authors and his colleagues on pig slurry (Robertson *et al.*, 1975). With external heating the digesting sludge is pumped from the digester through tubes in a hot-water heat exchanger and returned to the digester. The heat-exchanger water is provided by a biogas boiler or a CHP unit running on the digester gas, or in some older systems the burning digester gas directly heated the water through which the sludge pipes ran. The heat-exchanger pipes and hot water can be tube-in-shell, concentric straight tubes, or spiral tubes in a larger tube. External heat-exchangers can be relatively easily stripped for cleaning and maintenance compared with heat-exchangers in the digester, but they are more expensive (Brade *et al.*, 1982). They need hot-water pumps, and sludge pumps to circulate the sludge from the digester. The exchangers can be of such capacity that running can be intermittent.

Another form of external heating, so far as is known applied only to some sewage digesters, is by direct combustion. The raw sludge is pumped through the flame of a gas or other burner in a tank prior to its passage into the digester. Special equipment is required to pump the sludge and control the temperature and to avoid corrosion (Kidson & Ray, 1984, Hudson *et al.*, 1988). This method could pasteurise the raw sludge to reduce or eliminate pathogens.

Internal heating is most generally by hot-water heat-exchangers of different designs. Both flat-plate and coiled-tube types are in use. Draft tubes for mixing with a double skin through which hot water can circulate for heating are another option. A 680 m^3 digester for pig slurry with three large draft tubes for heating and mixing was run for some years by Summers *et al.* (1984). The upward movement of slurry through these draft tubes during the intermittent mixing cycle not only transferred heat but kept at least one of the heating surfaces of the heat-exchanger clean. Sludge can build up and dryout on the surfaces of internal heat-exchangers and so coils should be built around mechanical mixers or coils and plates fixed over gas-mixing jets to distribute the heated sludge and to keep the heat-exchanger surfaces clean. A maximum water temperature of about 75°C is used in external or internal heat-exchangers to help prevent drying of sludge on the surfaces.

Another form of internal heating has, so far as is known, been used only on a few sewage digesters. This is steam injection. Steam is injected directly into the digester tank through a jet and it requires good mixing to prevent the sludge overheating and heat-shocking the bacteria or even killing them. Like the direct combustion method (above), steam heating obviates problems of fouling of heat-exchanger surfaces. With softened and cleaned water for steam generation and a modulating steam valve controlled by digester temperature, steam heating is said to be cheap and reliable (Noone & Brade, 1985). However, as mentioned elsewhere, what may be cheap to the sewage works may be dear to the farm or industrial concern and apart from expense the use of live steam or flames would seem to be inapplicable for farm or similar use.

Loading rates
Loading rates are defined by the rates of breakdown of the feed constituents, the concentration of the solids in the feed sludge and the growth rates of the anaerobic bacteria (see Chapters 2,3). These factors can be used to fix the retention time of the digester. Many older sewage digesters were designed with retention times between 20 and 30 days (Ministry of Housing, 1954). This was to achieve a reduction in feed organic solids of 50–60% and included a significant contingency for poor mixing and dilute feeds. Improvements in the design of ancillary equipment and better understanding of anaerobic digestion have enabled retention times to be reduced to 12–15 days (Brade & Noone, 1981; Noone & Brade, 1985). The critical retention time to avoid washout of the digester bacteria has been suggested by laboratory experiments to be 5 days (Henze & Harremoes, 1983; Stronach *et al.*, 1986. See also Chapter 3). Most standard designs for retention times assume a feed solids content of 4–5%, but there are difficulties in consistently achieving this concentration in practice. Dilute sludges are common and it is now usual practice to thicken sewage sludges prior to feeding to the digester (Chapter 6). Typical running parameters of sewage-sludge digesters are given in Table 1 along with those from some farm slurries for comparison. The data for the farm digesters are for unseparated slurries (see Chapter 6).

Table 1
Typical operating conditions of a domestic sludge digester and farm slurry digesters.

	Parameter	Units	Typical value
Sewage	Biomass VSS	mg/l	2–6000
	Loading rate	kgVS/kgVSS/d	1
	Loading rate	kgVS/m^3*/d	2–6
	Retention time	days	15
	Removal of VS added	percent	50–60
	Removal of TS added	percent	35–40
Piggery	Loading rate	kgVS/m^3*/d	2.9
	Retention time	days	10
	Removal of VS added	percent	51
	Removal of TS added	percent	36
Dairy cattle	Loading rate	kgVS/m^3/d	2.9
	Retention time	days	20
	Removal of VS added	percent	38
	Removal of TS added	percent	32

*m^3 digester volume
VS as % of TS: sewage sludge 70; pig slurry 70; cattle slurry 85. TS in sludges generally used: sewage, 4–8%; piggery 4–6%; cattle 4–14%. The piggery and cattle results are from experiments by one of the authors, in practice farm wastes vary considerably with diet of animal and housing, etc.

The Contact digester

The Contact digester is essentially a stirred-tank and the mixing and heating systems are the same as those described above, but much of the biomass in the effluent stream is separated and returned to the digester to attain a high bacterial concentration and a long bacterial retention time (Chapters 3, 7). This allows the digester to be used for low-solids wastewaters. The equipment for separation of the biomass is in the effluent stream outside the digester. As explained in Chapter 7, separation of the bacteria is hindered by the gassing of the effluent and so the effluent is usually degassed before the biomass is settled; or the bacteria are retained in the tank by baffles, in which case the digester becomes akin to the UASB and other, similar, digesters (see Chapter 7). Three types of degassing unit are shown in Fig. 8. These have a vacuum to assist degassing and the gas is collected and added to the gas from the digester tank. The retention time in the degassing units is about 10–15 minutes. The high concentration of inorganic materials in some wastes, from sugar processing for example, can improve rates of settlement, but other wastes, e.g. from polysaccharide or protein processing, generate a poorly-settling biomass.

A common commercial design is a Scandinavian system originally applied to sugar-industry wastes. It is currently sold by Purac and known as the 'Anamet' process. The digester is mechanically mixed and biomass separation is assisted by a tower-shaped, stirred, flocculator prior to a lamella-plate settlement tank. Another widely-used system is the French Degremont process. The digester is mixed by gas recirculation and the effluent passes through a vacuum, stirred-

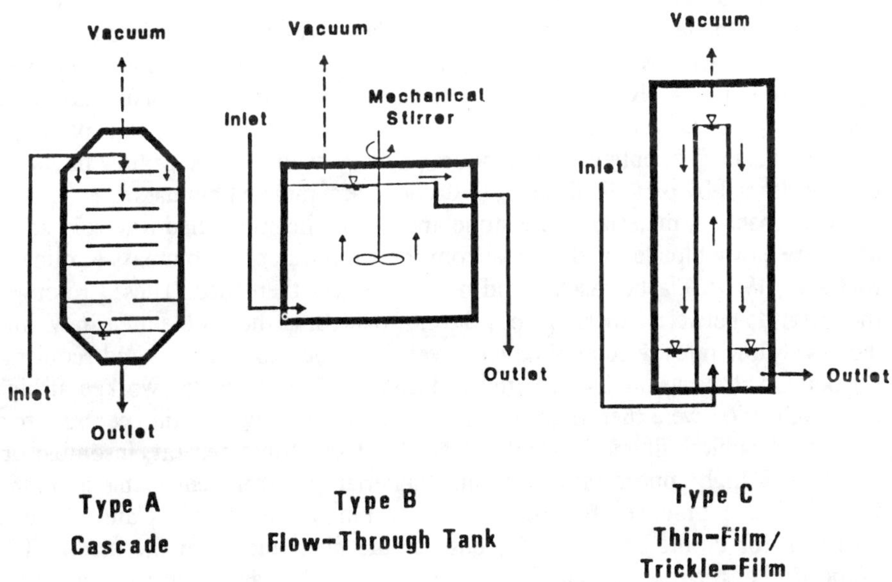

Fig. 8. Types of degassing units used in Contact digesters.

tank, degassing unit. Biomass separation is by a conventional stirred and scraped settlement tank. A third type is that sold by Sulzer of Switzerland. The digester is mechanically stirred with a settlement tank incorporated into the top of the digester. Settlement and recycle of the biomass is assisted by lamella plates within the clarifier. The UK Biomechanics system uses a combination of a falling film degasser and a heat-exchanger to cool the biomass prior to conventional settlement.

Typical running parameters for a mesophilic Contact digester are: loading rate 0·5–2·5 kg COD/m^3 digester/day; retention time 1–5 days; COD removal 80–90%; solids concentration 5–10000 mg/l. Because Contact digesters have a simple tank, like sludge digesters, they are not so much affected by suspended solids in the feed as are the retained-biomass types of digester. Contact digesters are currently used for treating wastes from sugar-processing, distilleries, citric acid and yeast productions, and factories producing sauerkraut, canned vegetables, pectin, starch and meat products (Nahle, 1991). There is published information on several Contact digesters in the UK: for instance on one of the earliest to be built for starch-processing wastewaters (Morgan, 1980; Butler, 1984; Butcher, 1988) and for a sugar-processing wastewater (Shore *et al.*, 1984; Smith, 1984). This is a 6600 m^3 tank with a 50 m^3 degasser and flocculator which works during a 6 month's beet sugar-processing season. COD removal is 86%.

Anaerobic filters

As described in Chapter 7, the anaerobic filter was one of the earliest types of retained-biomass digesters with the bacteria immobilised on a support to obtain high biomass concentrations and long bacterial retention times. Both upflow- and downflow-filters have been used in full-scale plants and in both the media are fully submerged. Downflow filters have counter-current gas and liquid flows and mixing is, therefore, intense. Recycle is also used to enhance horizontal mixing and minimise clogging. Downflow filters usually use orientated or tubular packings. Upflow filters use orientated and random packings with recycle. Upflow filters are usually cleaned by periodic flushing to dislodge accumulated biomass.

Earlier packing materials were stone and gravel, but these had low void space and were easily blocked with solids from the wastewater and biomass (Young & McCarty, 1969). Plastic raschig and pall rings were then tried. These prolonged the intervals between blockages but did not overcome the problem entirely, and increased flow rates were applied with recycle to increase flushing and scouring. Blockages still occasionally occurred and downflow filters, which worked at high liquid velocities, were then tried. Blockage and channelling has still been reported in some downflow filters, but only when the waste contained fat, insoluble organics or a high concentration of solid material (Environment Canada, 1986). Present design practice anticipates some accumulation of solids and includes provision for desludging from the base of the filter and from the media. The proportion of the digester containing the filter pack has also been reduced and is now typically 60–70% (Oleskiewicz *et al.*, 1986; Oleskiewicz & Thadari, 1988).

Solids may be detached from the media by temporary increases in upflow velocity induced by effluent and gas recycle (Camilleri, 1987; Ehlinger *et al.*, 1987). Once the solids have been removed performance recovers.

Most downflow filters have used the modular sheet- or block-media, while most upflow filters have used the random-ring type. Stone is now used only for low-rate, rural, filters, but mineral blocks and woodchips have been used, at least in small digesters, while unglazed porcelain rods are also advocated (see Chapter 7). The type of plastic material does not seem to greatly affect reactor performance as most surfaces can be colonised by bacteria (although see later). Nylon, polypropylene and polyvinylchloride have all been used, as well as ceramics.

The media used must be mechanically stable and able to tolerate some long-term compression. Attachment of bacteria to hydrophylic surfaces is very rapid. If the surface is hydrophobic, like the plastics, then the extracellular polymeric secretions of the bacteria plus the general debris act as a bridge between the cell and the plastic. The function of the media is to retain bacteria within the filter, but the predominant mechanisms of retention are different in the different types of filter. In upflow filters most of the active biomass is held as flocculated, granular, solids in the interstices of the packing. Fine, loose, solids are also present and held in place by the upward liquid flow. The former type of accumulated solids is the cause of clogging problems (Wheatley *et al.*, 1988; Young & Dahab, 1983). The specific surface area available within upflow filters is not as important as that in downflow filters. In downflow filters there is little solids in the interstices of the packing, the important biomass is a thin film firmly attached to, and closely following the contours of, the packing. Cross-flow, ordered packings have been shown to perform slightly better than regular or tubular media, and significantly better than random media even when the surface area of the latter is larger (Adams & McKinney, 1989).

There has been a number of studies on the ideal characteristics of media for downflow filters. Rough surfaces have been shown to promote more rapid start-up, and some ceramic materials, such as red tile clay, which leached trace nutrients, performed better than those which did not leach nutrients (Murray & van den Berg, 1981). This effect, though, may depend on feedstock composition. In practice, however, cross-flow, modular media in combination with a simple prefabricated digester design usually offers the best value for money.

Athough many materials have been used, apparently successfully, for filter media, detailed studies have suggested that the support material may influence the running of the filter in ways other than physical ones connected with porosity, blocking and flow rates. As mentioned in the last chapter, Fiestas *et al.* (1990) found that the support materials saponite, montmorillonite, sepiolite (Pangel and Pansil) and polyvinyl chloride influenced the kinetics and biogas production with olive-mill wastewater even when the support materials were free in a batch digester.

There have been fewer investigations on the ideal media for upflow filters. Larger blocks (> 100 mm) should resist clogging better than smaller (25–50

Table 2.
Some properties of typical anaerobic filter media.

Media name and description	Material	Weight (kg/m^3)	Specific surface area (m^2/m^3)	Void volume (%)
Flocor R. Random corrugated tube. 50 mm × 50 mm	PVC	90	250	95
ETA pack. Pall rings. 75 mm × 75 mm	Polypropylene	85	160	97
Furnace slag, porous stone	Expanded ceramic *	975	100	55
Flocor E. Regular sheets 1200 mm × 600 mm	PVC	90	100	94
Cloisonyle. Vertical vaned tubes	PVC	200	80	95

* with significant iron content.

mm), but they may be more difficult to clean. More research is needed on this topic. Examples of some typical media are given in Table 2.

There is now a significant amount of operating experience with anaerobic filters. Some typical operating parameters are given in Table 3. Further tests are required to provide more fundamental data but there is much emperical information that indicates that reactor hydrodynamics are the most important feature of design. Some comprehensive surveys that have been conducted by Young & Young (1989) and Young (1990) suggest that retention time is the most useful single indicator of performance. Typically, design difficulties arise from dead zones and short-circuiting of the liquid within the digester, and from different specific activities of the biomass. The degree of mixing has an important influence on the retention time: the greater the mixing the less will be the amount of short-circuiting and dead space. Upflow liquid velocity and gas production are important influences on the degree of mixing. New upflow filters behave as plug-flow reactors, but highly-active filters are mixed by the gas production. Research has only recently begun to consider critical upflow velocities produced by recycle and gas production. Early filter designs running at low loading-rates without recycle, and producing, for instance, less than one volume of gas per reactor volumeper day, did show plug-flow behaviour (Dahab & Young, 1982; Wheatley & Cassell, 1985). Reactors with higher gas productions (1–2 volumes gas/reactor

Table 3
Some typical operating parameters of an anaerobic filter.

Parameter	Value
Loading rate	2–10 kg $COD/m^3/d$
Retention time	10–50 h
COD removal	70–80%
Critical solids conc. in feed	1000 mg/l

volume/day) show mixed behaviour (Young & Young, 1988; Wheatley *et al.*, 1988). Downflow filters behave differently because of the counter-current gas and liquid flows and they are nearly always completely mixed (Samson & Kennedy, 1985; Hall, 1983). The amount of dead space or degree of clogging are measured by the standard retention time techniques (Levenspiel, 1972; Wheatley *et al.*, 1988).

In most cases effluent recycle is beneficial to the running of filters. It provides better mixing and improved effluent quality, and it can also assist with media washing. Recycle also provides extra alkalinity and nutrients, and will dilute the concentrations of very strong wastes. Providing for the most appropriate recycle flow rate depends on reactor shape, gas production and media, but it is usual to provide ratios of between 2:1 and 6:1. Upflow velocities with satisfactory performance vary widely; rates of between 0·2 and 0·7 m/h have been reported. Upflow velocities need to be set to avoid washout of the biomass and this will depend on the degree of mixing and the type of biomas, so some versatility in pump rate should be provided for.

Uniform distribution of the inflowing waste is also necessary to obtain optimum retention time and performance. Typical distributors for downflow filters are like cart wheels, with the number of 'spokes' depending on the diameter of the reactor. The distribution holes in the spokes point downwards into the filter bed, and the diameter of the holes, or number of holes, is profiled away from the pumped input. The recycle liquid is introduced through a similar, but larger and separate unit slightly above the feed distributor. The distributor spokes should protrude through the reactor walls to provide some possibility of rodding-out. The treated effluent, in the case of the upflow filter (which also needs a feed system which distributes the input) can be collected in a launder weir.

Overall, the simplicity and robustness of the anaerobic filter make it ideally suitable for treatment of dissolved industrial wastes. Loading rates are much higher than those of Contact digesters. However, despite these apparent advantages the filter is not as popular as the Contact or UASB digester and only about 15% of the industrial anaerobic treatment plants in Europe are of this type (Wheatley & Coombs, 1991). The major worry is excessive biomass accumulation and blockage of the packed bed. At the time of writing, there were 32 anaerobic filters in Europe, with about equal numbers using upflow or downflow operation. Most were in France, and all but two treated wastes from the agro-food industries, and nine of these were built for distillery wastes. Detailed information had been published on two, and there were no reports of blockages (Camilleri, 1988; Racault, 1990). Young (1991) published a list of upflow filters in the USA and Canada. Most were for the treatment of food and distillery wastes, although some were used for treatment of alcohols, acids and esters from the chemical industry (Witt *et al.*, 1979). Organic loadings were reported to vary between 0·2 and 16 kg COD/m^3/d, with hydraulic retention times between 12 and 96 hours. There were fewer downflow filters in the USA and Canada (Kennedy & Droste, 1991), but detailed information had been published on two. The design and performance of the filter at the Bacardi distillery has been widely publicised (Szendry, 1983, 1986; see also Chapter 7) and the details of a dairy plant were also reported (Environment Canada, 1986). The dairy filter used ceramic modular media and did block with fat after two years of operation. Bonastre & Paris (1989) have published an extensive survey of laboratory- and pilot-plants treating various kinds of wastes.

The upflow sludge blanket (UASB) digester

The problems caused by clogging of filters, particularly the earlier ones with mineral packings, resulted in investigations into methods of flocculating bacteria so they could be retained within the digester without support media and this resulted in the UASB digester for treatment of wastewaters. (The relation of the UASB to earlier, similar, digesters is described in Chapter 7). The development of the UASB has been very successful and it has applications for a wide range of industrial wastewaters and there has been some experience with sewage waters (see Chapter 7). There are about two hundred operating plants worldwide and it has become the most popular type of high-rate, wastewater, digester due largely to the research and development efforts of Lettinga *et al.*, (1979) and two Dutch companies marketing the process.

The UASB digester requires the active bacteria in the form of a high-density granular sludge which may then be retained within the digester tank despite the gassing and upflow velocity of the wastewater. Sludge granulation is a complex and not-fully-understood process. Comprehensive and detailed studies have been made in a number of countries, but principally in The Netherlands where the UASB process was devised (Lettinga *et al.*, 1987). To avoid problems in start-up, it is the usual commercial practice to inoculate the digester with a large volume of granular sludge from a working plant. Inocula of granular sludge of between 10 and 15% of the digester volume are used, if available, but these inocula can be expensive if they have to be purchased: e.g. US$6000 per tonne dry solids (Weiland & Rozzi, 1991).

The hydrodynamics of the digester, created by the feed distribution and the digester shape, are important in retaining the correct granular form of biomass. To maintain process efficiency, settling velocities of 10 m/h are required. Measured velocities of up to 90 m/h have been reported (Lettinga & Hulshoff-Pol, 1986). The quality or history of the inoculum also has an influence and as little as a 1% inoculum of a well-adapted, granulated, dense sludge has been successfully used (Souza, 1986).

Granulation is a natural process and is caused by a combination of the microbial morphology, the nature of the substrate and the accumulation of inorganic salts. Formation of rapidly-settling granules from an inoculum of ordinary biomass sludge from a stirred-tank or other digester may take only 50 days (Alibhai & Forster, 1986). The type of wastewater being treated has an important role in the formation of granules. A satisfactory sludge blanket was formed from yeast processing, sugar beet processing and potato wastewaters, but not from distillery, corn-starch production, dairy or abattoir wastewaters (Stronach *et al.*, 1986). The process is similar to scale formation. Key elements in the feed substrate for successful granule formation are calcium, phosphorus, ammonia, magnesium, aluminium and silicon (Kissel *et al.*, 1988). These materials are readily available in root-vegetable-processing wastewaters which contain some soil and clay. A large population of filamentous microorganisms, e.g. *Methanothrix* spp., is also essential and these may be associated with a high concentration of soluble components, mainly acetic acid, in the feed. In general, granular sludge will develop with mostly-dissolved wastes (Lettinga & Hulshoff-Pol, 1991). If an ideal start-up inoculum is available then full process-efficiency may be obtained within a month, but if a suitable inoculum is not available then the start could take six months (Souza, 1986). The overall start-up procedures used are similar to those for other types of digester in that the loading rate is increased step-wise (increments of 0·05 to 0·1 kg COD/kgVSS/d) provided that the digester pH and concentrations of carboxylic acids remain satisfactory.

The shape of the UASB digester, the upward liquid velocity and the baffles inside the digester are used to promote gas-solid separation. The most important features governing digester shape are the treatability of the waste and the efficiency of granule formation. If the waste is soluble, and a granular sludge has formed, then upflow velocities of 3 m/h are allowable. Digester heights can then be up to 10m (Lettinga & Hulshoff-Pol, 1991). If the wastewater contains significant amounts of suspended COD (and during start-up when granulation may not be complete), then the biomass is usually present as a voluminous, flocculant, sludge. Under these conditions, upflow velocities need to be more modest; e.g. 1–1.5 m/h. In practice, nearly all wastes are heterogeneous in composition and digesters are usually 3–5 m in height with up-flow velocities intermediate in the range of figures given.

The feed distribution system also exerts an important influence on the liquid upflow velocity and can be used to promote good contact between the bacterial sludge and the wastewater constituents. The large plan area of many UASB digesters allows the installation of a large number of small distribution points (usually one for each 1–2·5 m^2) which keeps upflow velocity to a minimum. A particular problem with UASB digesters with a large number of feed distribution points is the possibility of clogging of the nozzles. Each influent system should be easy to clean and monitor. Lettinga & Hulshoff-Pol (1991) have provided a design guide for the appropriate numbers of feed distributers according to the gas production and organic loading rate.

The effect of gas production on mixing is complex. Higher than optimum production will cause gulping and open paths through the sludge blanket, while a lower production will not encourage enough mixing within the blanket to avoid short circuiting through the dense mass of sludge granules. It has been shown (Bolle *et al.*, 1986) that at typical gas and liquid velocities (1–1·5 m/h) a sludge-blanket height of 3·4–4·0 m in a 6·7 m high digester is optimal.

Various designs of baffles inside the digester are used to promote gas/solid separation and, therefore, to retain the granules. The baffles provide surface area which reduces up-flow velocity and promotes biomass flocculation. The baffles are usually made of reinforced plastic so as to be light-weight and corrosion resistant. Solids which collect in the cones should settle back into the digester liquid. Lettinga & Hulshoff-Pol (1991) have provided a design guide based on wastewater characteristics. The gas-solids separator usually occupies 25–30% of the digester height and the baffles are set at an angle of between 45 and 60°, with an overlap of the plates of between 200 and 300 mm. There may be special problems with wastes that produce foam. Many effluents which contain fat and protein, such as dairy wastewaters, will foam when vigorously aerated in activated sludge, or in a gas-mixed anaerobic digester. Precautions against these foams blocking gas lines and effluent pipes are necessary. It is usually sufficient to install spray nozzles at the top of the digester above the baffles, but in very tall digesters it may also be necessary to add a chemical antifoam to avoid clogging the UASB gas separator. In any event, maintenance of the top and bottom of the digester to remove scale, scum and grit will be periodically necessary.

There is a good deal of design and operating experience with UASB digesters. UASB have been extensively applied to food-processing wastes and there are about fifty plants operating in Holland with sizes between 100 and 1500 m^3. There are five in the UK treating papermill waste, citric acid waste, dairy waste, and food and soft-drink preparation wastes. Problems have been encountered with dairy and slaughter-house wastewaters: losses of fine solids and periodic washout of bacteria (Sixt & Sahm, 1987).

The main operating problems with UASB digesters are in start-up and in maintaining the flocculated granules. Wastes with a high proportion of insoluble COD (i.e. > 15%) can give rise to poor granulation. Lettinga & Hulshoff-Pol (1991) give further details on recommended design loadings, and these vary according to the amount and solubility of the materials in the waste. High concentrations of suspended solids cause problems by reducing the activity of the suspended sludge, or lead to scumlayers if the solids are fat. Prolonged and continuous entrapment of feed solids in expanded flocculant sludge can lead to a spontaneous and sudden wash-out of the sludge bed. In order to avoid build-up of feed solids, periodic washing of the sludge by increases in hydraulic flow, lasting about one hour, are recommended (Lettinga & Hulshoff-Pol, 1991). If the waste does contain significant amounts of degradable suspended solids then a two-stage system with separate solids hydrolysis would normally be used in the design of a UASB system. Difficulties with foam and scale formation have also been reported (Lettinga & Hulshoff-Pol, 1991), but these are common problems with wastes containing protein, fat and calcium and it is unlikely that the UASB is any more susceptible than other types of digester.

The fluidised-bed digester

The difficulties of biomass separation in the Contact digester and loss of granulation or blockages in the sludge-bed or filter digesters can be overcome by what is really a combination of the two latter types. This is the fluidised-bed where the bacteria are attached to fine support particles which can be kept suspended in the digester tank (see also Chapter 7). Fluidisation of the mass of carrier particles occurs at a critical upflow velocity of the wastewater and this depends on the density of the particles and other sources of pressure loss within the digester. This fluidisation point can be defined by standard equations for pressure loss (Levenspiel, 1972) and is usually between 10 and 30m/h. The performance of fluidised-beds depends on the carrier materials and the hydrodynamics of the digester.

Sand has been the usual support medium for most fluidised-beds, but other natural materials such as anthracite and gravels have also been used. To reduce fluidisation velocities, lighter natural materials have been tried, for example, activated carbon, baked clays and porous ceramics. Synthetic materials can also be fabricated into porous structures with a low density, and plastic carriers have included resin beads, reticulated foams, PVC and other types of synthetic beads. Diffusion into porous particles may become limiting and the carrier may break down with use.

Much of the effectiveness of a fluidised-bed compared with an anaerobic filter can be attributed to the smaller support particles which can be used without risk of clogging. There has been little systematic research on the optimum size of carrier, but below 0·5 mm there is little advantage in the upflow velocity required for fluidisation (Iza, 1991). However, start-up has been shown to be much quicker with smaller particles (Enger *et al.*, 1988) and this was attributed to the lower liquid shear velocities with smaller particles (0·35 compared to 0·7 mm). Particle sizes are usually between 0·3 and 1 mm; smaller sizes are difficult to control and retain within the digester, larger sizes are difficult to fluidise. Surface roughness, specific surface area, strength and stability will affect media performance, as in anaerobic filters. The ideal physico-chemical characteristics of fluidised-bed particles have been listed for both aerobic (Walker & Austin, 1981) and anaerobic (Switzenbaum, 1983; Iza, 1991) processes: the particles must be uniform in shape and size to obtain an even fluidisation or bed expansion throughout the digester height. Some of these properties do, however, change with the amount of biomass, and that makes fluidised-bed digesters difficult to control.

The support particles are expanded and mixed by the gas production and effluent recycle. Sufficient recycle is needed to cause some lateral mixing of the particles and to avoid channelling. The amount of bed expansion depends on whether the digester is being run as an 'expanded' or 'fluidised' bed (see Chapter 7) but is typically about 20–25% of the digester volume. Obtaining uniform bed expansion is relatively easy while the carrier particles are homogeneous, but it becomes more difficult as the particles change shape during bacterial colonisation.

The design of the feed distribution system, as with other high-rate digesters, is a critical feature of the mixing characteristica. Many of the details of full-scale distribution systems are included in proprietary designs or are covered by

Table 4
Some design and operating parameters of fluidised-bed digesters.

Parameter	Design	Operating
Loading rate	2–15	2–50*
Retention time	10–50	0.5–24†
COD removal	70–90	70–80‡

Units: * kgCOD/m^3/d; † hours; ‡ percent.

confidentiality agreements, and are thus not freely available. Small-diameter (1–2 m) digesters can utilise an inverted cone with a downward pipe at the apex of the cone. Larger-diameter digesters have used perforated plates, nozzles, or square, annular or star-shaped pipework.

To improve the distribution and reduce the volume of recycled effluent, fluidised-beds are usually long, thin tanks with a 5 or 6:1 height to diameter ratio. A second, distinct zone is usually incorporated within the digester for clarification of the effluent. This can be an increase in the diameter of the top 20–25% of the vessel. Foaming can be a problem with certain protein-containing wastes or if the fluidisation energy required is high. This may be counteracted by an anti-foam agent or an upper grid to avoid particle loss. Designs in the USA usually use a separate vessel as the clarifier and particle separator and this also contains a filter grid. Table 4 shows some typical operating parameters of fluidised-bed digesters.

Biological fluidised-bed technology has been extensively studied at laboratory and pilot-plant scale, but there are as yet few full-scale examples. Most of the usual wastes have been tested in the laboratory (e.g. Jeris, 1983; Rudd *et al.*, 1985; Stronach *et al.*, 1986 and references in Chapter 7). The potential feedstocks are much the same as those treated by the otherretained-biomass digesters described previously. The fluidised-bed system has the important potential advantage that very large concentrations of biomass can be developed within the digester without problems of blockage and channeling (40–100 kg VSS/m^3; Enger & Heijnen, 1988). Potential performance is therefore high: typical loading rates are shown in Table 4.

There were three designs of fluidised-beds being built commercially with one type in Holland, France and Germany described by Enger *et al.* (1988), another in France and Spain (Iza, 1991; Oliva *et al.*, 1990), and 4–5 plants in the USA (Heijnen *et al.*, 1989). No anaerobic fluidised-beds had been built in the UK at the time of writing. Thus, there is very little full-scale design and operating experience and fluidised-beds are said to be difficult to scale-up and operate (Sixt & Sahm, 1987).

Other types

There are now many hundreds of large, mechanised, heated, anaerobic digesters operating in different parts of the world, as well as some millions of small, non-mechanised digesters operating on peasant farms in the warmer countries. The majority of the mechanised digesters, and all the non-mechanised ones, are stirred-tank sludge digesters. These treat domestic wastes from a village or a town, sometimes with municipal garbage or other wastes added; or farm wastes, with, in the case of the small peasant-farm digesters, house wastes or vegetation. Anaerobic digestion under controlled conditions is the only feasible biological treatment for wet, sludgy, wastes: uncontrolled biological systems such as lagoons or land-spreading are inefficient and pose many environmental problems. Non-biological treatments such as incineration are expensive and need plant which can be run only on a large scale with skilled workers.

Digestion plants can be on any scale and the degree of automation is governed by size and availability of labour. Small, peasant-farm digesters are stirred-tanks but rely on hand and gravity-assisted loading and unloading and on hand-paddles or gas production for mixing. Some of these small digesters are run on a batch system and batch digestion has been used in larger plants in temperate countries, although, so far as is known, no batch digester of bigger volume than the smaller stirred-tank farm or sewage digester has yet been built (see Chapter 6). The batch digesters have been used for more or less solid farm manures and vegetation. Loading and unloading is mechanical for the larger digesters; they may or may not be heated, but are not mixed. The digesters for solid municipal wastes mentioned in Chapter 6 were, at the time of writing, mainly pilot-plants. The one or two full-scale plants were automated, and these were on a continuous-flow basis, although batch digestions were being tried on a small scale.

The hybrid digesters for treating wastewaters and the two-stage, two-phase and other designs for sludgy wastes were mostly laboratory- or pilot-plants at the time of writing. However, large-scale plants of these types should develop in the next few years.

Apart from the digester tank, all large digesters need ancillary equipment for storing feedstocks and balancing flows or pretreating feeds. They also need tanks for containing effluents until they can be disposed of, and the effluent tanks in a sewage works are part of the process. Continuing slow digestion of the stored effluent at ambient temperature reduces remaining pollutants and pathogenic organisms, and storage allows the sludge to consolidate for easier removal. Automated digesters also require buildings for control equipment, pumps, boilers, engines and, possibly, automated equipment recording performance. There may also be secondary plant for further treatment of the digester effluent by biological or physico-chemical methods before the effluent is discharged.

CONSTRUCTION

Concrete

The first digesters were, as previously noted, large stirred tanks for the treatment of municipal sewage sludges. For many years these were built of concrete.

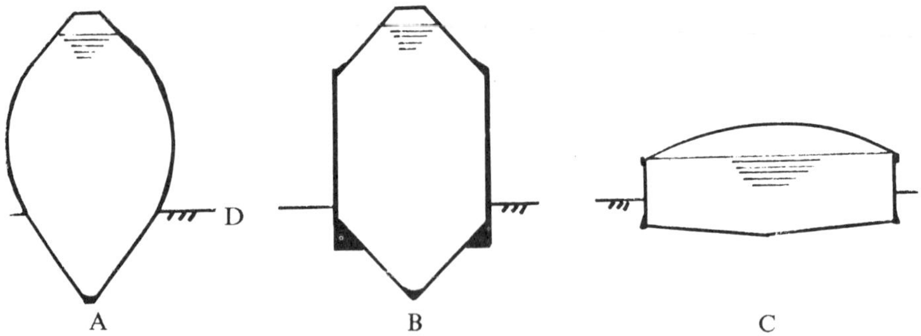

Fig. 9. Shapes of concrete sewage digesters. A, pear shape. B, tall cylindrical with conical floor for grit collection. C, older type of low aspect-ratio tank with flat, or almost flat, floor. D, possible ground level, but this can be higher or lower.

Capital costs were high, but the plant was expected to be operational for over fifty years. This was appropriate at a time when standards and technology were changing slowly and inflation was low.

One of the effects of the high capital costs of concrete digesters was to restrict the application of sludge digestion to the largest sewage works. There are 75 sewage works in the UK which serve populations greater than 100 000 (equivalent to a 2000 m^3 digester) and 95% of these use anaerobic treatment in concrete digesters. These large digesters are tanks of small height to diameter ratio and have flat, or very slightly sloped bottoms (Fig. 9). For large digesters over 2000 m^3 in size concrete is still the cheapest method of construction but with 80% of the UK population already served by sewage treatment there will be few digesters of this size built in the UK in the future. A number of concrete digesters were being built in Europe at the time of writing. Continental digesters are much taller than most UK concrete digesters (Fig. 9) and this tall thin shape is intended to make the digester contents much easier to mix by vertical motion. A steep conical bottom (45°) is also added to keep grit and scale in suspension for periodic removal. This type of tall digester needs to be well insulated because of the large surface area above ground and it is also very expensive to build; two or three times more than flat-bottomed digesters. This has led to a change in profile to an egg shape which can be constructed from successive ring segments and so reduces the amount of shuttering and scaffolding, which was said to reduce construction costs by 30% (Greeman, 1990). Six of this type of digester were built or being built in Germany at the time of writing. Two were under construction at Dortmund, for example, and were 12000 m^3 in volume and 24 m in diameter at the widest point. They were designed to treat screened, consolidated, primary- and secondary-sewage sludges at a retention time of 30 days. The digesters were mechanically mixed and sludge was drawn up a central draft tube and splashed over the surface of the digester liquid. The design and construction methods were based on experience of building four similar, but even larger, digesters (15 000 m^3) at nearby Bottrop.

Total costs for the whole sewage-treatment plant were to be £54M of which digester construction costs were £4M. This is equivalent to about £160/m^3: quoted UK costs for concrete digesters were £70/m^3. These figures were doubled by the costs of ancillary equipment.

Although some of the earlier farm digesters were made of concrete, and the Chinese-type digesters are still made of brick or stone with a concrete lining, large concrete digesters were too expensive and restricting for most industrial and farm use and so the standard prefabricated tanks from other areas of farm and chemical work were adopted.

Enamelled steel

The smaller sizes of prefabricated tanks made from enamelled-steel plates bolted together are 70% cheaper than concrete tanks. The glass-enamel coating is carried out at the works and holes for piping, as well as the holding bolts, can be made during fabrication of the plates. This reduces on-site work and associated delays. Such tanks have been used for many years as grain silos, storage tanks for farm slurries and other liquids, etc., so there is much experience in their construction. The tank is built up on site from the shaped plates. The edges of the plates are sealed with mastic as they are bolted into the rings which form the tank. The bottom ring is grouted into a flat concrete base. The roof is of the same enamelled steel plates or plastic. Fixed, or floating gas-holder, roofs can be fabricated. The flat floors can accumulate grit, but manholes near the base can allow easy access and degritting. The tanks require insulation and this can be internal or, more generally, external. Fifty to 100 mm layers of polyurethane or polyisocyanurate foams can be used, or glass-wool (external only). All external insulations need a cladding or painting to waterproof the materials and to prevent birds or rodents attacking, or nesting in, the foam. The largest digester size at the time of writing was 1000 m^3, the limit being set by the maximum size of the steel-plate roof, but larger sizes could be built by reinforcing the roof, which may be subject to significant flexing with the varying gas production. Multiple tank systems provide for greater capacity, and multiple tanks have some advantages over large single tanks in servicing and dealing with breakdowns. Increases in waste flow can be dealt with by adding extra tanks. Guaranteed plant life is 20 years compared to 40 for concrete.

The prefabricated stirred-tank digester has been very successful with sewage and farm wastes. There is no reason why the tanks should not be used for other wastes or other digester forms: some have been, but there is not as much experience in these areas as in the former.

Steel Tanks

Welded-steel tanks have been used for larger industrial digesters greater than 2000 m^3 in volume. They are about 10% cheaper than smaller enamelled tanks and 20–30% cheaper at the larger sizes where multiple enamelled tanks would be needed. It is usual to coat the insides of mild-steel tanks with epoxy-rubber. Such coatings inside digesters can become detached or broken and liquid will get behind them. There are doubts as to whether coating is needed. One of the authors and colleagues ran a 13·5 m^3 pilot-plant mesophilic digester on farm wastes for many years. This was made of welded mild steel and no corrosion was evident; on opening, the inside surface was found to be covered with a layer of black sulphide polished with the abrasion of the mechanically-mixed slurry. The outside of the tank was covered with a glass-fibre insulation under a riveted galvanised-iron-sheet jacket. Mild-steel digesters need insulation if they are running at above ambient temperature.

Stainless-steel has been proposed for digesters but its cost is too high for all except small, or pilot-plant tanks. Aluminium is unsuitable because of the high chloride concentration in the warm, digesting sludge. Galvanising provides satisfactory protection of steel tanks from the weather, but the rate of corrosion of galvanised steel in the digester liquid or the moist atmosphere above the liquid makes galvanising unsuitable for internal surfaces of digesters. Joining of pipes of different metals or alloys can lead to corrosion.

Plastics

Glass reinforced plastic (GRP) is corrosion-resistant and has been used for small digesters. There are difficulties with larger sizes because of its low strength, and the material is too flexible for large tanks without extra support. This can be overcome to a certain extent by excavation to sink the tank and backfilling, or the tank can be made into a horizontal tubular digester partly buried in the ground. GRP tanks are fabricated off-site and transported as a whole. This leads to an upper size-limit for complete tanks based on ease of transport, but larger tanks can be made up from moulded rings bolted together on-site. To provide sufficient self-supporting strength vertical digesters have to be tall and narrow and digesters up to 500 cubic metres have been built in this way. However, GRP digesters are no cheaper than enamelled-steel digesters at sizes above 200 cubic metres. The costs of moulds for plastic digesters also means that 'one-off' tanks are hardly viable.

Rural digesters

In Chapter 6 mention is made of farm digesters formed from slurry pits under animal houses. Some digesters for use on larger farms in temperate countries have been made of plastic-lined troughs dug into the ground (see Chapter 3, tubular digesters) and farm-slurry lagoons have been covered in various ways to form digesters. These can be relatively cheap ways of building digesters, but if the digesters are to be run at mesophilic temperatures similar to those of the tank digesters mentioned above, and gas is to be used, then the cost of the heating plant and other ancillary equipment may make savings on digester construction rather negligible: rubber or plastic linings for excavations are not cheap.

In warmer countries digesters for small peasant farms are most commonly constructed below ground. They may be troughs lined with cement or plastic sheet, but the most numerous are either approximately spherical (Chinese) or tubular (Indian Gobar) excavations lined with bricks, stones or concrete blocks, and finished with a coating of cement plaster to make them gas- and liquid-tight (see Chapter 6). Cubical excavated and brick-lined tanks have also been used, as have wooden tanks; the latter even in very large factory digesters (Taylor, 1972).

Gas-holders

Most appliances use digester gas at a pressure of a few inches water gauge and gas is stored at this pressure. In the Chinese digesters mentioned above the gas is stored in the top of the rigid digester and so the pressure, balanced by the slurry columns in the feed and effluent pipes, varies as gas is produced and used. This is not critical for running simple lights and gas cookers. However, most gas appliances run better with a constant supply pressure and the majority of gas-holders in use are designed to produce constant pressure.

The gas-holder of the Indian digester is a cylindrical metal tank of smaller diameter than the digester tank and rising and falling in the digesting slurry as the gas volume varies. Such 'floating-roof' gas holders are used in many large concrete sewage digesters. The gas holder is stabilised as it rises and falls by guide rails and rollers. One of the authors and colleagues ran for some years a 680 m^3 farm digester with a floating-roof gas holder of 190 m^3 capacity (Summers *et al.*, 1984). The digester was enamelled steel and the gas holder essentially a smaller-diameter enamelled tank with an enamelled roof. This was sealed by the slurry in the digester and the movement was stabilised by guide rails and rollers with a system of steel ropes to even out side pressures from wind. The wind caused no problems with the gas holder, but there was a tendency in hot weather for slurry to dry out in the annular space between the digester tank and the gas-holder tank and the resultant solids tended to impede the vertical motion of the gas holder. The solids were cleared at intervals.

If the digester has a rigid roof then a separate gas-holder is needed. The most usual type is the floating-dome holder working on the same principle as the floating-top holder just described. The bottom is a tank filled with water and gas is piped in above the water level to accumulate in the top tank which rises and falls with the gas volume. The bottom of the gas-tank is sealed in the water and the tank is stabilised by guide rails and rollers. The gas tank may need counter-balancing by weights to produce the gas pressure required. A water-sealed gas-holder used on a pilot-plant run by one of the authors is illustrated in the paper by Robertson *et al.* (1975). These gas-holders are usually made of mild steel, but some small ones in GRP are available. The separate water-sealed gas-holder or floating-roof digester have the advantage of acting as a safety valve if gas pressure is increased by failure in some other part of the system as excess gas will bubble out through the liquid seal if the gas production exceeds storage capacity. No major problems should be found in temperate climates with the separate gas-holder, although in cold weather the water can freeze. A small holder used by one of the authors was filled with water and glycol (car antifreeze) but this is too expensive for large holders; a water-oil emulsion can protect these. An electric heater, switched on if the temperature fell much below freezing point, was used on one of the authors' plants but this would be impracticable for large holders. Very hot weather might cause excessive evaporation of the water, so topping-up would be needed.

Plastic- or rubberised-fabric digester tops are used as gas holders. One large, cylindrical-tank farm digester was fitted with a rubberised-fabric top that inflated to a hemispherical shape as it filled with gas. However, so far as is known large digester-top gas holders have not become common, most plastic or rubber tops are used on small digesters; usually the-trough-in-the-ground type. The top may be sealed by fixing it mechanically to the edge of the digester trough, or the bottom edge may drop below the sludge level to give a water seal. Butyl-rubber or plastic bags are also used as separate gas holders. The bag or digester top is not usually inflated to stretch the fabric but to take up slack in a loose 'bag'. A uniform pressure can be obtained by having weights directly on top of the bag or operating through a system of levers. A number of designs were reviewed by Dohne (1980). Bags need separate safety valves to avoid overpressure and care has to be taken to see that fabrics are not torn or otherwise damaged. In hot weather dark-coloured gas bags can get very warm and so gas pressure variations can be considerable.

All these systems store gas at low (*c.* 2–6 in [5–15 cm] WG) pressure and the gas does not need to be specially cleaned before storage. The water in a water-sealed holder will scrub some impurities from the gas, so it is best changed after maybe a few years. The capacity of the gas-holder depends very much on the use made of the digester gas. If the gas is running a boiler or an engine virtually non-stop and this uses the available gas, then a gas-holder of capacity only a few cubic metres will act as a gas ballast volume for a large digester. However, if the gas usage is intermittent, say during the working day, then gas produced at night might be stored for use during the day and the gas-holder will have to be sized on the nightly gas production. (This assumes that the total day and night production can be used). In general, gas storage capacity should be as small as possible as steel water-sealed gas holders are as, or more, expensive than the digester tanks, and butyl rubber or other gas-impermeable fabric holders are of the same order of price unless they can be made in large numbers to one specification.

Digester gas can also be stored at medium (100–200 psi) pressure in commercially-available steel tanks. These tanks are not expensive for the volume of gas which they contain, but added to the capital cost must be added the cost of the compressor and associated valves and pipework and the running costs of the compressor. The tanks need to be more rigorously protected from external corrosion than do the low-pressure gas holders and they can be subject to internal corrosion from traces of hydrogen sulphide and ammonia, or even from the carbon dioxide if the gas is wet. The tanks need to be well painted internally and externally or enamelled as are the commercial tanks used for storage of LPG.

Pressurised gas tanks will be subject to local regulations for pressure vessels and may be inspected at intervals. Reducing valves will be needed to reduce the gas to the low working pressure of the gas appliancecs. Gas can also be stored at high pressure (2000 or more psi), but this needs special gas cylinders of the type used for commercial gas storage and the gas needs to be dried and purified before compression. Such storage would seem to be feasible only when pure methane was being sold in relatively small quantities for factory or other use.

The medium-pressure gas lines need mechanical safety valves; low-pressure lines can rely on blow-off from water-sealed gas holders or U-tube water seals on the digester or even from condensate pots in the gas lines which remove water from the gas. However, the U-tube water seals can freeze in very cold weather so oil may be used in some cases or a strong glycol solution. The bigger digesters usually have an automatic flare which burns off large excesses of unused gas and is started when the gas-holder is filled. Gas pipe runs are preferably lagged to reduce the possibilities of water freezing out of wet gas and blocking pipes or condensate traps.

Digester overflow and secondary tanks

Effluent removal is often by a weir or similar system which allows gravity overflow while retaining a few inches WG gas pressure in the digester. Gravity overflows should pose few problems with wastewaters of low suspended-solids content, but they can be subject to blockage from particulate solids in sludges; so, the effluent may be pumped out. This, theoretically requires balanced flows in the feed and effluent pumps for automated running. However, feed and effluent sludges differ in consistency, particle size and other attributes, and balancing of flows by similar on-off times of the pumps is difficult. Unbalance can produce either over-pressure or under-pressure in the digester head-space, as can blockages in feed or effluent pipes. To overcome unbalance the pumps can be timer-controlled but over-ridden by a level-sensor for the digesting sludge. Level-sensors have been float switches, conductivity probes or load-sensors in the liquid: the latter obviate the problems of foam and floating solids encountered by apparatus working at the liquid-gas interface. In a farm digester run by one of the authors and colleagues (Summers *et al.*, 1984) the digester effluent was pumped out to a pipeline which ran for some hundreds of yards around the farm buildings to the storage tank. To prevent blockages in this pipe causing blow-outs from the continued operation of the effluent pump the pump was fitted with a cut-off which stopped it when more than normal back-pressure developed in the pipe: in turn the digester feed-pump was cut off.

As mentioned before, digesters generally need feed-balancing tanks and storage tanks for effluent. These can be of concrete, or the same enamelled-steel as the digester, or the effluent tank can be a lagoon. Feed and effluent pumps and compressors and distribution pipes for gas mixing require protection from the weather. This may be provided by small wooden or other 'boxes' over the apparatus, or the pumps may be housed in the building holding the digester controls, boiler, engine, or other ancillary equipment for distributing hot water for site heating, etc. Such buildings may be prefabricated units or brick or concrete buildings. These last remarks apply to farm, factory or small sewage works in particular; the larger sewage works will run a number of large (maybe 1000 HP) engines and generators, or air-compressors for the activated-sludge plant, and these will be housed in large engine houses. Similarly, pumps will be on a large scale. At the other end of the scale, the small peasant-farm digester in a warm country requires only a rubber and metal pipe (and perhaps a condensate pot) joining the digester and the household gas rings and lights: with, perhaps, a small, open, brick or concrete tank to store effluent before land-spreading. Feed inflow is by gravity in open channels from the animal house. But gravity-flow feed is used in some large automated farm digesters – the permutations of design are many and do not depend entirely on digester size or siting; so only some short notes on digester design and construction can be given.

CONCLUSION

A brief description of the types of commercial digesters and their construction has been given. In sewage-sludge digestion there has been, overall, a move away from the large concrete digester to the smaller steel tank. However, the most rapidly-expanding field in western countries is probably that of the digesters for industrial wastewaters. Farm digesters were the leaders in the increase in interest in digestion in western countries in the early nineteen seventies, but there was some disillusion with early digesters which were often experimental as well as working digesters owing to lack of design information. This, combined with decrease in oil prices and lack of pressure from environmental factors, led to the present decrease in use and construction of farm digesters. In China and India and other warm countries with largely peasant farming there has been some disillusionment with small farm digesters, but new initiatives are being taken to increase numbers. The problems have been partly poor performance of digesters, and this has been largely due to poor management of the systems, although there have been some constructional problems. The former can be overcome by increased training of technicians in servicing of the digesters and of farmers in the running of the plants. Work to remedy construction problems continues. But a big factor here is still the price of even simple digesters compared with the money earned by the farmers. It is difficult to see how this problem can be overcome without government subsidies and loans; the construction of the digesters cannot be much simplified or cheapened. Another problem of the climatically-heated small digester in many areas is lack of gas production in winter or at nights. Various systems of solar heating or compost heating are being suggested and used, but a cheap, simple and reliable heating system for peasant-farm digesters would be of immense benefit.

As environmental concern grows then the market for digesters in western countries will increase, but at the present this market is limited. This has meant that although 'do-it-yourself' construction of digesters has probably virtually stopped many of the firms active in construction of digesters some years ago have ceased trading or have amalgamated. While the remaining firms have solved most of the problems of construction and use many standardised components, there is still a lack of knowledge about the digestion of many factory or other wastes and of the treatment of bimass residues that could form the sources of fuels of tomorrow. There is still a place for the laboratory and pilot-plant experiments on digestion.

REFERENCES

Adams, C.D. & McKinney, R.E. (1989). In *Proc. 44th. Purdue Ind. Waste Conf.*, Ann Arbor Science, Michigan, p. 1.

Alibhai, K.R.K. & Forster, C.F. (1986). *Environ. Tech. Letters* **7 (4)**, 193.

Baumann, P.G. & Huibregtse, G.L. (1981). *J. Wat. Poll. Cont. Fed.* **54**, 1194.

Bolle, W.L., van Breagel, J., van Eyeberger, G.C., Kosser, N.W.F. & Zoetemeyer, R. J. (1986). *Biotechnol. Bioeng.* **28**, 1615.

Bonastre, N. & Paris, J.M. (1989). *Process Biochem.* **24(1)**, 15.

Brade, C.E. & Noone, G.P. (1981). *Wat. Pollut. Cont.* **80**, 70.

Brade, C.E., Noone, G.P., Powell, E., Rundell, H. & Whyley, J. (1982). *Wat. Pollut. Cont.* **81**, 200.

Butcher, G.J. (1988). *Internat. Biodeterioration* **25(2)**, 71.

Butler, G.A. (1984). In *Anaerobic Treatment of Industrial Wastes*, ed. J. Coombs, BABA, Reading, U.K. p.

Camilleri, C. (1987). In *Biomass for Energy and Industry, 4th. EC Conference*, ed. G. Grassi, B. Delmon, J. F. Molle & H. Zibetta, Elsevier Applied Science Publishers, London, p. 1338.

Camilleri, C. (1988). In *Fifth International Symposium on Anaerobic Digestion: Poster Papers*, ed. A. Tilche & A. Rozzi, Monduzzi Editore, Bologna. p. 473.

Dahab, M.F. & Young, J.C. (1982). *Proc. 1st. Int. Conf. Fixed-Film Biological Processes,* King Island, Ohio, April 1982. p. 1337.

Department of Scientific and Industrial Research (1960). *Water Pollution Research.* The Report of the Water Pollution Research Laboratory. 1955–1959. H.M.S.O., London.

Dohne, E. (1980). In *Anaerobic Digestion*, ed. D.A. Stafford, B. I. Wheatley & D.A. Hughes, Applied Science Publishers, London. p. 429.

Ehlinger, F., Audic, J. M., Vernier, D. & Fraup, G.M. (1987). *Wat. Sci. Technol.* **19**, 261.

Enger, W.A., de Ram, R.A., Cignoli, A. & Gatli, G. (1988). In *Fifth Int. Symp. on Anaerobic Digestion: Poster Papers*, ed. A. Tilche & A. Rozzi, Monduzzi Editors, Bologna. p. 495.

Environment Canada (1986). Report EPS 3/FP/1 Environment Canada, Ottawa.

Fiestas, J.A., Martin, A. & Borja, R. (1990). *Biol. Wastes* **33**, 131.

Greeman, (1990). *New Civil Engineer*, **10**, 14–17.

Hall, E.R. (1983). *Wat. Sci. and Technol.* **15(8/9)**, 371.

Heijnen, J.J., Mulder, A., Enger, W. & Hoeks, F. (1989). *Chem Engng. J.* **43**, 3, B37.

Henze, M. & Harremoes, P. (1983). *Wat. Sci. and Technol.* **15(1)**, 1.

Hudson, J.A., Bruce, A.M., Oliver, B.T. & Auty, D. (1988). *J. Inst. Wat. Environ. Management.* **2**, 420.

Iza, J. (1991). *Wat. Sci. and Technol.* **24(8)**, 109.

Jeris, J. S. (1983). *Wat. Sci. and Technol.* **15**, 167.

Kennedy, K.J. & Droste, R.L. (1986). *Wat. Research* **20**, 685.

Kennedy, K.J. & Droste, R.L. (1991). *Wat. Sci. and Technol.* **24(8)**, 157.

Kidson, R.J. & Ray, D.L. (1984). In *Sewage Sludge Stabilisation and Disinfection*, ed. A.M. Bruce, Ellis Horwood, Chichester, UK. p. 399.
Kissel, J.C., Grotenhuis, J.T.C. & Zehnder, A.J.B. (1988). In *Fifth Int. Symp. on Anaerobic Digestion: Poster Papers*, ed. A. Tilche & A. Rozzi, Monduzzi Editore, Bologna. p. 207.
Lettinga, G. & Hulshoff-Pol, L.W. (1986). *Wat. Sci. and Technol.* **18**, 12, 99–108.
Lettinga, G. & Hulshoff-Pol, L.W. (1991). *Wat. Sci. and Technol.* **24(8)**, 87.
Lettinga, G., Man, A.D., de Grin, & Hulshoff-Pol, L.W. (1987). *Tribune Cebedeau* **40(519)**, 21.
Lettinga, G., van Velsen, A.F.M., de Zeeuw, W. J. & Hobma, S.W. (1979). In *Anaerobic Digestion*, ed. D.A. Stafford, B.I. Wheatley & D.E. Hughes, Applied Science Publishers, London. p. 167.
Levenspiel, O. (1972). *Chemical Reaction Engineering*, Wiley, Chichester, U.K.
Ministry of Housing and Local Government (1954). Report of an Informal Working Party on the Treatment and Disposal of Sewage Sludge, H.M.S.O., London.
Morgan, H. (1980). In, *Food Industry Wastes: Disposal and Recovery*, ed. A. Herzka & R.G. Booth, Elsevier Applied Science Publishers, London. p. 92.
Murray, W.D. & Van den Berg, L. (1981). *J. Appl. Bacteriol.* **51**, 257.
Nahle, C. (1991). *Wat. Sci. and Technol.* **24(8)**, 179.
Noone, G.P. & Brade, C.D. (1985). *J. Inst. Wat. Contr. UK.* **84(3)**, 309.
Oleszkiewicz, J.A., Hall, E.R. & Oziembio, J.Z. (1986). *Env. Tech. Letters* **7**, 445.
Oleszkiewicz, J.A., & Thadari, V.J. (1988). *Env. Tech. Letters* **9**, 89.
Oliva, E., Jacquark, J.C. & Prevot, C. (1990). *Wat. Sci. Technol.* **22**, 2, 483.
Racault, Y. (1990). *Wat. Sci. and Technol.* **22(1/2)**, 361.
Robertson, A.M., Burnett, G., Bousfield, S., Hobson, P.N. & Summers, S. (1975). In, *Managing Livestock Wastes*, ASAE, St. Joseph, Michigan, USA. p. 544.
Rudd, T., Hicks, S.J. & Lester, J.N. (1985). *Env. Tech. Letters* **6**, 209.
Rundle, H. & Whyley, J. (1981). *J. Wat. Pollut. Contr. UK.* **80**, 643.
Samson, R. & Kennedy, K.J. (1985). *J. Biotechnol.* **2**, 95.
Souza, M.E. (1986). *Wat. Sci. and Technol.* **18(12)**, 55.
Shore, M., Broughton, N.W. & Bumstead, N. (1984). *J. Wat. Pollut. Contr. UK.* **83**, 499.
Sidwick, J.M. (1986). *Public Health Engineer* **14(1)**, 47.
Sixt, H. & Sahm, H. (1987). In *Biotechnology of Waste Treatment and Exploitation*, ed. J.M. Sidwick & R.S. Holdom, Ellis Horwood, Chichester, U.K. p. 149.
Smith, N. (1984). In *Anaerobic Treatment of Industrial Waste*, ed. J. Coombs, BABA, Reading, England. p. 27.
Stronach, S.M., Rudd, T. & Lester, J.N. (1986). *Anaerobic Digestion Processes in Industrial Waste Treatment*, Springer-Verlag, Berlin.
Summers, R., Hobson, P. N., Harries, C. & Fielden, N.E.H. (1984). *Process Biochem.* **19**, 77.
Swanwick, J.D., Shurben, D.G. & Jackson, S. (1969). *J. Wat. Pollut. Contr. UK.* **68**, 639.
Switzenbaum, M. (1983). *Water Sci and Technol.* **15**, 399.
Szendry, L.M. (1983). In *Proc. 3rd. Int. Symp. Anaerobic Digestion*, Boston. p. 365.
Szendry, L.M. (1986). In *Energy from Biomass and Wastes VII*, IGT, Chicago, USA. p. 767.
Taylor, D.W. (1972). In *Proc. 3rd. Nat. Symp. Food Proc. Wastes*, New Orleans, USA. p. 172.
Tilche, A. & Vieira, S.M.M. (1991). Reactor design of anaerobic filters. *Wat. Sci. & Technol.* **24**, 8, 193–206.
Walker, I. & Austin, E.P. (1981). In *Biological Fluidised Bed Treatment of Water and Wastewater*, ed. P.F. Cooper & B. Atkinson, Ellis Horwood, Chichester, U.K. p. 272.
Wheatley, A.D. (1990). *Crit. Reports on Appl. Chem.* **31**, 171.
Wheatley, A.D. & Cassel, L. (1985). *Wat. Pollut. Contr.* **84(1)**, 10.
Wheatley, A.D. & Coombs, J. (1991). *The Potential of Anaerobic Digestion.* CPL Scientific, Reading, U.K.

Wheatley, A.D., Johnson, K.A. & Winstanley, C.I. (1988). In *Anaerobic Digestion 1988*, ed. E.R. Hall & P.N. Hobson, Pergamon, London. p. 135.
Weiland, P. & Rozzi, A. (1991). *Wat. Sci. and Technol.* **24(8)**, 257.
Witt, E.R., Humfrey, W.J. & Roberts, T.E. (1979). Proc. 34th. Purdue Ind. Waste Conf., Ann Arbor Science, Michigan. p. 299.
Young, J.C. (1991). *Wat. Sci. and Technol.* **24(8)**, 135.
Young, J.C. & Dahab, M.F. (1983). *Wat. Sci. and Technol.* **15(8/9)**, 369.
Young, J.C. & McCarty, P.L. (1969). *J. Wat. Pollut. Cont. Fed.* **41(5)**, R160.
Young, H.W. & Young, J.C. (1988). *J. Environ. Eng.* **114**, 621.

Chapter 5

PRETREATMENT OF FEEDSTOCKS

INTRODUCTION

The feedstocks for digestion can range in physical form from damp solids through slurries of particles suspended in liquid to virtually particle-free liquid. The liquid is water, but the suspended or dissolved materials contain components with a wide range of chemical compositions. These may be naturally-occurring polymers and simpler molecules such as carbohydrates, fats, sugars and amino acids, or they may be factory-produced organic chemicals. The complex mixture of chemical structures may be in one physical form, say ground vegetable material, and this is most likely to occur in feedstocks for energy production from biomass. However, most digesters are run for waste treatment and the mixture of chemical structures is most likely to be found in a mixture of particles of different shapes and sizes as well as materials in solution. Besides these completely- or partially-biodegradable materials which form the 'active' part of the digester feedstock there can be, particularly with wastes, amounts of other matter (metals, stones, grit, plastics) which is undegradable. Except for the very smallest digesters, these feedstocks have to be mechanically loaded into the digester. From the purely mechanical point of view, then, it may be necessary to pretreat the feedstock to remove, or change the physical form (usually a reduction in the particle size) of, some of these particles, degradable or undegradable, otherwise they would destroy or block pumps, pipes and the internal structures of the digester. From the biological point of view the bacterial attack on solid particles is from the surface (Chapters 2,3) and the bigger the surface area per unit weight, i.e. the smaller the particle, in general the faster and more complete will be the bacterial degradation.

The mixed bacterial population of a digester requires a mixture of sources of energy, cell carbon- and nitrogen-compounds, and growth factors such as trace elements (Chapter 2). In many feedstocks these bacterial requirements are more or less completely provided and although a slight deficiency in one component may lead to a slightly incomplete digestion it may not be economically worthwhile to try to remedy this deficiency. Some factory wastes, however, may contain only a few components and be virtually lacking in some important bacterial growth compound. If digestion is to take place at all then the missing factor must be added to the feed and this addition may be economically viable in the context of the waste treatment. The pH of the digestion must be about neutral for a complete digestion. The pH is usually balanced around this value by the bacterial reactions and the composition of the feedstock, not by an external control and addition of acid or alkali. This biological control can very often overcome adverse pH of the feedstock, but sometimes the feed is too acid and predigestion adjustment of the pH is required. For instance, Yan *et al.* (1989) added sodium bicarbonate as a buffer to cheese whey, and NaOH to adjust the pH from 4–6 to 7. Marshall & Timbers (1982) also found that whey required NaOH for pH control.

It is apparent, then, that although digesters may be fed with a virtually untreated feedstock, for the reasons given above, and others, some physical or chemical treatments of the feed can be required. These are considered below. Apart from the treatments mentioned later most liquid feedstocks containing suspended particles need mixing to produce a homogeneous suspension before the feed enters the digester. Digester feeding is generally semi-continuous and the waste is supplied from some collecting tank. During periods between feeds suspended solids in the tank may float or sink and later not be taken in by the digester feed pump. This not only leads to irregular solids input to the digester but may lead to blockages as the feed pump tries to take from a dense sludge at the bottom of the tank or from floating particles. In these cases it is usual to fit a mechanical or an air-bubble mixer or arrange for the feed pump to circulate the slurry round the tank before it pumps to the digester.

Some pretreatments of feedstocks have already been mentioned; for instance, the combining of wastes to provide a mixture of better C:N ratio (Chapter 2). However, if there is excess N in a waste in the form of ammonia then while combination with a waste of high C content may provide a theoretically more optimum C:N ratio, if the ammonia concentration is high this will not prevent ammonia inhibition of digestion (Summers *et al.*, 1987). A rather round-about way of pretreating the feedstock is suggested by the effects of animal feed constituents on the digestion of the animal excreta (Chapter 2). Economic factors and animal-production factors being equal, it could be possible to feed the animals to provide excreta of optimum digestibility. Athough the animal feed may induce faeces of different degradabilties in a digester, the reactions of digestion may lead to digested sludges of fairly uniform composition. For instance, in India many of the cattle on village farms are fed on poor fodders and the excreta from these animals is used as feedstock for digesters. The effluents are usually used as fertilizer, but it has been suggested that the effluents could be used as animal feed protein additives. Saxena *et al.* (1989) found that digestion could upgrade the protein content of slurries. When excreta from cattle fed on high-concentrate, low-concentrate or poor-roughage rations of 10, 7 and 6% crude protein were digested in a climatically-heated, Gobar, digester of RT 30 days, the digester effluents were all of uniform protein content (15%). The effluent protein was about 50% digestible in rations for buffalos when mixed with straw and molasses and could at least maintain weight of the animals. The effluent could be used as a protein supplement to poor rations when conventional concentrates were too expensive or not available. Plant geneticists are now trying to confer on plants greater resistance to pests and climate and to improve yields and digestibilities by animals of feed crops. Such changes may also change the faecal residues which form the anaerobic digester feedstocks, although increased digestibility in the animal would suggest that the residues will be less in volume, and of lower degradability in the digester. However, when it comes to crops grown for energy production genetic manipulation could increase anaerobic degradability and gas production. In the long series of experiments carried out on the growing and anaerobic digestion of seaweeds for energy production genetic improvements in growth rates of kelps is part of the work (Neushul & Harger, 1983). Such improvements would not make the material more digestible, but they would increase gas production per area of kelp per unit time. On the other hand Hiler (1986) reported genetic work in Texas designed to improve the composition of sorghum to make it more digestible as an energy crop. In most cases, though, the primary materials of the feedstock cannot be manipulated, but the feedstock itself may be treated in various ways to enhance, or allow, digestion.

Physical pretreatments

Removal of digestible solids, or of liquids

Removal of solids from a feedstock may involve degradable or non-degradable particles. Degradable particles may be removed for various reasons. With a dilute suspension of particles it may be economically more worth-while to allow the particles to settle to form a thicker sludge that can be digested with profitable formation of biogas in a stirred-tank while the supernatant, relatively free of particles, can be treated in a retained- biomass digester. Such removal of solids may be brought about by allowing the solids to settle in a holding tank and the supernatant to pass to a digester. Examples of this in farm-waste treatment are given in other chapters. In some farm systems the settled sludge is not digested but is allowed to accumulate in the tank or lagoon. It does undergo some degradation under these conditions and is removed at intervals for fertilizer use.

Cattle wastes can contain large particles of partly-degraded grass or other feed fibres as well as straw from bedding. These large particles cause problems in blockage of pumps and pipes and they require a long retention time for only partial digestion and contribute little to biogas production in a digester. If they are removed by a mechanical separator the slurry of finer particles will give slightly less biogas per unit volume than the original slurry, but it is more easily handled and will digest at a shorter RT and so need a smaller (stirred-tank) digester. Overall, the economics of separating some solids can be favourable, even with the addition of a separator and another pump to the digester plant (but see Chapter 6). Lo *et al.*(1983), Pain *et al.* (1984) and Summers *et al.* (1987) have described small- and large-scale application of such separation. The solids can be removed on the usual kinds of rotary screen press, rotary screen, brushed screen, vibrating screen, brushed screen with roller press, produced for farm use (see, for instance, Hobson & Robertson, 1977). Such separators require a uniform, pumped, slurry feed, but they can be programmed to work automatically with the slurry flow. The size of particle removed is controlled by the mesh aperture of the screen fitted. Centrifugal and similar separators are too expensive for farm use and would generally remove most of the solids. However, if all except the finest, mainly bacterial, suspended solids are to be removed to allow of treatment of the liquid in a retained-biomass digester as mentioned above, then the separator can be fitted with a very fine-mesh screen. The large, and small, separated solids are usually stacked and allowed to compost to form a fertilizer and soil conditioner.

Static or mechnically operated mesh and bar (see below) screens and/or settling, can also be used on food-industry wastewaters to remove the relatively large particles of meat and vegetables, starch, etc., from the waters. The particles may then be used for animal feeds or incinerated, or, in some cases, they might be digested in a sludge digester while the waters are treated in a retained-biomass digester.

However, it may be that what is removed is not a proportion of the solids but a proportion of the liquids. Sludge thickening is commonly practised in municipal sewage works. As mentioned elsewhere, the productivity of a stirred-tank digester depends on the solids content of the feed sludge. If the solids content is too low, gas production will be low and perhaps insufficient to heat the digester or to produce energy for other purposes. The feed sludges for a sewage digester are composed of primary sludge and excess microbial sludge from the aerobic treatment system: this latter, in particular, may be quite thin. The sludge needs to be thickened to 4 to 8% TS for digestion. The sludge is thicked by gravity settling in fairly tall tanks, typically about 3–4 m deep. The sludge reaches maximum settlement in about two days, but may be kept longer. After about five days storage there is a deterioration in the performance: settling sludge can begin to go anaerobic and start gassing. Stirring is needed to free gas and water from the thickening sludge layer and this is done by a slowly-revolving picket-fence stirrer. This is an arm, with vertical bars (about 3 or 4 cm apart) descending into the sludge, and rotating at about 0·1 to 1 rpm. The thickened sludge layer is run off to the digester while the supernatant can be returned to the sewage water for aerobic treatment. This treatment is viable for sewage works but would only in some exceptional circumstances be economically-viable for farm wastes. Also, the straw and other long fibres in farm wastes which were mentioned above would quickly block up a picket-fence stirrer.

Priestley (1990) described a rather novel method of thickening sewage sludge. Fine magnetite particles (5–20 μ) were washed in dilute caustic soda and then mixed into sewage water for 5–7 minutes with ferric sulphate or polyelectrolyte coagulants. The coagulated particles and magnetite were separated from the water by magnetism and the magnetite could be recovered and reused. The process was being tested in a 100 cubic metre pilot-plant. The use of coagulants for preparing feedstocks for anaerobic digestion is mentioned in another chapter, but the coagulants may affect digestion.

One aspect of separation of a feedstock stream can be, as previously mentioned, to allow of different treatments of the solids and liquids or larger and smaller solids. Kimchie *et al.* (1988) describe a scheme for digestion of piggery wastes in Israel. The waste is mechanically separated into larger-solids and liquid streams: the larger solids are digested in a two-stage thermophilic-mesophilic tank system while the smaller solids are sedimented from the liquid and passed to the solids digester and the liquid is treated in a sludge-blanket digester. Lo & Liao (1989) mixed cheese-whey (2 volumes) and dairy manure (1 volume). The mixture was passed through a separator to remove the largest particles and then fed to an anaerobic rotating-biological-contactor digester at 6·6% TS. An aerobic sequencing-batch reactor was used for final treatment of the digested effluent. Oleszkiewicz & Koziarski (1981) and Oleszkiewicz (1985) have described systems for total treatment of piggery wastes involving different types of anaerobic digesters and aerobic polishing systems in which, in some cases, the dilute slurry from the piggeries is thickened to provide a sludge treated in a tank digester and a low-solids supernatant treated in an anaerobic filter. In some piggeries much water is used for washing down the houses and so the waste produced is very dilute (*c.* 1–2% TS). A gravity thickener can give a sludge of 4–6% TS which can be digested in a stirred-tank type digester while the supernatant can be digested in a retained-biomass digester.

Separation may not involve mechanical screens, etc., but it can be a matter of separating various waste streams either to allow different methods of digestion or to allow digestion at all. The former could be exemplified in a factory waste where peels and starch granules from vegetable processing could be separated from sugar-containing waters either by diverting the streams at source or by allowing sedimentation. The sludge containing the solid wastes could then be digested in a stirred-tank digester while the low-solids waters could be treated in a retained-biomass digester. Conversely, a high-solids waste could be mixed with a low-solids wastewater stream to produce a slurry of solids content suitable for stirred-tank digestion. Waters containing detergents and biocides which could inhibit digestion could be entirely diverted from the waste streams going to the digesters. On the other hand, it may be possible to allow digestion of wastewaters containing detergents, etc., by mixing them at controlled rates with a main stream of degradable feedstock so that the antimicrobial substances are diluted to non-toxic concentrations.

Removal of non-biodegradable solids

Many wastes pick up non-degradable solids during collection. Examples are grit from animal-pen floors or sand used as a bedding. Stones and earth from cattle feed-lots have also caused problems. These additions are really unavoidable, as are animal hairs, feathers, grit used in poultry feeds, broken egg shells, etc. Avoidable (in theory) additions are baling twine, pieces of wood and floor slats from farm animal pens, lost tools, rubber gloves and even boots, rags and other such debris. This debris can block or break pumps and pipes and tangle with mechanical mixers and it can be very difficult to get rid of.

If the waste is a water with a low concentration of mainly small-particle solids, then screens of different sizes, settling tanks and cyclones can remove the debris, particularly if the flow-rate of the water is high. However, if the waste is a slurry, as, for instance, farm-animal excreta, and it is to be treated as a slurry, then removal of floating or denser debris is almost impossible. Settling tanks allow degradable, as well as undegradable, solids in the slurries to settle out; screens remove, and become blocked with, the desirable slurry solids. Even if the slurry is to be separated into two fractions for treatment, as above, debris is undesirable as it can damage the kinds of mechanical separators mentioned earlier. There seems to be little that can be done about debris in slurries except to remove it by hand when the debris is noticed or to try to prevent it getting into the slurry in the first place. Where this is impossible then occasional removal and cleaning of pumps and valves, etc., will become part of the routine jobs about the digester. Grit will pass into the digester, but some, or most, depending on size, will be kept in suspension by the digester mixers and will pass out with the effluent. That which settles may require the digester to be degritted after some years running.

In the case of sewage sludges prevention of debris getting into the sludge is brought about by removing, as far as possible, the debris from the sewage before it reaches the primary settling tanks where the sludge is produced. The debris in this case is in low concentration and has to be removed from a large volume of fast-flowing water; so coarse and fine screens can remove floating debris while cyclones, settling tanks or channels can remove grit and stones. The residual grit is usually removed mechanically and continuously from the cyclones or tanks. Similar methods can be used with food-factory wastewaters where bits of packaging materials are likely to be the undegradable debris in the wastes from the cooking side of the factory, but large amounts of soil and grit can be generated in the primary cleaning and peeling of root vegetables.

Bar screens for removing larger debris from sewage or factory waters are placed across the water flow at an angle of between 30 and 60° to the flow. Fine screens have bar spacings of about 10 mm and coarse screens of about 25–50 mm. The debris can be removed from the bars by hand or by mechanical rakes which move up the grid bars and deposit the debris at the top.

Mesh separators for factory use are generally similar in principle to the farm separators mentioned previously but may need to be constructed to higher standards for cleaning and general maintenance. Disposable filter bags have been introduced for low-solids waters. Various mesh sizes are available and the bags are effective for removing small amounts of cardboard, plastics, string and fat balls.

Wedgewire screens are popular for industrial wastewaters because they are largely self-cleaning. As suggested by the name, the parallel bars are wedge-shaped, the wider space being on the screened-water side. There are two basic types of screen, parabolic and cylindrical. The parabolic wedgewire is 2 m high with a top feed and weir box to distribute the water across the width of the screen. The water cascades down the parabolic curved surface and the solids move to the bottom while the water passes through the screen. The other types are half cylinders of wedgewire designed to fit into open channels. The screen is inclined to the water flow like a bar screen. The screenings are removed and drained by an archimedian-screw conveyer.

Solid wastes, as well as liquids, may need removal of non-degradable materials. Municipal refuse (MSW) needs to be sorted before it is digested. Sorting removes recyclable materials such as metal and glass, but it also removes plastics and other non-digestible materials which cannot be recycled but would lower the value of the digested waste as compost for commercial and private gardens. The composition and digestibility of the MSW can depend very much on whether the waste is sorted at the house or by mechanical or hand methods after collection (Mata-Alvarez *et al.*, 1990). All these undegradable materials could also damage, or interfere with the operation of, screw conveyers and other plant for macerating the waste and loading and unloading the digester.

Although digesters will take a feed varying in volume and pollutant loading without necessarily failing, the digester will work better if the feed is comparatively constant, and, of course, a constant feed means constant output of both properly-treated material and biogas: this makes it much easier to plan the disposal and use of these products. As mentioned elsewhere, the digester will in most cases require a feed-tank which will hold perhaps a day or two's feed volume and from which the feed pump can take liquids at intervals. This tank will also act as a balance tank to even out an intermittent flow of wastes to the digester plant. In some cases it may also act as a settling tank for the digestible or indigestible solids previously mentioned.

A balancing tank may also act as a fat separator. Fats can give the greatest gas production of any substrate (see Chapter 2) but in large amount and free in a liquid waste they can cause problems of solidification, with blocking of reactors and ancillary equipment, or, if liquid, of forming films over process instruments and other surfaces. In these cases it is usual to separate and remove the fat. Often, the recovered fat can have a commercial value. The fat is usually recovered by allowing the feed liquid to stand in a tank, when the fat rises and can be skimmed off the liquid surface either manually or mechanically.

Mainly biological pretreatments

As mentioned elsewhere, bacteria need carbon, nitrogen, sulphur compounds and trace elements for growth. If these elements were available and in the correct ratios, then a waste could theoretically be converted almost entirely to gas and microbial cells. As the elements are never available to the bacteria in the correct ratios then wastes will generally be only partially gasified, although the residue may be resistant to biodegradation and virtually non-polluting. To improve the content of available carbon or nitrogen sources it may be desirable (or necessary to induce digestion at all) to add some carbohydrate or nitrogen compounds to a waste. The additions should be of lowest possible cost and, preferably, another waste which has to be treated. Farm wastes are generally high in nitrogen. To improve the C:N ratio Summers & Bousfield (1980) added rotten potatoes (high in available C) to piggery waste for co-digestion. Potatoes might be available on a British farm. On the other hand where piggeries are near cities, as in Singapore, more industrial-type carbohydrates might be available and cardboard, sawdust, newspaper, sugar-cane waste and rice straw have all been tested as additives to piggery wastes (Wong & Cheung, 1989; Pathak *et al.*, 1985). Cheese whey and screened dairy manure were mixed by Lo & Liao (1989) to give a better liquid for digestion. These wastes could be available on the same farm and both require treatment. Urea or ammonia might have to be added to a sugary food-processing waste to allow digestion. Calcium, by forming insoluble salts, can reduce long-chain fatty acid toxicity in some food wastes and help granule formation in UASB digesters.

Many experiments have been done on the biological treatment of cellulosic wastes to enhance the availability of the cellulose as a constituent of animal feedstuffs (for review see Van Nevel & Demeyer, 1987) The usual technique is to allow growth of a cellulolytic and lignolytic fungus, possibly with addition of urea or ammonia as nitrogen source, to 'open up' the cellulose structure and to remove some lignin. The fibres can then be more easily attacked by rumen bacteria. The problem is that digestibility is usually increased by only a few percent and some of the cellulose is metabolised by the fungi. The product is then of lower weight than the original material and, overall, there is little, if any, increase, on the original weight, of digestible fibre. The protein content is increased, but this is only in proportion to the added nitrogen or the decrease in weight of other constituents. If an edible mushroom is used as the organism for breaking down the fibre then a cash return on the mushroom can be obtained and the process can become economic. The growth of mushrooms on water weeds and plant wastes has been suggested as a village industry in India and the spent substrates are then anaerobically digested along with cattle excreta in a system for total use of the materials. However, in this case the primary objective is the growth of the mushrooms and anaerobic digestion is a secondary process for disposal of the residues with the maximum benefit to the farmer (e.g., Bisaria *et al.*, 1983). Rolz *et al.* (1988) treated coffee pulp with 26 different white rot fungi to improve the material as an animal feedstuff. They found that both hemicellulose and lignin were degraded, but that the fungi also reduced the concentrations of polyphenols and caffeine. As these latter can be antagonistic to anaerobic digestion their removal could, along with lignin removal, improve the material as a substrate for anaerobic digestion; but against this must be set the loss in digestible carbohydrate in the fungal growth. Experiments by Richards *et al.* (1989) confirmed this loss. Various white-rot fungi did delignify barley straw to various extents, but they also used the cellulose and/or hemicellulose. Straw treated with *Pleurotus ostreatus* gave 20% more methane per unit weight of treated straw than did untreated straw, but because of the loss of carbohydrates from the straw the methane production calculated per unit weight of the initial straw was much less than from the untreated straw. Overall, it seems unlikely that pretreatment of digester feedstocks with fungi to enhance biogas production will have any large-scale application as it is confined to solid feedstocks and the operations demand specialised equipment and manpower. Aerobic bacteria have also been reported to degrade lignin in lignocellulosic materials, but the degradation does not seem to be as good as that produced by fungi (Kamra & Zadrazil, 1988). The remarks previously made about growth of fungi for delignification would apply to bacteria. If the ligninolytic enzyme system could be isolated and produced on a large scale then it could be theoretically possible to degrade lignin and fibres in large quantities. The aromatic monomers of lignin would then be anaerobically digestible, and the cellulose and hemicellulose made more degradable in digesters, without the loss of useful material through fungal metabolism. However, such a reliable and cheap process seems to be very far from production at the moment. Khazaal *et al.* (1990) attempted small-scale (10 g) treatment of barley straw with a fungal ligninase, but the results, measured both in terms of fibre analysis and an *in vitro* method for assessing digestibility in ruminants, were inconsistent and no definite improvement was found. The objectives of some of the biological and chemical treatments is to make the cellulose more hydrolysable to sugars by removal of lignin, but cellulase or hemicellulase enzymes themselves would not (one would suppose) hydrolyse any more of a lignified fibrous material than would digester bacteria, so use of such enzymes (as is suggested for production of fermentable-sugar feedstocks from cellulosics) would not seem to be a possible pretreatment system for cellulosic digester substrates. In any case, use of purified enzymes in pretreatment of lignocellulosic materials for digestion for pollution control or biofuel production would be likely in most cases to be ruled out on grounds of cost. A recent review of the use of fungi to increase feedstuff values of fibrous materials or to produce substrates for saccharification is the book edited by Zadrazil & Reiniger (1988).

However, in spite of the adverse considerations outlined above some experiments have been made on the addition of enzymes to digester feedstocks so that they can work before or after the feedstock enters the digester. So, addition of enzymes to help degradation of recalcitrant compounds may be a pretreatment or it may be done to enhance degradation in the digester to improve digestion and prevent some physical or mechanical problems. The main consideration above has been increased fibre digestibility. Gritzali *et al.* (1988) pretreated ground plant materials with *Trichoderma reesei* cellulase and obtained a small, but variable, increase in methane production in small batch digestions inoculated with contents of a working digester. However, the enzyme was shown to be degraded in the digestions (as might be expected) and had a half life of about 9 days. Different components of the enzyme system (cellobiohydrolase, endoglucanase and β-glucosidase) disappeared at different rates. The large amounts of enzyme used (about 50% of the weight of plant VS) and the small results suggest that the process would not be commercially viable.

One of the authors some years ago tested a commercial product said to degrade solid crusts formed on farm slurries, at a time when formation of crusts from grain husks and animal hairs was causing problems in a small farm digester. The material was not very effective at liquifying crusts and used at recommended doses would have been prohibitively expensive. In sewage sludges insoluble fats may be a major substrate and a major source of physical problems in digestion. Various commercial biocatalytic preparations, said to contain enzymes or specialised bacteria, have been produced over the years. Reports on their efficacies in sewage digesters have varied. Many users have reported no improvements, although a few have reported decreases in odours and greasy scums or increased rates of digestion (Rudolfs, W., 1932; Heukelekian & Berger, 1953; Wells & McKinney, 1955; Grune & Sload, 1954; Pearson, 1957; Bryan, 1952; Robinson, 1954; Corder, 1955).

Composting is an aerobic treatment of solid cellulosic wastes by naturally-developing consortia of mesophilic and thermophilic bacteria, actinomycetes and fungi to break down the plant fibres and to metabolise easily-degradable material to give a stable fibrous product that can be used as a soil conditioner and fertilizer. Microbial metabolism and use of substrate initially heats up the compost and microbial death allows later cooling. Unlike the fungal growth mentioned above, it uses a natural inoculum of organisms in the material. The aeration can be by natural ventilation or forced-air. Either process is probably cheaper in manpower and apparatus than a defined culture of a fungus. If the composting is not carried to completion then the fibre structure can be partially degraded and the product is more easily attacked by the digester bacteria. The composting of cattle manure before digestion to heat the manure and to make it more digestible is mentioned elsewhere (Chapter 6) and a similar batch digester system, for 25%TS cattle manure, with an aerobic pretreatment was described by Meszaros & Matyas (1988).

Under the usual conditions, continuous digestion can be started up at a low loading rate and the rate can be increased as the various bacterial populations develop without the digestion failing due to excess acid production. A properly-running continuous digester will then take feedstocks of high volatile acid content or high content in relatively easily-fermentable carbohydrates (Chapter 2). However, unless a batch digestion has a very large inoculum of a balanced digester flora it is possible that startup with a feedstock containing large amounts of easily-fermentable carbohydrate, or acids, will fail because the acids are not metabolised to methane. The provision of a sufficiently large inoculum to prevent this may be uneconomic as the inoculum can be perhaps 40% of the digester volume. Ten Brummeler & Koster (1989) found startup of batch digestions of municipal solid wastes was subject to acid inhibition due to the rapid fermentation of some easily-degradable fractions of the waste such as starch and sugars. Composting of a waste before digestion could remove such easily-degradable materials by aerobic metabolism without the production of acids and the residue could then be slowly degraded anaerobically and acid production limited to the amount which could be metabolised to methane. Ten Brummeler & Koster (1990) tested composting as a pretreatment for municipal wastes and found that composting for two weeks gave a product that started anaerobic digestion without becoming acid. However, 23·5% of the VS was degraded during the composting, which meant a loss of 40% of the potential methane production, so the economic prospects for this pretreatment did not seem good.

A French process also uses aerobic composting of straw, manure, weeds and vegetable wastes (not wood or vine shoots) as a pretreatment for digestion. Here, the advantages of the process are seen as a temperature rise which makes further heating of the feed unneccessary, and a partial degradation of the fibre to enhance digestion; but, again, there will be some loss of potential methane production (Dehaye, 1987), and another precomposted batch digestion of farm wastes is described in Chapter 6. If the composting is carried out around the digester it is possible to transfer the composting heat directly to the digesting sludge. This has been suggested by various workers and has been carried out, for instance, in Korea where small, climatically heated, digesters have been kept at 35°C during the winter by being surrounded by composting rice straw and pig manure. However, the composting is only used to heat the digester, a separate lot of straw and manure being digested. Both batch and continuous composting and digestion seem to have been tested (Anaerobic Digestion, MIRCEN 1986).

Water hyacinth is one of the many aquatic plants being tested as a digester feedstock both from the point of view of 'alternative' energy production and to get rid of a pest which clogs and stagnates water courses. A problem with the aquatic plants is scum formation in the digester. To prevent this with *Alvinia* the plant was shredded to burst the air bladders in the leaves and to produce a denser slurry. Maceration of water hyacinth, followed by one-day drying and seven days of composting had similar physical effects and also increased the digestibility of the plant (Eng Leon-Foo, 1986; Tentscher & Zin, 1986).

Boopathy (1988) found that aerobic treatment of a 25% TS sludge of coffee pulp for 15 days enhanced the subsequent anaerobic digestion, but that pulp given aerobic treatment for 25 days produced less gas than did the untreated pulp. The oils in the peels of citrus fruits are toxic to digestion (McNary *et al.*, 1951; Lane, 1980) and in laboratory experiments Mizuki *et al.* (1990) found that *Citrus unshu* peel could not be digested at loadings above 2 g/l/d, which corresponded to about 65 μl of peel oil per litre/d. Vigorous air sparging of the homogenised peel at 30°C or steam sparging in an open pan, vaporised the peel oil and the resultant peel was non-toxic in digestions. Wolford *et al.* (1952) and Lane (1983) also used aeration to remove toxic oil from citrus wastes. Glancer & Bann (1988) used aerobic bacteria and yeasts in a microbiological pretreatment step to remove toxic aromatic compounds, antibiotics and quaternary ammonimium salts in wastewaters from cellulose-treatment, pharmaceutical and chemical industries. Hamdi (1991) used an aerobic prefermentation of olive-mill wastewaters with *Aspergillus niger* to remove toxic materials before anaerobic digestion.

Traditionally composting is a solid-state aerobic fermentation in which the fermentation heat is confined and allowed to heat up the substrates so that the fermentation passes from a mesophilic to a thermophilic temperature range. In the usual aerobic treatment of liquid wastes the rate of aeration is relatively low and the heat of fermentation is allowed to dissipate in uncovered and uninsulated tanks. If aeration rate is high and the tank is covered, and insulated from the ground, then the heat of fermentation is conserved and the temperature rises (it can rise to about 50–60°C) and the reactions become thermophilic. Some years ago two pilot plants were described for the thermophilic treatment of farm slurries and the process was called 'liquid composting' or 'aerobic digestion' (Hobson & Robertson, 1977). The mechanical aeration and the temperature produced problems in heavy foaming, which had to be controlled by foam-cutting rotors, and the power inputs were high. The process did not seem to develop further, but some research continued, particularly on sewage-sludge treatment. The aeration reduces odours, and, like all thermophilic processes, reduces pathogens in wastes at a higher rate than do mesophilic processes; but biodegradable sludge components are left. Overall, much less is known about the basic microbiology and biochemistry of the process than is known about anaerobic digestion. However, some pilot and larger-scale plants have been built for sewage-sludge treatment. When used alone the aerobic digestion needs RT of 5 to 7 days. The aerobic digestion can be combined with anaerobic digestion with the aerobic system either preceding or following the anaerobic digestion. When preceding anaerobic digestion the aeration acts in a similar manner to pretreatment of solid wastes by composting and more easily-degradable components of the sludge are digested (and pathogens are at least reduced) in a retention time of about half to one day. The heat generated then reduces the heat input required for the mesophilic anaerobic digestion where degradation of the more recalcitrant sludge components takes place. When the aerobic process follows the anaerobic process, the aerobic metabolism removes the VFA and other residues of the anaerobic digestion and, if the retention time and temperature rise are sufficient, residual pathogens are destroyed. Like the anaerobic digestion, a sludge TS of about 8% is suitable for an efficient aerobic process. Some recent advances in thermophilic aerobic treatment of sewage sludges are discussed in the book edited by Bruce *et al.* (1989).

Singh *et al.* (1983) allowed a 15–16% TS slurry of cattle wastes to stand in open containers for 1 to 5 days before being fed to a digester. The solids- and cellulose-digestions, and specific gas productions, were increased by this 'predigestion'. However, although the predigestion was done in open containers the reaction was primarily an anaerobic fermentation as the slurries went acid (pH *c.* 6·4) during the predigestion. Although it is difficult to see why, if the reaction was only the acidification step of anaerobic digestion, it should result in increased substrate degradation in the subsequent digestion.

The making of vegetation into silage as a means of preserving digester feedstocks and increasing fibre digestibility is mentioned in Chapter 6. The silage acids are degraded in the digester. However, acid production during storage of feedstocks may have deleterious effects on digestion, particularly if the buffering capacity of the feedstock is low. Wastewaters containing easily-fermented sugars may develop a lactic acid fermentation while standing in tanks and pipes, particularly if the weather, or the factory surroundings, are warm. In the authors' experience grass juice prepared as an animal feedstuff fermented in tanks in a matter of an hour or two on hot days. Whey and other sugar-containing solutions may similarly ferment in holding tanks. Proteinaceous animal excreta may degrade to produce high concentrations of acids and ammonia if stored too long. The acids may usually be degradable when the liquids are fed to a continuous digester, but ammonia can cause toxicity. The acids can be neutralised before digestion to keep the digester pH correct, but overcoming these problems is best done by feeding the materials to the digester as quickly as possible. But it may be that the feedstock as produced is either too acid or alkaline for digestion and the buffering action will not allow of pH adjustment in the digester, so the pH has to be adjusted to near 7 before the feed enters the digester. In some cases it may be necessary, or advantageous, to also add nutrients (e.g. N as ammonia or urea, phosphate, trace elements) to the feedstock to improve or allow digestion.

Particle-size reduction

There are two reasons for pretreatments of digester feedstocks to reduce the size of particles in slurries. One is to make a freer-flowing slurry with less tendency to block pipes and pumps and which can be used at a higher solids content than the untreated material. The other is to increase the rate and extent of digestion of the particles. The latter is mentioned in previous sections and will be mentioned later, but the problems of pumping slurries of small straw-particles are also mentioned. Whether it will be a viable proposition to chop, mill or grind a feed material to improve the physical characteristics will depend on the problems of using the untreated material and the monetary and energy costs of the treatment. A proposed chopper for water hyacinth consumed about 20% of the biogas produced by the digester (Tentscher & Zin, 1986). It will also depend on the digestion process. Obviously, long straw and other dry plant materials can be loaded into a batch digester by grab but cannot be pumped into a continuous digester as a constituent of a slurry. If fed by pump, or as a dry feed loaded by an auger or similar system, the straw will need a certain amount of chopping or grinding. Such size reduction will be most needed with feeds of plant materials for energy production; excreta feedstocks have already been subject to particle-size reduction in the preparation of human or animal feeds and in the digestive tracts. Milling or grinding is then mainly needed for reduction of animal bedding materials and debris getting into the final digester feedstock.

However, digestion of cellulosic plant fibres is carried out by bacteria attached to the fibres and in the case of grass stems and similar materials research on the rumen system has shown that the bacteria gain access to the fibres through the broken ends of stems and damaged surfaces of fibres. Thus, milling plant materials increases the damaged areas for attack by the rumen or digester bacteria and so will increase the rate, and possibly the extent, of degradation. Jerger *et al.* (1983) milled wood to particle sizes between 0·003 and 8 mm and used the resultant particles at 5% TS in a batch-digester feed slurry. They found that particles of 0·8 mm diameter gave the highest ultimate methane yield (0·33 m^3/kgVS) but that there was little difference at 0·003 mm (0·31 m^3) and 8 mm (0·28 m^3). Solids degradations were in the same proportions, 63·6, 59·7 and 57·6%. The particle size made more difference to the rates of degradation, 95% of the methane yield in a 60 day digestion of the 8 mm particles could be obtained in 20 days with the smallest particles. This might be expected, as although the milling would increase areas of bacterial attack it would not necessarily change the structure of the particles. The results of Sharma *et al.* (1989) were similar in that they found that the highest gas yield from the stem of *Ipomoea fistulosa*, a woody weed of India, was obtained from particles 0·4 mm diameter and that reducing the particle size further gave rather lower gas yields per unit weight of stem. On the other hand, reducing the particle size from 6·0 to 0·4 mm considerably increased gas yield. Further work on a number of agricultural residues again suggested that reducing the particle size below 0·4 mm was not worthwhile in terms of gas production (Sharma *et al.*, 1988). Llabres-Luengo & Mata-Alvarez (1988) found that straw particle size had no significant effect on the ultimate biogas yield.

When the solids of a feedstock are mechanically reduced in size, the effects of particle size on the extent of degradation of the particles and the production of gas will depend on the shape of the particles as well as the size (assuming that no chemical change or physical change in microstructure is brought about by the reduction). Hobson (1987) considered the surface degradation of particles of different shapes and sizes and showed that the rate and extent of degradation in a fixed RT was affected by these two parameters. Tomato wastes of different particle sizes digested at the same RT were degraded to different extents. Below 4mm size degradation increased with decrease in size, but degradation was little different for particles from 4 to 20 mm size. This could be explained by the 4 mm and under particles being roughly spherical while the larger particles were plate-like (Hobson, 1987; data from Hills & Nakano, 1984). The effects of size reduction will thus depend, to some extent, on the methods used. Various choppers, mills and grinders will have different physical effects on the particle structure and these effects will also be influenced by the initial particle size and structure. For instance, a clean cut dividing a piece of plant stem doubles the number of ends exposed to microbial attack, but a shearing cut or hammering could split the stem and thus expose much more of the degradable inner stem to the bacteria. Badger *et al.* (1979) found that ground straw gave higher gas productions than chopped straw. Extensive mechanical treatment may also affect the molecular structure of the material. Long-time ball-milling, as practised in the preparation of filter-paper substrates for cellulolytic bacteria, has been said to cause depolymerisation of the cellulose.

Straws and other plant wastes and whole plants grown for energy production are dry-milled to reduce the particle size for anaerobic digestion or other biological treatment; for ethanol production for instance. The energy used in the process is, as previously mentioned, a vital factor in determining whether the milling is worth while when compared with any increased or more rapid degradation. Cadoche & Lopez (1989) found that hammer mills or knife mills were most usually used for dry particle reduction and they compared the energies consumed by such mills in reducing hardwood, straw or corn stover to different particle sizes. The relationship between energy consumption by the mill and particle size reduction was not linear and varied with different substrates, but if the final particle size was kept in the range 3–6 mm then energy consumption was less than 30 kWh per ton of processed material.

The sizes of particles in feed residues in animal excreta are smaller than those of the original feeds. However, because the particle size has been reduced by digestive enzymes and bacteria in the gut as well as by the physical actions of chewing by the animal, the relations between particle size and anaerobic digestibility are different from those of mechanically-processed digester feedstocks. Chang & Rible (1974) separated cattle and poultry faeces into particles of over 1 mm and various sizes under 1 mm. The protein content of the smallest particles was highest, suggesting bacterial attachment, but the ash content was also highest, suggesting that these particles had been almost entirely biodegraded: the fibre content of the biggest particles was highest, presumably because they were less degraded. Peck *et al.* (1988) showed that the smaller particles from cattle excreta contained more lignin than the larger and were less digester-degradable and gave lower specific gas productions than did the larger particles. They attributed the high mineral content of the smaller particles to grit from the animal-house floors. In the previous experiments the ash contents showed the same trends in freshly excreted wastes and those collected from floors.

In the previous paragraphs dry milling for particle-size reduction has been considered. Laboratory ball-milling of filter paper is often carried out on wet paper. Szikriszt *et al.* (1988) described a pilot-plant digestion of the organic fraction of municipal solid waste which had been ground with sewage sludge and water, at a solids content of 7–10% in a hydro-pulper, to the consistency of paper-mill pulp. The biogas yield was high, but it was not compared with unpulped waste and because of variations in the composition of MSW and of conditions for digestion it is difficult to compare these results with others.

Plant materials can be fragmented by using ammonia freeze explosion. This process was developed by Dale & Moreira (1982) and used for treatment of straw and forestry (aspen) wastes (Abdullah *et al.*, 1985; Tengerdy & Nagy, 1988). The solids are treated with liquid ammonia for an hour at 35°C and 300 psi and then the pressure is suddenly released volatilising the ammonia. The product contains a small amount of the ammonia, but the lignin content is decreased and the digestibility of the cellulose by enzymes or fungi is increased. This process could not be used with many wastes as the feedstock has to contain only about 3% moisture and the wood wastes were also milled before treatment. The liquid ammonia used was twice the weight of the wood. While this process could produce a relatively high-priced animal feed or fermentation product, it does not look as if it would be economically or practically worthwhile as a pretreatment for anaerobic digestion feedstocks.

In producing 'puffed' breakfast cereals the water in the grain is heated under pressure and the pressure suddenly released to result in an explosive release of the water as steam. A similar process, but with added superheated steam under high pressure, has been used for making woods and straws more biodegradable. The high temperature produces complex chemical and physical changes in the fibre lignin and cellulose and hemicelluloses. These become liquid or paste-like at different temperatures and when the pressure is suddenly released the material cools with the structure disintegrated and the cellulose more accessible to enzymes, although the hemicellulose may be less degradable. Again, the process seems to be much too complicated for treating anaerobic digestion feedstocks, unless used in the future in energy production by anaerobic digestion of wood and other 'energy crops' with the lignin used for production of adhesives and other chemical products.

Chemical, with or without physical, pretreatments

Chemical pretreatments to increase the digestibility of cellulosic materials in animal feedstuffs have been the subject of much research, particularly in the use of waste fibres as feeds. Treatments which increase animal digestibility of fibres will also increase their degradability in anaerobic digesters. However, a treatment which may be economically possible for production of a relatively high-priced feedstuff may be uneconomic when applied to a waste of no real value. In addition, feedstuffs are solids and the treatments are designed for solids not the slurries or liquids which comprise most digester feedstocks.

Although lipid is the principal substrate for the production of methane in all, or most, sewage sludges, in digestions of animal wastes and terrestrial vegetation of energy-crops plant structural polysaccharides are the principal sources of methane. There has been much discussion about the factors limiting the degradation of plant structural polysccharides in the rumen; the physical structure of cellulose fibrils, side chains on hemicelluloses, waxes, silica, etc., but it would seem that although these factors may have some effects the principal determinant of the rate and extent of polysaccharide digestion is lignin (see review by Chesson & Forsberg, 1988). The extent of lignification of plant fibres varies with the type of fibre (wood, vegetable, etc.), and the age of the plant. Thus, various types of plants used for energy crops will vary in degradability, and the degradability of any particular crop will depend on the age of the plants; straws and residues of mature plants being most indigestible. Some comparisons of the rates of digestion of straws and pure cellulose or delignified fibre were given in Chapter 2. The residues of plant feedstuffs in animal wastes will be more indigestible than the original plant material as the more-easily degradable polysaccharides will have been digested in the animal gut, and the surfaces of the plant particles, which is where bacterial attack occurs, will be more lignified than those in the fresh vegetation. The degree of lignification and its relationship to digestibility of fibres has been measured in various ways. A review of modern chemical and physical methods of fibre analysis and relationships to the evaluation of the fibres for feedstuff and industrial uses is the book edited by Chesson & Orskov (1989). Rivers & Emert (1988) investigated the effects of a number of physical and chemical pretreatments on the enzymatic degradability, crystallinity of the cellulose and other properties of rice straw and bagasse. They concluded that the cellulose crystalinity was not a major factor in degradability, but that the overall nature of the lignin-polysaccharide commplex was. Some effects, shown by electron microscopy, on the ultrastructure of a lignocellulosic waste (rice husks) of treatments by chemical and physical methods were described by Lhoneux *et al.* (1988) and, although advances are continually being made in research on lignin and lignified fibres, the book by Crawford (1981) contains much useful information on lignin and lignin biodegradation. To increase the extent and rate of degradation of plants and plant residues the lignin should be removed and the cell wall structure 'opened up' to allow better access of the bacteria and bacterial enzymes. Although not always specifically mentioned, the chemical treatments discussed below are often preceded by mechanical grinding or shredding as this allows of better access of the chemicals to the fibres and in many cases improves the handling of the material as a feedstock for enzymatic treatment or anaerobic digestion.

Various chemical treatments have been used to increase the digestibility in the rumen of ligno-cellulosic feedstuffs. A review of the subject is the book edited by Sundstøl & Owen (1984). The most usual chemical treatment to bring about these changes in vegetation used as animal feeds is by treatment with alkali, usually sodium hydroxide. The sodium hydroxide, as a solution (usually about 4% by weight, although from about 2 to 12% has been used) is sprayed on to the solid fibres and mixed in. The solution is usually used in volume of one to one by weight of the fibres. The mixture is allowed to stand at ambient temperature for from about half a day to four or five days and then the material is air dried. The times and alkali concentration depend on the nature of the fibrous material and only experiment can determine the optima for a particular material. The material is usually treated in bulk and stored for animal feeding, so further degradation of the fibres may take place during storage. Sodium hydroxide is a dangerous chemical to use in bulk, especially on farms, and the less dangerous calcium hydroxide has been suggested. Since the calcium salt is a weaker alkali than the sodium salt many people concluded that it would not be as effective and so little work has been done on the calcium hydroxide compared to that on sodium hydroxide. However, Winugroho *et al.* (1984) concluded that calcium hydroxide could be as effective as sodium hydroxide in increasing the digestibility by cellulase of straw but it had to be used in twice the amount (10 or 20% as against 5%) of the sodium salt when the time of reaction (24 h) was the same. Calcium oxide and a commercial calcium hydroxide (91% hydroxide), in similar amounts, were also effective. In this case a soaking method was used where the calcium salts were mixed in various amounts of water (400, 800 or 1200 ml) and 100 g of straw held in a nylon bag were suspended in the water. The amount of calcium salt in the water was 5, 10 or 20 g per 100 g straw. The residual salts solution was squeezed out of the straw after the treatment. In the spray treatments the straw is only dampened by the alkali solution, and a limit of about 5% alkali is set by the effects of excess sodium on the rumen digestion. This also limits the improvement in digestibility that can be obtained. In the soaking method excess alkali is squeezed out of the treated straw and a higher alkali concentration can be used for treatment (Ibrahim & Pearce, 1983). Hadjipanayiotou (1984) mixed solid calcium hydroxide or oxide with wet barley straw (2 or 4% by weight on the straw) and kept the mixtures sealed up for up to 150 days. The oxide was ineffective in improving digestibility.

Sodium carbonate, chlorite and hypochlorite are amongst other chemicals which have been used for pretreatment of straws for feedstuff production and chlorine and sulphur dioxide gases have also been used. However, an alkali as widely (or perhaps more widely) recommended as sodium hydroxide for improving the digestibility of lignocellulosic materials in feedstuffs is ammonia. The ammonia can be added as liquid anhydrous ammonia, ammonium hydroxide solution or as urea which is hydrolysed by plant enzymes to give ammonia. Urine has also been suggested as a low-cost source of urea and ammonia. The ammonia, whatever way it is applied, is used at about the same weight concentration on the damp straw as sodium hydroxide and the treated material is kept sealed up for some days. Recommended details of treatments have varied for the same fibre and vary with source of fibre. Besides the physical and chemical attack on the fibre structure the ammonia has a useful bonus for ruminant feeding as the residual alkali improves the nitrogen content of the feed, and is unlikely to be a possible source of microbial toxicity as are the metal alkalies. Residual ammonia would also improve the C:N ratio of low-nitrogen plant residues for anaerobic digestion. There are many hundreds of papers on treatments of feedstuffs with ammonia and other alkalies, a lead into the literature and some results are provided by the papers of Vaccarino *et al.*, 1987; Ibrahim & Pearce, 1983; Naim *et al*, 1986; Jayasuriya & Perera, 1982; Kumarasuntharam *et al.*, 1984; Ibrahim *et al.*, 1985).

Autoclaving of straws at about 14 to 28 lb/sq in can result in increased digestibility, and this treatment can be used alone or added either before or after alkali treatment (Nesse *et al.*, 1977; Garrett *et al.*, 1981; Shah & Zia-Ur-Rehman, 1987; Oji & Mowat, 1978; Rangnekar *et al.*, 1982; Rai & Mudgal, 1988) and the effects of the treatments can be synergistic.

Pfeffer (1979) did small-scale experiments on thermophilic digestion of sodium hydroxide-treated, ground, wheat straw, but maximum practical digestion was only about 38%, similar to that of untreated barley straw (Hobson, 1979). Ngian *et al.* (1977) mixed equal weights of sodium hydroxide solution and dried pig faeces and allowed the mixture to stand at ambient temperature for 14 days. Up to 7% NaOH increased gas production in batch digesters by up to 33% more than that from untreated faeces: 9% NaOH was no different and 12% decreased gas production. The faeces were used at only 1·2% TS in the digesters. Van Velsen (1979) increased digestions of pig waste by additions of alkali or acid to pH 13 or 1 for 1 hour at ambient temperature, or, better, at 100°C. Gossett & McCarty (1975) found that treating municipal garbage at 133°C for 180 minutes with acid or alkali at very low or very high pH increased digestibility (similar treatment at neutral pH had little or no effect). Treatment at pH 13 and 250°C further increased digestion (McCarty *et al.*, 1976) and in a two-stage system where digested sludge from the first stage was treated at pH 13 and 200°C before the second stage there was an additional 73% VS reduction and methane production. However, inhibition of digestion occurred when 1–3 g heat-treated lignin was present, which suggests the process might not be suitable for straws or animal excreta of high lignin content. Pfeffer & Khan (1976) found that municipal refuse treated with 3 g NaOH (in 300 ml water) per 100 g refuse at 130°C for 1 hour gave 20% more gas in a thermophilic (60°C) digestion than did untreated refuse.

Boopathy & Larsen (1988) got slightly better digestion of a whisky-distillery waste (containing yeast and other particles) after it was kept at pH 1 or pH 13 (or both in sequence) for 1 hour before digestion. Datta (1981), and Beccari *et al.* (1989, 1991) all used dilute alkali to pretreat lignocellulosic wastes (MSW, corn stover, etc.) before digestion. A typical treatment was 0·5% NaOH (16:1 waste:alkali) for 3 days at 25°C.

The proportions of waste activated-sludge and primary sludge in the feed-sludge for digesters can vary in different sewage works, and activated sludge can be about 70% of the digester feedstock. A method which made cell walls in the zoogloeal microbial sludge more easily degraded in the digester would be useful. High- (60–100°C) and very high- (100–225°C) temperatures, alone or combined with acid (Haug, 1978; Haug *et al.*, 1983; Hiroaka *et al.*, 1984; Stuckey & McCarty, 1978) have been used to try to improve sludge digestibility. Heat generally improves digestibility, although temperatures above 200°C produced compounds inhibitory to digestion. However, any increase in digestibility (and thus decrease in weight and pollution in the sludge solids to be handled) and increase in biogas production has to be balanced against the costs of the pretreatments. Heating large volumes of activated sludge, especially to over 100°C, involves large amounts of energy and heavy and costly apparatus. Any large concentrations of chemicals (*c.* 300 meq/l) are also expensive when thousands or millions of gallons of sludge have to be treated. Rajan *et al.* (1989) tested low (5–100 meq/l) of NaOH and $Ca(OH)_2$ at ambient temperature for their ability to partially hydrolyse activated sludge and so increase digestibility. The alkali concentrations were used with different sludge solids concentrations (0·5–4% TS) to give from 4 to 24 g NaOH/100 g TS and 7·4–29·6 g $Ca(OH)_2$/100g TS. NaOH gave better results than the calcium salt, and 4g NaOH/100 g TS, with a 4% TS sludge at 20°C, for 4 or 5 hours, was best. The authors pointed out that the treated sludge did not need neutralising before digestion and Na concentrations were below toxic levels.

These results might be expected, but the methods seemed at the time barely applicable on a large scale and, so far as is known, no large-scale process has since been attempted.

Tests have also been made on the alkali and heat treatments of plant materials to give solid substrates for biogas production, and gas production is enhanced over that from the untreated materials (e.g., hot acid-sodium chlorite on vine shoots, Jimenez *et al.*, 1989; sodium hydroxide on various plant residues, Ghosh *et al.*, 1980; Hasan Dar & Tandon, 1987; Annachhatre *et al.*, 1987).

The problems of chemical treatments of slurries were mentioned above, and Hashimoto (1982) described laboratory experiments related to a pilot plant in which manure and crop residues were allowed to ferment to acids in a heated vessel at either mesophilic or thermophilic temperatures. The effluent was passed over a separator screen which removed large solids and the liquid passed on to a methane fermenter. The separated solids were then treated with alkali in a heated reactor to produce a material that would be further degraded in the hydrolysis reactor, to which the alkali-treated solids were returned. The small-scale experiments described were on the effects of alkaline treatment at 120°C for 1h on digestion of manure-straw mixtures, but whether this treatment was to be eventually used in the larger-scale plant is not known. Peroxides have also been used to delignify fibrous materials, but Malik *et al.* (1989) found that peroxide treatment of cattle-waste slurries did not improve anaerobic digestion. They used 30 ml of 30% hydrogen peroxide to 2·75 litres of cattle slurrry, and, besides the dangerous nature of peroxide solutions, such a treatment, even if successful, would need very large volumes of peroxide to treat digester feedstocks measured in hundreds of cubic metres. Oxidation by alkaline permanganate of the cattle slurry also proved ineffective in increasing biogas production. Ando *et al.* (1988) increased the digestibility of cedar wood chips by cellulase with a treatment with peracetic acid followed by steam explosion. This could be used for production of biogas from woods, but the economics would have to be considered.

The aromatic compounds produced as wastes in various industrial processes are difficult to degrade anaerobically or can be toxic to the bacteria. Ozone treatment can degrade these compounds to products that will digest comparatively easily (Stover *et al.*, 1982; Medley & Stover, 1983; Gilbert, 1987). Continuing this work, Wang (1990) found that ozonation converted o-cresol, benzenesulfonic acid, 2,5-dichlorophenol and 2,4-dinitrophenol to organic acids and other products and these were biodegradable. The reactions were complicated and products depended on time and amount of ozone and pH of the reaction mixtures. However, the doses of ozone were large (many grams of ozone per gram of substrate) and it would seem unlikely that such treatment would ever be possible or economical on a large scale. Yang suggested that it might be possible to use small doses of ozone to decrease toxicity of aromatic compounds rather than large amounts to try to increase biodegradability.

The treatment with gaseous sulphur dioxide mentioned before is an acid treatment, and other acidifications of wastes have also been described. Hydrochloric, sulphuric and acetic acids have been used, either alone or in combination with other reagents (e.g. zinc chloride) or physical treatments, the latter including radiation in an electron-beam accelerator (Kumakura & Kaetsu, 1979, 1984; Chen & Gong, 1982; Bhattacharya *et al.*, 1987; Rolz *et al.*,1986). These acid treatments were akin to the alkali treatments, the objective being to partially hydrolyse the hemicellulose and open-up the fibre structure. Treatments at pH 1 or 13 were described earlier, but Boopathy and Larsen (1988) found a combined acid and alkaline treatment improved gas production from a distillery waste sludge more than did acid or alkali separately. The sludge was adjusted to pH 1 with sulphuric acid for one hour at 30°C and then adjusted to pH 13 with NaOH for one hour before the pH was adjusted to 7 and batch digestion started by seeding. More drastic acid hydrolysis is being used to hydrolyse the cellulose and hemicelluloses in fibres completely to sugars which are used as substrates for fermentations by pure cultures of microorganisms to give chemical end-products such as ethanol and microbial proteins. Such treatments could give sugar solutions which could be converted to biogas in retained-biomass anaerobic digesters: whilst this could be a possibility for energy production it would seem to be an expensive method of biogas production and unsuitable for waste treatments. However, Yang & Chang, (1985) and Reig *et al.* (1989) have done small-scale tests on anaerobic digestion of the liquid produced by treatment of pure cellulose with alkali. A 10% suspension of the cellulose in 6% calcium hydroxide was heated at 250°C in an autoclave for 30 min. The latter authors used the calcium salt as the former authors found methane production from the hydrolysate was slow, probably due to inhibition by the residual sodium ions when sodium hydroxide was used. The resultant solution contained mainly lactic, acetic and succinic acids and could be gassified in a fluidised-bed digester with a few hour's RT. Calcium hydroxide cooking of cellulosic fibrous materials also produced organic acids that could be digested (Tsai & Tsao, 1985). The processes were aimed at large-scale digestion of municipal solid wastes. Acid hydrolysis, under pressure, of wood to produce sugars and other digestible products for biogas formation was being investigated by Baugh & McCarty (1988) In these cases the starting material is digestible, the chemical treatment produces other materials which it is hoped will be more degradable than the original and so the increased biogas production will more than pay for the treatment. Anaerobic digestion of lignites was previously mentioned and here the starting material is not biodegradable but is converted into products that can be digested to gas; reports of an international conference on this subject are given by Wise (1987) and some work on biological solubilisation of chemically-treated lignites was reported by Maka *et al.* (1988). It appeared that microbial attack on lignite was related to chemical oxidation of the constituents of the coal.

In the production of paper pulp, lignin and hemicellulose is removed from wood and other lignocellulosic materials by treatment with sulphites and alkalies, but pulp production by means of solution of lignin in organic solvents ('organosolv' pulping) is now being used. These solvents are also being tested for pretreatment of lignocellulosics for ethanol and chemicals productions. Solvents used have been aqueous methanol, ethanol, butanol, ethylene glycol, phenol, formic acid, acetic acid, usually with the addition of a catalyst. For instance, Rolz *et al.* (1986) treated Lemon Grass and Citronella bagasse with 6 volumes of 60% ethanol–40% water containing 40 g NaOH Na_2O per kg substrate and 0·05 g anthraquinone per 100 g substrate as catalyst. The mixture was autoclaved at 175°C for 4 h in a rotating cooker and the solid residue drained and washed with water and dried. The treatment removed lignin and hemicellulose and improved the cellulase hydrolysis of the residues to a similar,or rather greater, extent as did treatment with 8% (by weight of substrate) sodium hydroxide at ambient temperature for 10 days. The sodium hydroxide treatment is obviously a much cheaper and easier process. Bungay (1981) used aqueous butanol and Pannir *et al.* (1980) used ethylene glycol as lignin solvents to improve enzymic hydrolysis of agricultural crop residues. Fan *et al.* (1982) reported that organosolv treatments were expensive in comparison with other pretreatments. The dissolved lignin and hemicellulose may be used for other commercial processes, but it does not seem as if organosolvs would be a pretreatment for most digester substrates. A chapter by Puls in the book edited by Van der Meer *et al.* (1987)

reviews aspects of organosolv treatments, and this book also contains papers on other delignification processes that have been mentioned here.

CONCLUSION

In the foregoing sections a number of possible physical, chemical and biological treatments for anaerobic digester feedstocks have been discussed. Some of these have so far been used mainly for preparation of animal feedstuffs from lignocellulosic wastes, some are being tested with the aim eventually of providing substrates for enzymic hydrolysis, or the actual sugars, for fermentations to produce fuels, chemicals and microbial proteins. With the exception of alkali treatments for upgrading fibres for ruminant feeds, few of the processes have been used on a large scale in a commercial set-up.

Mechanical treatments such as separation, screening, mixing or diversion of wastes streams, are already standard practice for wastewaters and slurries in sewage works and in other digester plants. On the other hand, many of the chemical treatments designed to change the chemical and/or physical nature of components of the digester feedstocks have generally been tested only on a small scale and their application to regular use in full-scale plants is problematical. The tests do show that various chemical treatments can make solid fibrous materials more readily and extensively degradable by enzymes and bacteria, but the treatments in general need chemicals that are dangerous to handle and/or apparatus that is expensive and runs at high temperatures or pressures. Such considerations suggest that the processes would be of use only in a factory system where the chemicals and high temperatures and pressures could be used under controlled conditions, and that economics would confine their use to production of biogas as an alternative energy source; possibly in the future as prices of conventional fuels increase. The fact that relatively high concentrations of chemicals have to be used also adds to the expense and confines the processes to solid substrates: the making of thousands of gallons of farm slurry into a 5% alkali solution would seem to be beyond possibility and the alkali would have various deleterious effects on the subsequent digestion. Biological treatments are mainly designed for solid feedstocks but, again, the use of either defined whole organisms or separated enzymes would seem to be hardly cost-effective at the moment even for solids, and certainly not for slurries. If shortages of conventional fuels become apparent, and the prices of these increase, then production of biogas fuel from pretreated woods and other vegetation may become economically viable, although many processes need much more work before they can be transferred from the laboratory to the factory. Maceration is probably the only worthwhile physical pretreatment of any feedstocks at the moment, as this may be done for slurries with feed pumps equipped with chopper blades or well-tried choppers for solid feedstocks.

At the moment, the most important solid feedstock is probably municipal solid waste. Sorting, maceration or shredding is feasible and generally necessary;

but of the chemical and biological pretreatments, composting, because it can be done with low energy inputs and simple plant, is the most feasible. It also has the advantage that the heat energy generated in the process can offset some of the energy inputs to the digester thus lowering overall costs.

Even for sewage treatment, small works are virtually unmanned and many waste treatment plants on farms and factories have to be run with the minimum of attention. If pretreatment of the wastes before digestion is essential it needs to be kept as simple as possible.

REFERENCES

Abdullah, A.L., Tengerdy, R.P. & Murphey, V.G. (1985). *Biotechnol. Bioeng.* **27**, 20.

Anaerobic Digestion MIRCEN (1986). Microbial Resources Centre, Karolinska Institute, Stockholm. **3**, 121, 135.

Ando, S., Kakimoto, T., Itoh, K., Arai, I., Kiyoto, K. & Hanai, S. (1988). *Biotechnol. Bioeng.* **31**, 802.

Annachhatre, A.P. & Khanna, P. (1987). *Biotechnol. Bioeng.* **33**, 924.

Badger, D.M., Bogue, M.J. & Stewart, D.J. (1979). *N. Z. J. Sci.* **22**, 11.

Baugh, K.D. & McCarty, P.L. (1988). *Biotechnol. Bioeng.* **31**, 50.

Beccari, M., Campanella, L., Majone, C., Rolle, E. & Todini, O. (1989). Acidogenic fermentation of the organic fraction of municipal solid wastes. *7th. Int. Conf. Chemistry for protection of the environment*, Lublin, Poland, Sept. 7-9.

Becarri, M., Majone, C., Riccardi, E., Rolle, E. & Scarinci, A. (1991). *Biores. Technol.*

Bellamy, D. & Wheatley, B.I. (1988). In *Fifth Int. Symp. Anaerobic Digestion, Poster-Papers*, ed. A. Tilke & A. Rozzi, Monduzzi Editore, Bologna. p. 899.

Bhattacharya, S., Biswas, P.K. & Tolbert, M.E.M. (1987). *Biol. Wastes*, **19**, 215.

Bisaria, R., Gujral, G.S. & Bisaria, V.S. (1987). *C.R.C. Critical Rev. Biotech.* **7**, 17.

Boopathy, R. (1988). In *Fifth Int. Symp. An. Digestion, Poster-papers.* ed. A. Tilche & A. Rozzi, Monduzzi Editore, Bologna. p. 451.

Boopathy, R. & Larsen, V.F. (1988). Ibid. p. 455.

Bruce, A.M., Colin, F. & Newman, P.J. (1989). *Treatment of Sewage Sludge*, Elsevier Applied Science, London.

Bryan, A.C. (1952). *Public Works*, **83**, 1969.

Bungay, H.R. (1981). *Energy, the Biomass Option*, Wiley, New York.

Cadoche, L. & Lopez, G.D. (1989). *Biol. Wastes*, **30**, 153.

Chang. A.C. & Rible, J.M. (1974). In *Managing Livestock Wastes. Am. Soc. Agric. Eng.*, St. Joseph, Michigan, USA. p. 339.

Chen, L.F. & Gong, C.S. (1982). *Biotechnol. Bioeng. Symp.* **12**, 57.

Chesson, A. & Forsberg, C.W. (1988). In *The Rumen Microbial Ecosystem*, ed. P.N. Hobson. Elsvier Applied Science, London. p. 251.

Chesson, A. & Ørskov, E.R. (1989). In *Physiochemical Characterisation of Plant Residues for Industrial and Feed Use.* Elsevier Applied Science, London.

Chynoweth, D. & Jerger, D. (1988). In *Fifth Int. Symp. An. Digestion, Poster-Papers.* ed. A. Tilche & A. Rozzi, Monduzzi Editore, Bologna. p. 913.

Corder, W.A. (1955). *Water Sew. Works*, **102**, 42.

Crawford, R.L. (1981). *Lignin Biodegradations and Transformations*, John Wiley & Sons, New York.

Dale, B.E. & Moreira, M.J. (1982). *Biotechnol. Bioeng. Symp.* **12**, 31.

Datta, R. (1981). *Biotech. Bioeng.* **23**, 61.

Dehaye, J. (1987). *6th Biogas Computer Conference*, Anaerobic Digestion, MIRCEN, Microbiological Resources Centre, Stockholm, **4**, 161.

Eng Leon-Foo (1986). Anaerobic Digestion MIRCEN. Microbial Resources Centre, Karolinska Institute, Stockholm. **3**, 126.
Fan, L.T., Lee, Y.H. & Gharpuray, N.M. (1982). *Adv. Biochem. Eng.* **23**, 155.
Garrett, W.N., Walker, H.G., Kohler, G.O., Hart, M.R. & Graham, R.P. (1981). *J. Anim. Sci.* **51**, 409.
Ghosh, S., Henry, M.P. & Klass, D.L. (1980). *Biotech. Bioeng. Symp.* **10**, 165.
Gilbert, E. (1987). *Water Res.* **21**, 1273.
Glancer, M. & Ban, S.N. (1988). In, *Anaerobic digestion 1988*, ed. E.R. Hall & P.N. Hobson, Pergamon Press, Oxford. p. 465.
Gossett, J.M. & McCarty, P.L. (1975). 68th. An. Meeting Inst. Chem. Eng., Los Angeles.
Grethlein, H.E. (1984). *Biotechnol.* Adv. **2**, 43.
Gritzali, M., Shiralipour, A. & Brown, R.D. (1988). In *Methane from biomass, a systems approach*, ed. W.H. Smith & J.R. Frank. Elsevier Applied Science Publishers, London. p. 367.
Grune, W.N. & Sload, R. (1954). *Sew. Ind. Wastes*, **26**, 1245.
Hadjipanayiotou, M. (1984). *Ag. Wastes*, **10**, 187.
Hamdi, M. (1991). *Biores. Technol.* **36**, 173.
Hartley, R.D. & Jones, E.C. (1978). *J. Sci. Food Agric.*, **26**, 711.
Hasan Dar, Gh. & Tandon, S.M. (1987). **29**, 805.
Hashimoto, A.G. (1982). *Ag. Wastes*, **4**, 345.
Haug, R.T. (1978). *J. Wat. Pollut. Cont. Fed.* **50**, 73.
Haug, R.T., Lebrun, T.J. & Liberato. D.T. (1983). *J. Wat. Pollut. Cont. Fed.* **55**, 23.
Heukelekian, H. & Berger, M. (1953). *Sew. Ind. Wastes*, **25**, 1259.
Hiler, E.A. (1986). *Sorghums for Methane Production.* National Tech. Inf. Service, US Dept. Commerce, Port Royal Road, Springfield, VA22161, USA.
Hills, D. J. & Nakano, K. (1984). *Ag. Wastes*, **10**, 285.
Hiroaka, M. *et al.* (1984). *Sci. Technol.* **17**, 529.
Hobson, P.N. (1979). In *Straw Decay and its Effect on Disposal and Utilisation.* ed. E. Grossbard, John Wiley & Sons, Chichester, UK.
Hobson, P.N. (1987). *J. Ferment. Technol.* **65**, 431.
Hobson, P.N. & Robertson, A.M. (1977). *Waste Treatment in Agriculture.* Elsevier Applied Science, London.
Ibrahim, M.N.M. & Pearce, G.R. (1983). *Ag. Wastes*, **8**, 195.
Ibrahim, M.N.M., Wijeratne, A.M.U. & Costa, M.J.I. (1985). **13**, 197.
Jayasuriya, M.C.N. & G.R. (1983). *Anim. Food Sci. Technol.* **8**, 271.
Jerger, D.E., Chynoweth, D.P. & Dolenc, D.A. (1983). *In Energy from Biomass and Wastes vii.* Institute of Gas Technology, Chicago. p. 585.
Jiminez, S., Cartagena, M.C. & Arce, A. (1989). *Biol. Wastes*, **29**, 211.
Kamra, D.N. & Zadrazil, F. (1988). In *Treatment of Lignocelluloses with White Rot Fungi*, ed. F. Zadrazil & P. Reiniger, Elsevier Applied Science, London. p. 56.
Khazaal, K.A., Owen, E., Dodson, A.P., Harvey, P. & Palmer, J. (1990). *Biol. Wastes.*
Kimchie, S., Tarre, S., Lumbroso, E., Green, M. & Shelef, G. (1988). *Biol. Wastes.* **26**, 275.
Kumakura, M. & Kaetsu, I. (1979). *Int. J. Appl. Radiat. Isotopes*, **30**, 139.
Kamakura, M. & Kaetsu, I. (1984). *Ag. Wastes.* **9**, 279.
Lane, A.G. (1980). *J. Chem. Technol. Biotechnol.* **30**, 345.
Lane, A.G. (1983). *Environ. Technol. Lett.* **4**, 65.
de-Lhoneux, B., Gerlache, L., Clemente, A., Roda-Santos, M.L., Menaia, J.A.G. & Fernandes, T.H. (1988). *Biol. Wastes.* **23**, 163.
Llabres-Luengo, P. & Mata-Alvarez, J. (1988). *Resources Conserv. Recycl.* **1**, 27.
Lo, K.V., Bulley, N.R., Liao, P.H. & Whitehead, A.J. (1983). *Agric. Wastes* **8**, 155.
Lo, K.V. & Liao, P.H. (1989). *Biol. Wastes.* **28**, 91.
Mahoney, E.M., Varangu, L.K., Cairns, W.L., Kosasic, N. & Murray, R.G.E. (1987). *Water Sci. Technol.* **19**, 249.
Maka, A., Srivastava, V.J., Kilbane, J.J. & Akin, C. (1988). *10th. Symp. Biotech. for Fuels & Chemicals.* I.G.T. Document Center, Chicago.

Malik, R., Tauro, P. & Dahiya, D.S. (1989). *Biotech. Bioeng.* **33**, 924.
Marshall, D.R. & Timbers, G.E. (1982). Paper no. 82-6519, Winter Meeting ASAE, St. Joseph, Michigan, USA.
Mata-Alvarez, J., Cecchi, F., Pavan, P. & Llabres, P. (1990). *Biol. Wastes.* **33**, 181.
McCarty, P.L. (1964). *Public Works*, November. 91.
McCarty, P.L., Young, L.Y., Gossett, J.M., Stuckey, D.C. & Healy, J.B. (1976). In *Microbial Energy Conversion*, ed. H.G. Schleigel & J. Barnea, Verlag & Goltze, Göttingen. p. 179.
McNary, R.R., Woolford, R.W. & Patton, V.D. (1951). *Food Technol.* **5**, 319.
Melley, D.R. & Stover, E.L. (1983). *J. Water Pollut. Cotrol Fed.* **55**, 489.
Meszaros, G. & Matyas, L. (1988). In *Agricultural Waste Management and Environmental Protection, Proceedings, Vol. 2.* ed. E. Walte & I. Szabolcs. C.I.E.C. & FAL, Braunschweig, Fed. German Republic. p. 97.
Mizuki, E., Akao, T. & Saruwatari, T. (1990). *Biol. Wastes* **33**, 161.
Naim, N., Abdel-Mohsen, S.I., Lofty, A.S. & Abdel-Monem, H. (1986). *Biol. Wastes*, **16**, 103.
Nesse, N., Wallick, J. & Harper, J.M. (1977). *Biotech. Bioeng.* **19**, 323.
Neushul, M. & Harger, B.W. (1983). *Kelp Biomass Production; Yield, Genetics and Planting Technology*. Nat. Tech. Information Service, U.S. Dept. of Commerce, Port Royal Road, Springfield, VA 22161, USA.
Ngian, M.F., Ngian, K.F., Lin, S.H. & Pearce, R.G. (1977). *J. Environ. Eng. Div.* **103**, 1131.
Oji, U.I. & Mowat, D.N. (1978). *Can. J. Animal Sci.*, **58**, 177.
Oleszkiewicz, J.A. (1985). *Agric. Wastes*, **12**, 185.
Oleszkiewicz, J.A. & Kosiarski, S. (1981). *Agric. Wastes*, **3**, 123.
Pain, B.F., West, R., Oliver, B. & Hawkes, D.L. (1984). *J. Agric. Eng. Res.* **29**, 249.
Pannir, S.P.V. & Ghose, T.K. (1980). *2nd Int. Conf. on Bioconversion and Bioenergy, Eng. III*, Delhi, India.
Parkin, G.F. & Speece, R.E. (1983). *Water Sci. Technol.* **15**, 261.
Pathak, B.S., Jain, A.K. & Dev, D.S. (1985). *Agric. Wastes* **13**, 251.
Peck, M.W., Hawkes, F.R. & Hawkes, D.l. (1988). In *Fifth. Int. Symp. An. Dig. Posterpapers*, ed. A. Tilche & A. Rozzi. Monduzzi Editore, Bologna. p. 411.
Pearson, E.A. et al. (1957). *Sew. Ind. Wastes*, **29**, 1066.
Pfeffer, J.T. (1979). In *Fuel Gas Production from Agricultural Residues and Biomass*, Dynatech Co., 99 Erie Street, Cambridge, USA.
Pfeffer, J.T. & Khan, K.A. (1976). *Biotechnol. Bioeng.* **18**, 1179.
Priestley, A. (1990). *Water & Wastewater.* **5**, 31.
Puls, J. (1987). In *Degradation of Lignocellulosics in Ruminants and Industrial Processes.* ed. J.M. Van der Meer, B.A. Rijkens & M.P. Ferranti, Elsevier Applied Science, London. p. 75.
Rai, S.N. & Mudgal, V.D. (1988). *Biol. Wastes*, **24**, 105.
Rajan, R.V., Jir-Gaur, & Bill, T.R. (1989). *J. Wat. Pollut. Cont. Fed.* **61**, 1678.
Rangnekar, D.V., Badve, V.C., Kharat, S.T., Sobale, B.N. & Joshi, A.L. (1982). *Anim. Feed Sci. Technol.*, **7**, 61.
Reig. M., Toldra, F., Gow, J.T., Jansen, N.B. & Tsao, G.T. (1989). *Biol. Wastes*, **29**, 201.
Richards, S.R., Thomas, G.W. & Spendler, F.H. (1989). In *Energy from Biomass 4.* ed. G. Grassi, D. Pirrwitz & H. Zibetta, Elsevier Applied Science, London. p. 449.
Rivers, D.B. & Emert, G.H. (1988). *Biol. Wastes*, **26**, 199.
Robinson, R.R. (1954). *Public Works*, **85**, 116.
Rolz, C., de Arriola, M.C., Valladares, J. & de Cabrera, S. (1986). *Agric. Wastes*, **18**, 145.
Rolz, C., de Leon, R. & de Arriola, M.C. (1988). *Biol. Wastes*, **26**, 97.
Rudolfs, W. (1932). *Sew. Works J.* **4**, 782.
Saxena, K.K., Nath, K. & Srivastava, S.K. (1989). *Biol. Wastes*, **28**, 73.

Shah, F.H. & Zia-Ur-Rehman (1987). *Biol. Wastes*, **19**, 63.
Sharma, S.K., Mishra, I.M., Sharma, M.P. & Saini, J.S. (1988). *Biomass*, **17**, 251.
Sharma, S.K., Saini, J.S., Mishra, I.M. & Sharma, M.P. (1989). *Biol. Wastes*, **28**, 25.
Singh, R., Jain, M.K. & Tauro, P. (1983). *Agric. Wastes*, **6**, 167.
Summers, R. & Bousfield, S. (1980). *Agric. Wastes* **2**, 61.
Summers, R., Hobson, P.N., Harries, C. & Richardson, A.J. (1987). *Biol. Wastes* **20**, 43.
Sundstøl, F. & Coxworth, E.M. (1984). In *Straw and Other Fibrous By-products as Feed.* ed. F. Sundstøl & E. Owen. Elsevier, Amsterdam. p. 196.
Sundstøl, F. & Owen, E. (1984). In *Straw and Other Fibrous By-products as Feed.* Elsevier, Amsterdam.
Stover, E.L., Wang, L-W. & Medley, D.R. (1982). *Ozone Sci. Engng.* **4**, 177.
Stuckey, D.C. & McCarty, P.L. (1978). *Biotech. Bioeng. Symp.* **8**, 219.
Szikriszt, G., Frostell, B., Norrman, J. & Bergstöm, R. (1988). In *Anaerobic Digestion 1988*. ed. E.R. Hall & P.N. Hobson, Pergamon Press, Oxford. p. 375.
Ten Brummeler, E., Koster, I.W. & Zeevalkink, J.A. (1988). In *Anaerobic Digestion 1988*. ed. E.R. Hall & P.N. Hobson, Pergamon Press, Oxford. p. 335.
Ten Brummeler, E. & Koster, I.W. (1990). *Biol. Wastes*, **31**, 199.
Tengerdy, R.P. & Nagy, J.G. (1988). *Biol. Wastes*, **25**, 149.
Tentscher, W. & Zin, A. (1986). Anaerobic Digestion MIRCEN. Microbial Resources Centre, Karolinska Institute, Stockholm. **3**, 134.
Tsai, G.J. & Tsao, G.T. (1985). *Annual Rep. Ferm. Proc.* **8**, 187.
Vaccarino, C., Lo Curto, R.B., Tripodo, M.M., Belloco, E., Lagana, G. & Patane, R. (1987). *Biol. Wastes,* **20**, 79.
Van de Meer, J. M. et al. (1987). Eds. *Degradation of Lignocellulosics in Ruminants and in Industrial Processes*. Elsevier Applied Science, London.
Van Nevel, C.J. & Demeyer, D.I, (1987). In *The Rumen Microbial Ecosystem*. ed. P.N. Hobson, Elsevier Applied Science, London. p. 387.
Van Velsen, A.F.M. (1979). In *Energy Problems with Effluents from Livestock*. CEC, Brussels. p. 476.
Verma, M.L. (1983). In *The Utilisation of Fibrous Agricultural Residues*. ed. G.R. Pearce, Aust. Gov. Pub. Services, Canberra. p. 85.
Wang, Y-T. (1990). *Water Res.* **24**, 185.
Welander, T., Malmqvist, A. & Yu, P. (1988). In *Anaerobic Digestion 1988*. ed. E.R. Hall & P.N. Hobson, Pergamon Press, Oxford. p. 267.
Welander, T. & Andersson, P.E. (1985). *Wat. Sci. Technol.* **17**, 103.
Wells, W.N. & McKinney, R.E. (1955). *Sew. Ind. Wastes*, **27**, 871.
Winugroho, M., Ibrahim, M.N.M. & Pearce, G.R. (1984). *Agric. Wastes*, **9**, 87.
Wise, D.L. (1987). *Resources Conserv.* **15**, 229.
Wolford, R. W., Patton, V.D. & McNary, R.R. (1952). *Food Technol.* **6**, 418.
Wong, M.H. & Cheung, Y.H. (1989). *Biol. Wastes* **28**, 143.
Yan, Y.G., Lo, K.W. & Liao, P.H. (1989). *Biol. Wastes* **27**, 289.
Yang, S.T. & Chang, M. (1985). *Annual Rep. Ferm Proc.* **8**, 187.
Zadrazil, F. & Reiniger, P. (1988). eds. *Treatment of Lignocellulosics with White Rot Fungi*. Elsevier Applied Science, London.

Chapter 6

DIGESTION OF HIGH-SOLIDS FEEDSTOCKS

INTRODUCTION

The forerunners of the anaerobic digester were treatments for whole human excreta in the form of sludges of faeces and urine; thus, the digester developed as a method of treating wastes containing solid particles. It is only comparatively recently, as described in other sections, that effective methods of digesting wastewaters of low solids-content have been developed. The digestion of high-solids wastes is thus concerned with the original sewage digesters and these were of the stirred-tank type, although other digesters have been since produced especially for digestion of feedstocks of very high solids content.

In the context of digestion 'high solids-content' feedstocks might be denoted as slurries with 4% or more Total Solids, 4% being about the minimum TS at which digestions of sewage sludges are run and the minimum TS at which other digesters, such as those for farm wastes can be effectively run. 'Effectively' in this context means running without danger of washing out the bacteria in a continuous process and producing sufficient gas to heat the digester and/or have enough gas for other purposes. The upper limit to solids content is set by the nature of natural feedstocks (e.g. vegetable matter) and the need of bacteria for water for growth. These factors set the limit at about 30% TS or rather higher. Digestion of high-solids feedstocks can take place as a continuous or batch process, and some examples will be described in the next sections. The solids content of the feedstock dictates to some extent the type of digester and, particularly, the ancillary equipment needed to handle the feed. With some solids, for example those in pig wastes, a slurry of up to about 8% TS might be handled by pumps and pipes; with cattle slurries somewhat higher TS can sometimes be used. At about 14%, which is the TS content of cattle faeces-and-urine slurries, pump and pipe handling becomes increasingly problematical and recourse must be had to gravity or scraper feeding in open channels, and, as TS become higher, to handling of the wet solids with cranes and grabs or tractor shovels and similar equipment. About 15% TS, depending on the feedstock, is probably the limit to the slurry that can be used in a continuous-flow type digester. These figures are only for guidance as the composition of the solids makes a lot of difference to their handling and flow properties. For instance, animal hairs in farm slurries can cause great problems by inducing blockages in pipes and pumps even at slurry TS contents around 5% which would not be difficult to handle if the solids were entirely faecal particles. Similarly, long stems of grasses and straws can cause more problems than shorter, chopped, stems: but some materials such as slurries of finely ground straw can solidify in a pipe while pressure is being applied by a pump (Hobson, 1979; Pfeffer, 1979). Digesters for high-solids feedstocks can be run at mesophilic or thermophilic temperatures. Fielden (1981) calculated that for piggery slurries above 4% TS, digestion at the upper mesophilic temperatures (38–42°C) was most efficient in producing a positive energy balance for a digester heated by water from a biogas boiler and having no effluent/influent heat exchange. For 7·5% TS cattle slurries operation at thermophilic temperatures might be optimal.

Simple and modified batch digesters

The time-course of a batch digestion has been described in Chapter 2, Fig. 3, with its inoculation-, lag-, production- and stationary-phases. The curve for bacterial growth or product formation given there is of a laboratory pure culture and is rather idealised for a digestion: except, perhaps, for digestion of a pure substrate. Powell *et al.* (1983) found that production of methane from acetate in a laboratory batch culture by mixed bacteria enriched from a sewage digester followed approximately the pure-culture growth. There was an initial phase when added acetate was being digested by the inoculum before growth started and then an exponential increase in methane produced by exponentially-growing bacteria. The final phase was one of decline in methane production as the acetate was used up; but there was a period before this when gas production was at a constant rate which was ascribed by the authors to degradation of acetate uncoupled from bacterial growth for some unknown reason, but perhaps because of production of some toxic product or depletion of some nutrient other than acetate (sources of nitrogen and trace elements and salts were added with the acetate). The batch digestion of a heterogeneous substrate such as pig waste can show a number of peaks and troughs of methane production rate as different components of the waste are degraded (Hobson, 1985; Svendsen & Blackburn, 1986). Similar changes in methane production rate will be found as one feed input is degraded in a continuous-flow digester, although whether the variations in gas production will be noticeable will depend on the frequency of feed additions. Ferrara *et al.* (1984) added various pure substances (either found in or produced from animal-waste digester feedstocks) once a day to a digester running on pig waste at an RT of 14 days. Gas production from glucose peaked after about 30 minutes from glucose addition and then declined over the next 8 hours. Acetate accumulated after glucose addition indicating that conversion of glucose to acid was faster than conversion of the acetic acid to methane. Addition of acetate resulted in peak gas production after about 2 hours and then a decline. The rate of gas production from hemicellulose increased for about 3 hours and then became steady. Production from cellulose steadied after about 1 hour. In both cases the gas production rates were lower than the peak production rates from glucose and acetate. Acetate and reducing-sugar concentrations remained practically constant after addition of the cellulose or hemicellulose, indicating that the hydrolysis of the latter was slower than the fermentative or methanogenic steps. It may sometimes be necessary to add alkali during the early stages of a batch digestion of an easily-fermented feedstock to neutralise acids produced and to keep the pH in a region which will not inhibit the slower conversion of acids to gas.

Batch digesters have generally been used in two ways. On a small, laboratory, scale they are used for preliminary experiments on the feasibility of digestion of low- or high-TS feedstocks: on a full-scale they are generally used for digestion of solid feedstocks, that is those with TS above the 15% or so mentioned above.

Laboratory batch digesters have varied in size and complexity. To some extent the size depends on the particle size of the feedstock and the analyses needed (i.e. whether samples will have to be taken during the digestion and what size of sample is needed) and the sensitivity of the gas-measuring apparatus. Probably the most general size has been between one and three litres working volume. However, Cullimore *et al.* (1985) used test tubes filled with slurry and fitted with a rubber bung through which passed a needle fitted to a plastic syringe. The tubes were held in an incubator and gas production was measured from the rise of the syringe barrel. (The method was originally used to test gas production from rumen contents (El Shazly & Hungate, 1965).) The authors used the method to test the effects of temperature on rates of methanogenesis. Chen *et al.* (1988) used a similar syringe technique for measuring gas production but used 100 ml bottles as the digesters when comparing methane productions from fresh- and dried-cattle manures. Svendsen & Blackburn (1986) used similar bottles with syringes to measure gas in the batch digestions mentioned above. The bigger bottles (1l and above) used as batch digesters are connected to a liquid-displacement or mechanical gas meter system for measurement of the relatively large volumes of gas produced. Acidified water can be used in the gas-collection system if loss of CO_2 is to be avoided and total biogas production measured. Conversely, if the volume of methane alone is needed the gas can be passed through an alkaline solution to absorb CO_2. Some batch digesters are stirred more or less continuously; by a paddle on a shaft from the top, or from below by a magnetic stirrer, or by being placed on a shaker table. Others are mixed at intervals by hand or by mechanical means. As most batch digestions are long-term a useful precaution is to use dark-glass or covered bottles to avoid growth of photosynthetic organisms in the digesting material.

A common use of laboratory batch-digesters has been to determine if a substrate is digestible and, if it is, what is the maximum yield of gas to be expected and what is the rate of gas production, the latter being correlated to some extent with the RT which would be needed in a continuous-flow digester. Although some feedstocks such as animal excreta may contain bacteria to act as an inoculum, to ensure comparison between different digestions, and to make the time-scale similar to the RT of a continuous digestion, an inoculum is usually used. The inoculum size in relation to the amount of substrate solids and the digester volume has varied in different experiments, as has the nature of the inoculum. For experiments on digestion of new substrates the inocula have generally been taken from working continuous-flow digesters operating on sewage or animal wastes, Usually, the coarser particles are removed from the inoculum to leave as far as possible a suspension of bacteria with little substrate material which could complicate measurements of gas production from the test substrate. Or, the inoculum is allowed to ferment as a batch digestion for some days to use up the residual substrate. A survey of different experiments suggests that about 10% by volume is the usual size of inoculum, but there are variations in volume, and inocula which are the same volumetrically may contain different concentrations of active bacteria. It has been assumed that provided the digestion starts almost immediately growth of the bacteria will rapidly obviate any differences in inoculum between experiments and that the methane produced when gas production ceases represents the maximum degradation of the substrate. However, Hashimoto (1989) found that the inoculum size did affect the final gas yield. He used inocula from a cattle-waste digester in tests of methane yield from straw. The inoculum digester was not fed for three weeks before the inocula were taken, to reduce residual cattle-waste substrates in the inoculum. The volume of inoculum used ranged from 10 to 90% of the volume of straw slurry. The batch digestions were carried on for 150 days at 35°C and gas productions levelled off after about 30 to 60 days. Final gas productions were drastically reduced at inoculum : substrate ratios (on a VS in inoculum and substrate basis) of less than 0·25. At ratios above 0·25 gas production increased slightly up to a ratio of 2, after which it remained constant. The results suggested that in all batch tests of digestibility it would be wise to include digestions with different proportions of inoculum. On the other hand, Llabres-Luengo & Mata-Alvarez (1988) found, using a recirculating batch digestion, that although the initial rates of hydrolysis were increased by increasing inoculum (5 to 20% inoculum VS on substrate VS) the differences were not significant after six or seven days of fermentation.

Laboratory batch digesters have been used to determine the methane yields from many substrates; such as textile willow dust (Balasubramanya *et al.*, 1986) solids from slaughter-house and chocolate- and biscuit-manufacturing wastes (Sayed *et al.*, 1988; Ranade *et al.*, 1989), rumen contents from slaughter-houses (Tritt & Kang, 1991), energy crops and crop wastes (Badger *et al.*, 1979; Zubr, 1985); seaweeds (Brouard, Bories & Sauze, 1983; Yang, 1981) woody plants (Jerger *et al.*, 1982), etc. A large amount of testing of crop wastes and terrestrial and aquatic weeds has been done to determine their suitability as additions to cattle wastes for feeding small-farm digesters in India and elsewhere to increase biogas production and to use otherwise discarded or non-utilised materials (e.g. Dar & Tandon, 1987; Sharma *et al.*, 1989). Pig manure with sawdust, cardboard, newspaper or sugar-cane waste was tested in batch digesters by Wong & Cheung (1989). Batch digesters have also been used in testing the effects of pretreatments of digester feedstocks, for instance NaOH treatments of straws (Hashimoto, 1986); the effects of antibiotics and disinfectants on digestions (Hilpert, Winter & Kandler, 1983); and the death of pathogenic organisms during thermophilic and mesophilic digestions (McKain & Hobson, 1987; Olsen & Larsen, 1987; Olsen, Jorgenson & Nansen, 1985).

To obtain a constant supply of gas, and treat a continuous supply of waste, a number of batch digesters running out of phase is needed. As an example of multiple, full-size, batch digesters to provide a constant source of gas, the small-scale experiments of Balasubramanya *et al.* (1986) were scaled-up in a plant with six unstirred digesters around a common gas holder and feed tanks (Balasubramanya *et al.*, 1988). Each digester could hold two tonnes of cotton willow dust. The willow dust was mixed with sodium hydroxide in a pretreatment to increase digestion and digested at ambient temperature, either 'dry' (1 part willow dust to 1·5 parts alkaline liquid) or wet (1 part dust to 6 parts liquid), with an inoculum of cow dung, for 80 to 100 days. Ambient temperature was 25 to 30°C, depending on season. About 50% of the willow dust was digested to gas which contained 55–60% methane after the first week. (The composition of the gas from a batch digester will often vary with time, especially during the starting period).

Jewell (1980) and Jewell *et al.* (1985) on the basis of small-scale tests calculated that large (about 30 ft by 30 ft tanks) batch digesters containing packed crop residues at 30–40%, or higher, TS could be run on cycle times of 100 to 300 days and metabolic heat would be sufficient to heat the insulated digester tanks to mesophilic temperatures; but so far as is known such digesters have not been built. There have been some full-scale batch digesters built. Clausen & Gaddy (1983) described a plant with four 4500 gallon digesters operating on a 10% slurry of ground poor-quality grasses and hay. The digesters were operated on a 60-day cycle (at 35°C) so that overall gas production was continuous and the gas could be used in an engine generator. The digesters were horizontal tanks and the contents were agitated by an oscillating paddle for 5 minutes twice a day. Initial inoculation was from sewage sludge, but later inoculations came from the spent digested crops. A number of on-farm batch digesters for cattle manure (excreta plus straw bedding) have been described in France, particularly (e.g. Demuynck *et al.*, 1984; Wheatley, 1978). They have varied in complexity from a pile of manure inside a heated, gas-tight, plastic cover to steel tanks loaded by cranes and rail trucks. Some were designed to produce useful biogas and fertilizer, for others the residual fertilizer was the main product. A batch-digestion process for mixtures of crop stalks and pig-waste has been developed on laboratory and full-scale plants by Guo-Chao Sun *et al.* (1987) and Ke-Zin Liu *et al.* (1987). The plants are meant for small Chinese farms and the feedstock is a mixture of, say, rice straw and manure of 25–30% TS incubated at ambient temperature for about 4 months in summer and 5 months in winter. The size of inoculum was found to have an effect on the rate and final yield of biogas and an inoculum of at least 30% by volume of sludge from a working sewage or farm-waste digester was recommended. A main reason for development of the dry process was to supply the traditional 'dry' manure to farmers in certain regions of China as well as supplying gas for household use. Zeeman *et al.* (1988) found that an inoculum of 50% by volume of digesting cow manure adapted to lower temperatures was needed to start batch digestions of pig or cattle manure at low ambient temperatures of 5 to 20°C: without inoculum, acids but no methane, were formed. At 25 or 30°C methane production began after about 60 or 30 days.

Molnar and Bartha (1989) described a novel batch digester used in Hungary on an experimental farm. This consisted of a number of units each of which was a circular concrete base with vertical sides with an overall capacity of 85 m^3 and a working volume of 80 m^3 of wastes. The digester was loaded by tractor shovel with 25% beef-cattle manure and 75% corn stalks, husks and leaves chopped to 20–30 mm particles. The C:N ratio was adjusted to 25:1 by addition of slurry from the cattle pens. The Total Solids concentration was 27% and some lime was added to buffer the mixture. The mixture was allowed to compost under natural aeration for 5–10 days in some experiments, but this was not reproducible and it was found better to mechanically aerate the mixture for 2–3 days (summer or winter). The aeration raised the temperature of the mixture to about 55°C and then a flexible, insulated plastic 'tent' was lowered over the pile. The pile became anaerobic and was inoculated with manure/corn from a digester kept at 55°C. The mass then digested at a slowly decreasing temperature for about 30 days. The digested material could then be removed and the process started again. The gas-proof tent was sealed into the digester base by water troughs and the gas was collected in a separate water-sealed gas holder for use.

A batch process rather different from the conventional batch has been tested. On a small scale the solid substrate was a mixture of straw and cattle wastes (Hall *et al.*, 1985) which was placed in a heated tank and a liquid inoculum added. The inoculum was allowed to drain from the bottom of the digester and recirculated continuously. The recirculating, bacteria-containing liquid could be passed through other tanks before returning to the initial tank. After adding tanks in sequence exhausted tanks could be replaced in sequence, thus giving a semi-continuous process. Hall *et al.* operated the batch tanks over 40 to 70 day periods at 35°C with TS of the initial straw and cattle-waste mixture of 26·3%. A similar system was used on a pilot-plant scale by Chynoweth *et al.* (1990), with sorted municipal solid waste (MSW) as feedstock. Tanks were connected in series for leachate flow, as above. After the primary inoculation a fresh tank of MSW was inoculated by circulation of the liquid leachate from a working tank. After some 20 days of the 42-day cycle time of each tank digestion would occur without further inoculation and so the leachate recycle was turned to an 'internal' flow. As in the previous experiments, continuous gas production could be obtained by emptying and refilling tanks in sequence. ten Brumeller *et al.* (1991) also found start-up time of a batch digester for MSW was shortened by leachate recycling.

In the last process described all the reactions of digestion take place in the one tank, as in a conventional batch fermentation process, and unless the reactions of hydrolysis and fermentation and methanogenesis are balanced the reactor can go acid and fail, unless the pH is adjusted upwards as it begins to fall. A rapid fermentation and fall in pH can be put to use in an extension of the circulating batch digester to a batch two-phase system. In this extension of the system the solids-containing tank is used for acidification of the feedstock and the percolating, acid-containing liquid passes on to a second, 'methanising', tank where the acids are converted to gas. The pH of the liquid thus rises in the methaniser to about 7 or above and the effluent from the methaniser is returned to the acidifier tank. The methaniser can be a stirred-tank, but has more generally been a retained-biomass type of digester as the acid effluent contains little solids. Rijkens

(1981) described such a laboratory batch process in which a tank containing solid waste was inoculated and water was percolated down the tank. This water was then passed through a sludge-blanket digester and returned to the solids digester. A hydrolytic bacterial population developed in the first tank and a methanogenic population in the second. After some time when only slowly-degradable solids are left in the first tank of such a system the concentration of acids in that tank becomes low and a methanogenic flora will develop, so the methaniser can be disconnected and the digestion finished in the one tank. Rijkens tested municipal solid waste, straw, sugar-beet pulp, poppy-seed hulls and mixtures of some of these; the progress of the digestions was similar. Koster (1984) studied the liquefaction and acidification of tomatoes in laboratory batch digesters percolated with water, but did not couple this to a methanogenic reactor. Viturtia *et al.* (1989) coupled a hydrolyser for mixtures of fruit and vegetable wastes of about 6% TS with a recirculating methaniser in a laboratory-scale apparatus. Rijkens later (1983) ran a pilot-plant in which an acidifier for waste tomato plants was percolated with water and the percolate passed to a UASB methaniser. Coble & Egg (1987) built and operated successfully a system in which three 619 litre tanks containing sorghum were connected to an anaerobic filter. The sorghum was allowed to ensile (acidify) at ambient temperature for 6 weeks. Water was then added to one silo and the leachate passed through the filter and then back to the silo. The filter operated at 35°C. After about 100 days of recirculation of liquid, gas production in the methaniser started to decline and methane production started in the silo. At 126 days gas production in the silo was greater than in the filter. At 200 days of operation gas production was very low, but in a full-scale system another silo would have been brought into operation before this. Once the sorghum had ensiled its constitution should not change. No inoculum is usually needed in silage making, but apparently the methaniser in this system did not need inoculating either. Zauner (1988) described laboratory-scale acidification of ensiled sugar beet leaves (13·6% TS) in a type of stirred-tank reactor and conversion of the acid effluent to biogas in an upflow filter. Potato pulp (15% TS) was similarly digested.

Ensiling was also advocated by Stewart (1980) for conservation of energy crops before they were used as feedstock for a stirred-tank, continuous, digester. Mathisen & Thyselius (1985) found similar gas productions in batch digestions of fresh and ensiled crops. Barry & Colleran (1982) digested grass-silage liquid in an anaerobic filter as a method of controlling pollution from farm silage making and Summers & Bousfield (1980) had earlier digested grass silage liquid along with a slurry of pig excreta in a stirred-tank digester. In both these cases the silages were produced for animal feeding: the liquid seeping from silos or silage pits is intensely polluting, and digestion could be a method of reducing this pollution.

Pauss *et al.* (1984) in a discussion of some factors affecting the digestion of municipal wastes briefly described a two-phase digestion in which bacterial acidification of the solids took place in a percolating digester and the liquid was converted to methane in a fluidized-bed digester.

So far as the authors are aware no full-scale plant on this percolation of solid feedstock has been tested, but a digester with intermittent bottom loading and circulation of liquid to a sprinkler at the top of the tank was running in France some time ago (Demuynck *et al.*, 1984). However, this plant seemed to use a diluted manure feedstock, although solids were allowed to accumulate in the digester.

Verstraete *et al.* (1981) reviewed some aspects of two-phase digestions and the reasons for using the process.

Gijzen and co-workers (1986, 1988*a,b*) and Zwart (1988) have tested on a laboratory scale a stirred-tank digester in which an inoculum of rumen bacteria and protozoa hydrolyses cellulosic substrates (e.g., paper mill sludge: a fraction of municipal solid waste) to acids and the acids are converted to methane in a retained-biomass digester. The effluent from the methaniser is recirculated to the hydrolyser, but additional loading of solids to the hydrolyser converts this into a semi-continuous process rather than a batch process as the ones previously described.

A further extension of this hydrolyser and methaniser principle is the chemical hydrolysis of the solids and methanisation of the hydrolysate. Reig *et al.* (1989) treated 10% cellulose sludges with alkali at 250°C to produce a solution of mainly lactic, acetic and succinic acids which was methanised in a fluidised-bed reactor at 35°C. This was done on a laboratory scale but the authors suggested that the process could be a viable one for treating municipal solid waste.

Fed-batch digesters

The principle of the fed-batch digester has been explained previously. On a large-scale some farm lagoons can be regarded as fed-batch systems in that the lagoon is gradually filled up with slurry and the filled lagoon is left until a suitable time for emptying and use of the digested slurry as fertilizer. A lagoon that is continuously fed after it has filled up begins to overflow and becomes a type of stirred-tank digester. In warm climates the solids may actually be concentrated by evaporation of water in spite of removal by digestion. In wet climates the reverse may, of course, occur. The performance of anaerobic lagoons is much influenced by climatic conditions, and they can cause pollutional problems, so in many areas they are no longer used. However, in some cases their use as low-cost digesters has been advocated. This involves covering the lagoon with a plastic, or other, cover and fitting the cover with a gas-take-off system. Safley & Westerman (1989) described floating, rigid-framed, gas-collecting covers used on two farm lagoons. Balsari & Bozza (1988) also described a synthetic-rubber lagoon cover held up by floating polyurethane blocks. These and other workers have reported gas productions of, generally, about 0·05 to 0·35 m^3 per m^2 of lagoon surface with maxima about 0·9 m^3, for different lagoons with cattle, pig or poultry wastes at different times of year. Obviously, gas production will depend on ambient temperature (and hence location and season) and on slurry concentration and depth of the slurry under the cover. However, this work (and some covered lagoons made by farmers) has shown that usable quantities of gas can be recovered from large lagoons and used to run engines, etc. Safley & Waterman (1992) described the performance over about two and a half years of a covered lagoon in the USA fed with separated dairy-cattle waste of approximately 1% TS. Lagoon temperature followed ambient over long periods but was much less variable in the short term and was not lower than 10°C in winter. Gas production fluctuated with season, but was satisfactory overall. Cadwallader (1991) described a high-density polyethylene floating cover for a paper-mill waste lagoon which he said was very hard-wearing.

A more controlled fed-batch digester can be made either by deliberate heating of a tank or by utilising the heating and insulating effects of making the digester from under-floor excreta tanks in animal houses.

The heated fed-batch tank for animal wastes has been advocated by Hill and co-workers (1981, 1986) as a system suitable for small-farm use. They recommend digestion at 30 to 35°C as being commercially viable (gas production at lower temperatures being too slow) and cycle times of 60 or more days. About 30% of the digested sludge is left after emptying the digester to act as inoculum for the next run. Pilot-plant studies have been done, but how far the process has gone on a commercial scale is not known. However, a fed-batch digester using a 700 m^3 heated tank has recently started operating on a Dutch research pig farm (Zeeman, 1991).

The use of waste tanks under animal houses as digesters has been advocated for some years. The animal excreta is deposited in channels from which it is scraped or flows under gravity to a downpipe leading into the underfloor tank. The amount of excreta in the tank increases with time, as the animals are kept in the house for 24 hours a day for the whole year (with beef cattle or pigs) or for certain months of the year (usually for dairy cattle). Some time ago, Haga *et al.* (1979) described an underfloor digester for dairy cattle in Japan. This had two submersible pumps for stirring and was being operated on the fed-batch principle until the tank filled up, when it would become a continuous-flow digester. Some under-house digesters are operated on this dual principle, others are operated entirely as fed-batch systems while the animal houses are occupied in the autumn and winter months, and the digested slurry is removed during the summer months when the cattle are out at pasture. The feed slurry may be faeces and urine alone or with straw or other bedding.

Much work has been done on underfloor digesters in Switzerland where cattle are kept indoors during the winter and put out to pasture in the summer. The usual temperature of stirred-tank digesters used on farms and in treatment of sewage and other sludges is about 35°C: under-floor digesters usually run at temperatures of less than 25°C and sometimes below 10°C. These have been referred to as 'psychrophilic digesters'; anaerobic lagoons operate at similar temperatures. The rate of the bacterial reactions producing methane decreases with decrease in temperature in the mesophilic range, so that in practice gas production is a function of temperature and retention time of the substrate in the digester system. For instance, over the normal range of RT for stirred-tank pig-waste digesters of 10–20 days, effectively no gas is produced at digester temperatures of about 13°C: a model showed a minimum RT for solids digestion at 13°C of 25 days, while at 15 and 20°C the minimum RT were 17 and 8 days (Hobson, 1983; van Velsen, 1977). Cullimore *et al.* (1985) studied gas production by the contents of a pig-waste lagoon in laboratory experiments at different temperatures between 4 and 40°C. They observed a linear relation between temperature and gas production below 26°C and a continued increase in production between 26 and 40°C, but the increase in rate per unit temperature change was smaller than at the lower temperatures. This latter is in accord with other results (Summers & Bousfield, 1980). Cullimore *et al.* found that the minimum temperature for gas production was between 3 and 9°C, but over two years of lagoon operation there appeared to be a decline in the minimum temperature at which gas production started, suggesting that a flora adapted to low temperatures was gradually developing (see Chapter 2).

Wellinger & Kaufmann (1982) monitored two underfloor digesters over two years. In the operational temperature range, 15–21°C, gas production increased with temperature, but a flora adapted to low temperatures seemed to develop gradually, as specific gas production increased over the two years. These authors found that start-up of the digesters was very difficult, with 6 months running being needed before stability was obtained. At 20°C pig- and cattle-waste digestions could be started without an inoculum, but at 18°C or below an inoculum was needed. At 15°C or higher an inoculum of mesophilic sludge of 50% of the final volume of the digester allowed digestion to start. With large inocula of sludge adapted to low temperatures, digestion could be started at 5 or 10°C. However, a large inoculum obviously reduces the volume of manure that can be added to a fed-batch digester and gas production at 10°C or below was too low to be commercially useful, so the authors recommended 15°C as a lower temperature limit for running fed-batch digesters. Zeeman *et al.* (1986) found that at this temperature a 13% inoculum of well-adapted digesting slurry could start digestion within a reasonable time. However, Zeeman *et al.* (1988) recommended some form of heating system be installed in underfloor fed-batch digesters as start-up could be expedited by temporarily increasing the temperature to about 25°C, and, also, a short temperature increase could cure a digestion that was beginning to break down. Sutter & Wellinger (1988) reported on an underfloor digester in a cow house which was running initially as a fed-batch and later as a continuous-flow digester. In the latter mode the retention time was 40–50 days and the temperature of the water-heated- and mechanically-stirred-digester contents was 20°C. Although biogas production was comparable to a stirred-tank operating at 35°C and 25 days RT the net energy production was 30–40% better than from comparable-size mesophilic digesters because of the lower heating requirements of the under-floor digester. A Swiss fed-batch digester under pig housing and running at 15–20°C is described by Demuynck *et al.* (1984). The gas is collected in a flexible gas holder and used in a boiler for space heating. The floor of the animal house in under-house systems must, of course, be impermeable to gas and the manure must be channelled from the animal flooring to a drain which leads below the liquid level of the digesting waste or through a liquid trap which prevents escape of gas. The overflow, if there is one, must, similarly, be through a liquid weir or other gas trap. Intermittent stirring, usually by a paddle system, helps to keep the digesting mixture homogeneous and evens out the temperature of the material. The submerged-pump stirring of the Japanese digester previously mentioned also heated the slurry, but in the authors' experience submerged pumps wear rapidly in gritty piggery slurries.

Continuous-flow stirred tanks

The theoretical chemostat continuous bacterial culture is fed with medium containing dissolved nutrients in a continuous flow below the surface of the culture liquid, and the incoming medium is instantaneously mixed into the culture. In practice, because of the possibility of the bacteria growing back up a medium inlet below the surface of the culture liquid some form of gap is built into the inlet tube (e.g. Hobson, 1965). Because the flow rate of the feed is usually low the gap induces dropwise flow of medium into the culture. Detailed observations can show that even drops coming into the culture at intervals of seconds can induce oscillations in the bacterial metabolism (Brooks and Meers, 1973), but these usually have to be ignored. The retention times of stirred-tank digesters treating high-solids feedstocks are usually long, so the flow rate of feed into even large digesters is relatively low. High-solids feedstocks present problems in pumping the feed sludges through pipes and the pumps required have to be of relatively large bore and running continuously have a capacity much greater than the required hourly or daily flow rate. Thus, most high- solids digesters have to be fed intermittently. The feed pump may be programmed to work for a few minutes every hour or so, or for a longer time once or a few times a day. Because of the even greater difficulties of pumping high-solids slurries through pipes a few millimetres in diameter most laboratory (a litre or two capacity) digesters have to be fed once or twice a day by manual introduction of portions of the feed. If the feed is ground to produce fine particles then the slurry is no longer representative of the full-scale feedstock After the introduction of the feed, the same volume of effluent is removed, by hand from the laboratory digester or by pump or by overflow from the pilot- and full-scale digester. Some workers have made a distinction between the true continuously-loaded culture and digesters loaded, say, once or twice a day, calling the latter 'semi-continuous' or 'intermittent' loading. However, so long as equal volumes of feed and effluent are introduced and removed at the same time the digester can be considered as a continuous-flow culture. Some oscillations in gas production are induced by intermittent feeding, but as the general rate of conversion of the feed solids is low the overall gas production is relatively smoothed out even with comparatively long intervals between feeds. Bursts in gas production caused by activation of mixing systems are usually more noticable than increases in gas production due to introduction of feed.

The basic construction of the stirred-tank digester has been described in Chapter 4. Digesters to this design, or with minor variations in detail of feeding, mixing and heating systems, are being used for all kinds of substrates with solids contents of about 4 to 8%. These will not be described in detail here; a number of different designs are described by Demuynck *et al.* (1984). Some variations on the 'standard' stirred-tank will be discussed in more detail.

The climatically-heated, small-farm digester

The digesters just mentioned are heated, mechanically-operated plants suitable for use in any area where large volumes of sludge have to be digested and there is an infrastructure capable of delivering feedstock to the plant and using the gas generated. Although the plant will be automatic or semi-automatic, so needing few workers for everyday running, it will still need a backup of a skilled and semi-skilled workforce to attend to any problems arising. Thus, the mechanised plant is required on many farms in developed countries where large volumes of sludges have to be processed and labour for day to day running is not cheap. However, in many areas of developing countries labour is plentiful, but farms are small and farm incomes are low. On the other hand, such areas are often facing fuel shortages as wood, the main fuel, is used up, and they have a need for fertilizers, but chemical fertilizers are too expensive. For such areas, anaerobic digestion of farm wastes has been, and is continuing to be, developed as a method of producing a clean fuel and a useful fertilizer. The gas is a better fuel than wood or cowdung, and unlike the latter the biogas leaves the dung residue as a fertilizer. Over many years, small (2, 3 or so, cubic metres) continuous-flow, stirred digesters have been developed which can use the wastes from a few pigs or cattle plus kitchen residues, human excreta and crop wastes or weeds to generate sufficient gas for simple cooking and lighting for a peasant farm. Since these digesters are generally in warm countries they are climatically heated, and since labour is available, feed and effluent are manually maintained. There are many designs and construction methods of such digesters, depending in part on locally-available materials, but the Chinese and Indian (Gobar) digesters are the most widely used: in China there are many millions of digesters, and in India thousands. The the Indian digester is a vertical brick cylinder, partly buried in the ground and with inlet and outlet pipes running from near the bottom. In some cases, the cylinder has a dividing wall so that the slurry circulates from the inlet over the wall and to the outlet. The gas-holder is a floating metal drum over the digester and this can be provided with a mixing paddle which can be occasionally turned by rotating the gas-holder. The Chinese digester is an almost spherical underground tank made of concrete or concrete-lined brick or stone with inlet and outlet pipes. Gas collects in the top of the digester and so, unlike the Indian digester, the gas pressure varies as more or less gas is collected. The digesters are illustrated in Fig.10B and C. They run on sludges of about the 6–8% solids previously mentioned, at mesophilic temperatures. However, since only in certain areas and certain seasons do ambient temperatures approach the 30–35°C of the heated mesophilic digester the digesters generally run at low mesophilic temperatures and in consequence require retention times of ca. 50 days to attain reasonable digestion of the solids. Much research and develop-

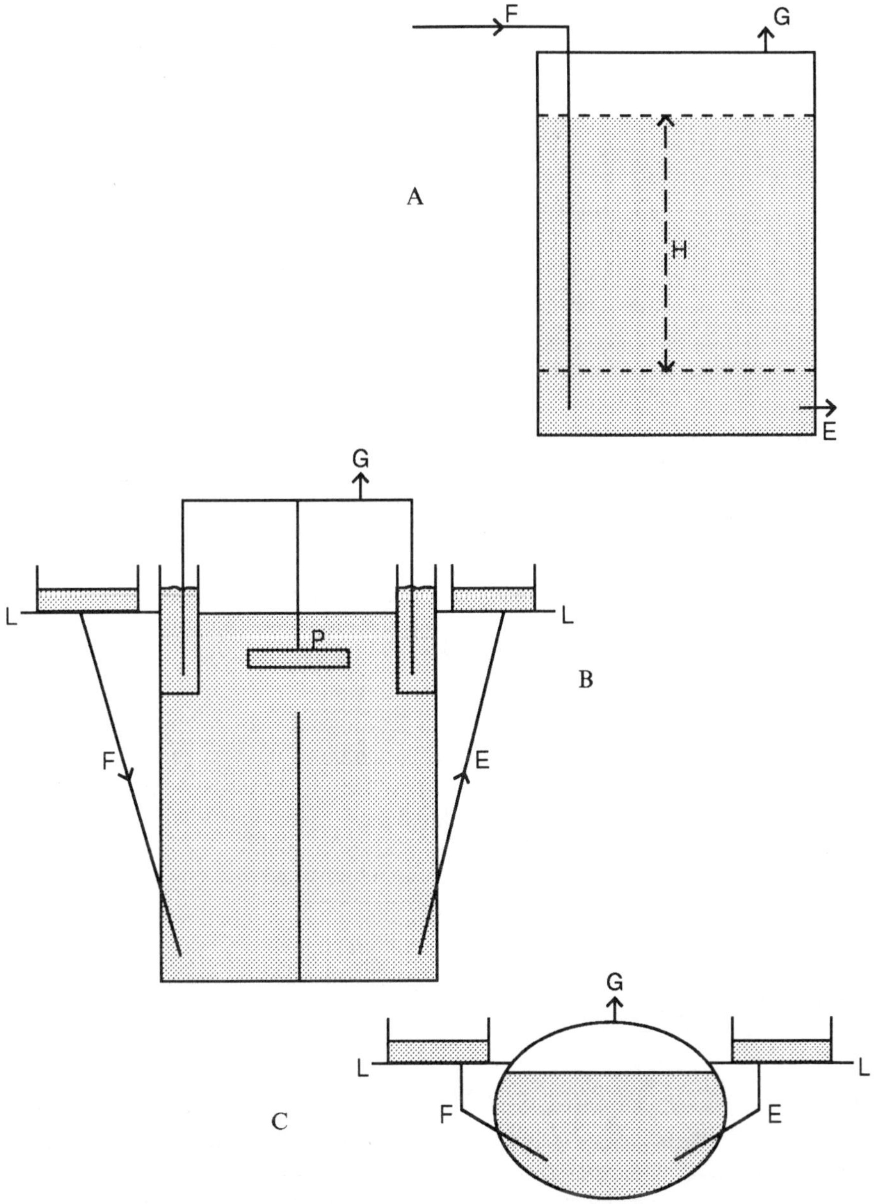

Fig. 10. Diagramatic representations of some digester systems. A, fed-batch tank digester. The digester is gradually filled up over the depth H and then emptied, retaining the sludge below H as inoculum for the next filling. B, Indian Gobar digester with floating gas-holder. The lower dividing wall is optional. C, Chinese concrete or stone and concrete digester. F, feed inlet; probably pumped for A but gravity fall from collecting tanks in B and C. E, effluent. G, gas flow. P, paddle stirrer attached to gas-holder top. L, ground level.

ment has gone into these and similar digesters and many papers on aspects of their construction and running have been published: some general references are: Biogas Technology, an information package (1985); A Chinese Biogas Manual (1979); Fixed Dome biogas Plants, A Design, Construction and Operation Manual (1987); Advances in Biogas Technology (1986). Nazir (1991) has described construction technology for biogas plants in rural areas of countries such as India. The digesters, although apparently simple, are not without problems. In some areas, of India for instance, winter and night-time temperatures can be very low and digester activity little (e.g., Kishore, 1989; Kalia & Kanwar, 1989; Kaul *et al.*, 1987). Solar heating panels to heat the feed or digester, insulation, compost heaps around the digester and other things have been suggested or tested to try to increase digester temperatures (e.g., Kishore, 1989; Dayal *et al.*, 1985; Yadav *et al.*, 1987). In the Chinese form of digester lined with cement, gas is contained at low pressure in the cemented dome. As ground movements can crack the lining, recementing at intervals is required to prevent gas escape. Sludge build-up may require emptying and desludging of the digester at intervals of about a year. Since in some areas, of India for instance, the cattle are on poor diets, excreta production is low and availability of cattle dung can become a limiting factor in digester running. Hence there has been a search for additional feed materials and the feed may consist of dung plus vegetation, etc., as mentioned previously. And, like all digesters, unless the designed programme of feeding and effluent removal is adhered to the digester will not function correctly. A further problem is that although the digesters are cheap by 'Western' standards they are expensive in terms of the income of peasant farmers, and this is a problem which is difficult to solve even with Government loans. Goswami (1991) has recently described a rectangular digester with floating gas-holder and manual stirring, for use in India, which is said to be of low-cost. The feedstock is vegetation. Gas productions from the digesters should be exactly as found in laboratory tests and in mechanised digesters when RT, operating temperature and composition of feedstock are taken into account. Well-digested sludge is comparatively free from pathogens and weed seeds and is generally used as a fertilizer, either liquid or sun-dried.

The 'standard' stirred-tank sewage digester

Until a few years ago all stirred-tank sewage-sludge digesters were large vessels of some thousands- to about two million-gallons capacity and such digesters are still in use or being constructed (Chapter 4). These digesters were built in concrete and were usually cylindrical in shape, although the ratio of height to diameter varied. There were other designs, some egg-shaped. The variations in shape were designed to promote better mixing or release of gas. Many of the bigger digesters were partly buried in the ground to improve insulation and also to make the plants as inconspicuous as possible to the general public. Pump and control rooms were very often below ground as well. The gas was collected in floating-top gasholders on the digester itself or in separate water-sealed gasholders. However, if the gas is not to be used there is actually no necessity to have a top to a large digester. The bacterial uptake of oxygen described in Chapter 2, combined with the low solubility of oxygen in sludge, results in only the top few centimetres at the air-liquid interface being aerobic or partially aerobic, the main mass of the sludge remains anaerobic and so the fermentative and methanogenic bacteria can metabolise as in a closed tank. An example of an open-tank digester is shown by Brade & Noone (1981). However, such digesters do give off odiferous and flammable gases and so cannot be used in a confined works or near buildings. Mixing and heating of all types of digesters was by a variety of methods, internal and external; the detailed design of the digester and these ancillary fittings was often particular to the firm building the plant. However, all worked basically on the stirred-tank, continuous-flow principle, although some combined this with some aspects of the fed-batch system. The retention time was usually about 30 days and the temperature about 30–35°C, although some digesters ran at lower temperatures or even ambient temperature and longer retention times. Ideally, all such digesters should have produced similar results in terms of reduction of sludge solids and pollution potential, as the composition of sewage sludges is fairly uniform. The sludge feeding a sewage digester is usually a mixture of the primary debris settled off from the incoming sewage water (large faecal particles, papers, rags, etc.) plus excess sludge from the activated-sludge or other acrobic treatment side of the sewage works (a mixture of slimy microbial flocs and entrapped small particles of debris). However, surveys showed that performance of digesters in different works, or even in the same works, varied greatly (Swanwick *et al.*, 1969; Brade & Noone, 1981). Some digesters gave the results expected from laboratory tests, but others fell very much short of this ideal performance. This was not due primarily to the microbiology of the process but to mechanical aspects. Digesters were found to be inefficiently heated, with temperature variations over the digester contents and temperature variations with time. Inefficient mixing, usually mechanical, produced sludging-up of the digesters with masses of grit and sludge particles. Such masses were not only inert as regards microbial action but also altered the effective volume of the digester thus upsetting estimates of retention time from the feed flow and the original tank volume. Inefficient mixing could also lead to variable retention times as the chances of a sludge particle being washed out of the digester were not covered by the statistics of a perfectly-mixed sludge. Irregular mixing as well as sludged-up heat-exchangers were causes of the previously-mentioned temperature irregularities. Feeding was often irregular and at long intervals. This obviously gave irregular microbial activity.

Programmes designed to refurbish (e.g., gas instead of mechanical mixing), and control the running of, such poorly-performing digesters produced great improvements (Brade & Noone, 1981) but the digesters were still large plants requiring extensive civil and mechanical engineering facilities for construction. They were also expensive, in absolute terms and per head of the population served by the sewage works, and were suitable only for sewage works serving large populations which required large digester volumes. In the late 1970's and 1980's small digesters, of a hundred, or a few hundred cubic metres (100 m^3 is approx. 22 000 gallons) capacity, were being built for farm digester plants in

Britain. These used glass-enamelled steel tanks which were commercially available in various prefabricated sizes and usually used as farm slurry tanks, grain silos, etc. Such tanks, when insulated, can be used as continuous, stirred-tank, digesters with gas mixing and hot-water heating. Digesters on this construction principle were adopted by the sewage industry, as their cost was about one tenth (per head of population) that of the large concrete digesters, and they could be used in small sewage works. For larger works a number of tanks could be built to run in parallel. As with the conventional large digesters, the biogas is used in combined heat and power units to produce electricity and heat for keeping the digesters at mesophilic temperatures and for building heating, or in boilers for heating alone. The digesters are efficient and retention times of near twelve days, rather than the thirty days or more previously considered as desirable, are generally used. These digesters are being used now in many countries, and there are also similar designs with steel tanks being produced by other firms for farm and sewage digesters, although concrete is still used for the largest digesters (Chapter 4).

A necessary ancillary for some sewage works is apparatus for sludge 'thickening' before the sludge is introduced into the anaerobic digester. As previously mentioned, digesting sludges of less than about 4% TS is inefficient, and sewage sludges, particularly those from the aerobic plants, can be lower in solids than this. The solids content must, then, be increased (the sludge 'thickened'). This can be done mechanically in tanks where the sludge is slowly stirred and gravity settled (see Chapter 4). Or it can be done by coagulating the solids by addition of chemicals. These are usually iron or aluminium compounds, but the coagulants, alum and ferric chloride, can depress digestion of the sludge on both laboratory- and large-scale (Gossett *et al.*, 1978). Dentel and Gossett (1982) showed that substrates insoluble in water and those forming complexes with Fe and Al (amino acids, proteins, long-chain fatty acids) were affected: thus digestion of glucose was not affected while digestion of cellulose was. The aluminium and iron salts form insoluble hydroxides which coagulate colloidal particles and entrap larger particles. Lime is also used as a coagulant as it forms insoluble calcium carbonate which entraps sludge particles; lime should not affect digestion unless present in excess when the bulk of calcium carbonate precipitate can cause mechanical problems. Synthetic organic compounds are available as coagulants for sludges, and use of some natural compounds such as chitin has also been advocated (Sandford, 1989). The synthetic compounds are polyelectrolytes and a large number of formulations are on the market from different manufacturers. Laboratory testing of the effectiveness of the different compounds in coagulating the sludge to be treated on the large scale is the usual procedure. The coagulants should also be tested for their effects on anaerobic digestion of the sludge. Chitin is a poly-hexosamine obtained in bulk from crab and similar shells. Its use as a coagulant and for other purposes is now being promoted, but it is not, so far, as much used as chemical coagulants. It should have no effect on digestion of sludges.

The 'standard' digester design, usually with gas mixing, has been refined for use with sludges of all kinds of about 4 to 8% TS, usually nearer the lower value. Problems in running should be few, but one thing that can cause mechanical difficulties is the presence of 'stringy' materials in the feed. With sewage sludges such materials are mostly rags, but some farm wastes and other feedstocks can contain long straws and other plant fibres which can cause problems in mixing and feeding digesters. To overcome such difficulties, and to digest feedstocks of greater than 8% TS some variations in design of the stirred-tank digester have been produced.

Thermophilic digestion

The majority of sewage digesters run at mesophilic temperatures, 35–39°C being the maximum. The rate of digestion increases with temperature, although the absolute limit of degradation of the substrates at 'infinite' retention time is not changed. Although the extent of digestion at long RT will not be affected by digestion at high temperatures, at RT where digestion is not complete (the usual RT for running digesters) the same percentage degradation of solids may be obtained at lower RT at high temperatures than at mesophilic temperatures. This could enable the use of smaller digesters for a particular feedstock flow. However, although theoretically better, very few thermophilic sewage, or other, digesters are in use. One reason is that since the main thermal energy input to the majority of digesters is to heat the feedstock from ambient to digester temperature much more heat input is required to run a digester at 60°C than at 30°C. Garber (1977) reported on a thermophilic sewage digester that, although solids reduction (at the same RT) was greater in a thermophilic than a mesophilic digestion, gas production per weight of Volatile Solids was lower. The percentage of methane in thermophilic gas was higher. However, the volatile acid content of the thermophilic digested sludge was higher and the sludge gave odour problems. The thermophilic sludge was easier to mechanically dewater than mesophilic sludge, but more of the nitrogen was solubilised and lost on dewatering. Thermophilic digestion was much more sensitive to temperature changes than was mesophilic. Starting up thermophilic digestion was not easy. Overall, the problems of thermophilic sewage digestion have outweighed the advantages and similar problems have been found in stirred-tank thermophilic digestion of other feedstocks. One of the authors found that it took more than a year to start up thermophilic (about 55°C) pilot-plant digestion of pig waste and the digestion was not then stable, failure occurring for no obvious reasons. The digested sludge was also of unpleasant odour. Others have found difficulty with thermophilic pig-waste digestions, one factor being the generally high ammonia content of pig waste, the high free ammonia being more toxic to the bacteria at higher temperatures. Hashimoto (1983) found that laboratory-scale pig-waste digestions at 55°C with a feedstock containing 6·25% Volatile Solids were unstable, while digestions at 35°C were stable. With feeds of lower solids content the bacteria seemed to be able to adapt over a long start-up time to the thermophilic conditions, although Van Velsen *et al.* (1979) found adaptation to high ammonia concentrations did not occur at thermophilic temperatures. Hill *et al.* (1985) reported that waste scraped from piggery floors and so containing only faeces, not the faeces and urine of the slurries used by others, could be digested in a pilot-plant at 55°C. The digester was brought up from 35 to 55°C in 5°C steps over 40 days; for the next 50 days gas production and quality varied greatly although loading remained constant. After this, the digester was taken to be in a steady-state after loading changes when gas production over 20 days had a coefficient of variation of less than 10%. The data were collected over the last ten days of the period. Over these short periods the digestion seemed to be reasonably stable.

The thermophilic digestion of cattle wastes seems, in general, to have been more successful, at least in small-scale tests. Converse *et al.* (1977) digested cattle faeces alone, faeces plus urine and faeces, urine and straw at 60°C and 35°C. In both cases the straw did not digest well, but digestions appeared to be stable. They pointed out, though, that while the net energy output of the digester averaged 69% of the gross energy output as gas under mesophilic conditions, under thermophilic conditions, because of the greater energy needed to heat the feed, the net energywas only 40·3% of the gross energy. Varel *et al.* (1977) found 60°C to be the optimum temperature for digestion of wastes from cattle fed on maize, oats and some silage. At 65°C efficiency was much less, which is in accord with the laboratory batch digestions of Hashimoto *et al.* (1981). Studies on digestion at 60°C were carried on by Wohlt *et al.* (1990) who found reduction in solids degraded and in methane production at higher feed-solids concentrations. However, Zeeman *et al.* (1985) found pilot-plant digestion at 50°C of waste from European cattle fed on a different ration to be extremely difficult. Their work suggested that total ammonia concentration was a more important cause of the poor digestions than was solids concentration. It is a general finding that there is an increase in gas production per volume of feed with increasing solids concentration in the feed up to some particular concentration, above which specific gas production declines (Chapter 3). However, as in most cases feedstocks of higher solids contain higher concentrations of other feedstock constituents it is generally difficult to ascribe the effects solely to the solids. The ammonia content of poultry wastes is high (Hobson *et al.*, 1980) and Hassan *et al.* (1975) found that methane production per kg poultry-waste solids was more than twice as high at 35°C than at 50°C.

Factory wastes are generally low in ammonia and in some cases are produced at high temperature: such wastes can be digested at thermophilic temperatures without the necessity for heating the feedstock. Pea-processing wastes leave the machines at near boiling point and Lunsford & Dunstan (1958) showed that these could be digested in stirred-tank digesters at 55°C at shorter RT than at 35°C (3·5 compared to 6 days), while the gas yield per weight of solids was rather higher at 55°C. Most of the solids in this waste were dissolved.

Thermophilic digestion of solid municipal wastes in tank digesters has been previously mentioned. Cooney & Wise (1975) digested 10% sewage sludge and 90% municipal waste as a 4·8% TS slurry at 65°C and found better gas production than at mesophilic temperatures, although, as reported for sewage sludge, the volatile acid content of the effluent was 4 times that from a mesophilic digestion. Marique *et al* (1989) digested 25% TS MSW in a pilot-plant stirred-tank digester at 10 days RT and 55°C with liquid recycle. Gas production was 20% better than at 35°C.

Marchaim (1988) has recently described a thermophilic digestion of slaughterhouse wastes (rumen contents, blood, faeces, urine, straw) at 10% Total Solids and 12 day RT. The effluent solids were used as horticultural peat substitute. One advantage of the thermophilic process for this waste was that the high temperature effectively reduced coliform and Salmonella numbers, although it did not affect Clostridia. It is a general finding that destruction of pathogens in thermophilic digestions is more rapid than in mesophilic systems.

Modifications to the stirred-tank digester

Greasy scums can form in sewage digesters and have to be mixed back into the bulk of the liquid. However, one of the problems of running digesters for animal wastes and vegetable matter is the formation of a scum of plant fragments and animal hairs floating on the top of the digesting liquid. If mixing is not adequate then once such a scum begins to form it can build up and then begin to dry out to form eventually a 'cover' to the liquid which prevents gas escaping and the liquid overflowing in the effluent outlet. Gas mixing, whether free-rising gas or gas directed through draft tubes, will prevent scum formation, but paddle mixers working in the middle, or near the bottom, of the tank do not necessarily provide enough surface turbulence to prevent scum formation. A 13·6 m^3 digester run for many years, without scum formation, on pig, cattle and other wastes, was fitted with a twin-rotor, flat-blade turbine stirrer rotating discontinuously at low speed (50 rpm for 3 minutes per hour) with one rotor just under the surface of the liquid (Robertson *et al.*, 1975; Hobson *et al.*, 1980). Other systems have recirculated the digester liquid over the top of the digester contents or fed the influent over the top of the contents, with varying success.

Stewart (1983) described a digester for slurries of ensiled vegetable matter of about 10% solids. This had an auger feed with the vegetable matter being mixed with digester effluent liquid. An Archimedian-screw pump working in a draft tube lifted digester contents from the bottom of the tank and spread them over the top of the liquid. A system used in many of the older, large, sewage digesters.

An Archimedian screw mixer in a draft tube (turning at 600 rpm for 10–15 minutes once per day) was reported by Baeten and Verstraete (1988) to prevent scum formatiom in a stirred-tank digester treating 6–7% TS pig waste.

A system for preventing scum formation that is different in principle, but rather similar in design, tested in a digester which has run for many years, is the 'completely-filled digester' (Baader, 1985). The digester is of 100 m^3 capacity running on cattle manure and waste vegetable matter. The tank has a conical top and bottom and is completely filled with digesting material. The digester contents are circulated down a central draft tube by an Archimedian screw and up the outside of the tube. The feed of vegetable silage and/or manure is of high-solids content (the silage being 42–77% TS) and it is pushed by an auger into the bottom half of the digester through a tube where it is mixed with liquid recirculated from the digester bottom to give digester contents of about 9% TS. Gas and liquid leave at the top of the digester and surplus solids are removed from the bottom. Gas (60% methane) yields have been about 400l/kg TS fed at RT of 17–20 days and 33°C. Maximum loading rate has been about 3·2 kg VS/m^3 digester volume/day.

The idea of digesting municipal solid wastes (MSW) as an alternative to landfill disposal has been a subject of research and discussion for some years. The exact composition of municipal solid wastes varies with the town of collection; a small country town in Britain might collect less factory and shop wastes than a large town, and the wastes from a tropical town could differ from those of a temperate-climate town because of the different lifestyles of the inhabitants. However, municipal wastes consist, in general, of: biodegradable organic materials; kitchen vegetable wastes and garden wastes, papers of different kinds: undegradable chemical organic matter; mainly plastics: undegradable inorganic materials; tins and other metal objects, glass, stones and ashes. The material is often sorted before dumping and metals and other recyclable materials are removed along with large debris such as bicycle frames, furniture, etc., for separate disposal. The remaining organic materials can form a digester feedstock and about half this material is digestible in a reasonable retention time.

The earliest discussions and experiments on municipal waste digestions considered mixing the solid material (about 40% TS) with sewage sludge or water to make a slurry that could be digested in a continuous-flow, stirred-tank digester in the same way as sewage sludges. One such system which has been studied in large-scale plant for many years and has been the subject of a number of reports is the RefCoM process (e.g. Walter, 1983; Pfeffer, 1987; Chynoweth & Legrand, 1989). Here, the shredded refuse was made into a slurry of 4 to 10% TS with recycled water obtained by dewatering the digester effluent. The slurry was digested in a thermophilic (60°C), stirred-tank. Various sizes of tanks have been used during the laboratory, pilot and large-scale tests; the most recent plant had two 350 000 gallon digesters. Mixing was by top-driven paddles which were designed to circulate the slurry up and down the tank. In earlier runs these mixers picked up rags and similar materials which altered the flow pattern of the mixing and allowed formation of surface scum which could solidify (as has previosly been mentioned, rags can cause similar problems in sewage-sludge digesters). But redesign of the paddles obviated this. Overall, the digestion was successful, the main problems being with the waste separation and production of the feed, but the plant was reported to be closed at the time of writing. The gas was about 50% methane and VS destruction about 50%. The digested sludge was dewatered and the liquid used to slurry the incoming refuse and sewage sludge.

A small pilot-plant digester, which has run on various feedstocks but has recently been running on a 7% TS mixture of sewage sludge and sorted municipal waste, is not mixed and is designed to retain solids for 2 to 3 times longer than the liquid retention time. The slurry is sprayed into the top of the vertical tank (of large height to diameter ratio) and effluent is removed from the bottom. The solids are concentrated by gravity and gas uplift, etc.,in the middle and top of the digester. Experiments are continuing. (Biljetina *et al.*, 1987, 1988).

A pilot-plant horizontal-cylinder continuous-flow digester, continuously mixed with a screw type of mixer rotating on a longitudinal axle, was described by Szikriszt *et al.* (1988). This was used to digest a mixture of the organic fraction of municipal waste mixed with sewage sludge, but the novel part of the process was that the feed mixture, of 7–10% TS, was finely disintegrated in a hydropulper before being fed to the digester. The liquid from the dewatered digester effluent was recycled to be mixed with the waste and sewage sludge. The RT was 20–30 days at 38°C and gas production was 250–3001 kg TS fed.

Unless there are some inhibiting factors in the feedstock, the higher the loading rate of a digester the higher the gas production rate per volume of digester and so the more efficient the digester in terms of size of digester for a particular purpose and economic returns in usable energy. But combined with this maximising of gas production must be a maximising of pollutant reduction and gas production per unit of feed. For optimum solids reduction a long RT is usually needed, but if the solids content of the feed slurry cannot be altered the only way of increasing the loading rate is by loading a bigger volume of slurry per unit time, and this decreases the RT and so efficiency of solids digestion. So, at a short RT it might be possible to get a larger volumetric gas production than at a long RT because of the greater amount of solids loaded, but only at the expense of a lower solids breakdown and lower pollutant destruction. Where this problem might be overcome by increasing the TS of the feedstock, the previously-mentioned problems of pumping thick slurries can arise. So there is the limit to slurry thickness in continuous-flow digesters previously mentioned. However, if the solids content of the feed can be increased so that the feed itself is solid then problems of pumping slurries may be overcome; at least in some cases. Such solid wastes can be loaded by large-bore augers or by grabs through self-sealing ports. With solid-state digestions large loading rates and gas productions can be obtained, while there is no liquid-gas interface at which a scum can form. Similarly, there is no large liquid phase through which solids can settle and solids compaction is reduced by lift of the gas formed.

A continuous unstirred-tank digestion of the sorted organic fraction of municipal solid wastes, sorted waste paper, paper-making sludge, etc., has been running in Belgium on a large scale for some years (Baeten & Verstraete, 1988). The solid material is mixed with sewage sludge or some other liquid waste, or with water or the liquid returned from the digester effluent, to give a feedstock of about 30–35% TS. The feed is loaded into the top of a cylindrical digester by an auger and removed from the bottom in a similar fashion. The process is a continuous digestion with an overall retention time of 18–21 days at 55°C. The effluent material is dewatered in a screw press and the liquid can be returned to make up the feed. (In any digester, apart from reduction in solids in the water originally present, water can be produced by the bacteria during fermentation and methanogenesis). Gas is collected in a separate gas holder and is used for electricity generation. The digester contents do not need mixing.

The biogas, of methane content 55%, is produced at the rate of 125 to 185 litres per kg organic matter added to the digester. The actual rate varies with the composition of the waste: as said before the compositions of municipal wastes vary with the lifestyle of the inhabitants of the cities and the industries in the city and the degree of sorting of the raw waste. The digestion process can be used with other solid organic wastes and these may be more degradable than MSW. About 52% of the organic fraction of MSW was digested under the conditions mentioned above and this was a 55% reduction in the COD of the waste. Fruit and vegetable market wastes were reduced in weight by 67%, whereas restaurant wastes, paper-making sludge, and brewery draff were reduced by over 90%. Loading rates varied from about 10 to 20 kg/ m^3 digester volume/ day (Baeten and Verstraete, 1988).

Van Meenen *et al.* (1988) investigated the effects of changes in the running parameters of the dry municipal waste digestion. The bacteria in a digester running at 55°C could withstand a temperature rise to 63°C for a maximum of 20 minutes, but, as with other digesters, the bacteria remained active for at least 3 months if the temperature was dropped to ambient, or as low as 4°C, without feeding. The digested residue could be dewatered to 55% TS and the bacteria remained active, so such a dewatered material could be used as inoculum for another digester. However, the bacteria were affected by pH changes from 8·3 to 6·9 or 9·0. More than 10 g sodium chloride per kg digester feed decreased gas production. This latter could be a factor affecting sewage sludge digesters when the sewage contains road drainage in winter when roads are treated with salt, although concentrated salt solutions from melting ice would most likely be diluted to non-toxic concentrations in the general sewage water. The rumen anaerobic bacteria and some digester bacteria have been shown to be mildly halophilic (Chapter 2).

A similar process using a continuous-flow tank digester with sorted MSW of 30–35% TS is in use on a large scale in France. In this case the digester contents are mixed at intervals by gas injected into the bottom of the tank, as in gas-mixed sludge digesters. This can, again, be a mesophilic or thermophilic process, but is generally run at 37°C. Results very similar to those previously mentioned are obtained at retention times of 15 to 20 days (Zaoui, 1989). Because of the high solids-content of the feed, gas-production rates in both these solid-digestion systems are high; over 3 m^3 methane/m^3 digester volume/day compared with about 1 to 1·5 biogas volumes from the usual sludge digesters.The digested solids, after dewatering, can be sold as fertilizer ('anaerobic compost') or combusted for heat production, or used as inert landfill, etc. Advantages over conventional land-fill disposal of MSW are that the biogas can be generated near its site of use, the gas does not contain the halogenated hydrocarbons and other toxic or corrosive gases found in land-fill gas, the residue is inert and there are not the problems of leachates and smells associated with normal land-fills.

Although the ancillary equipment for utilising the biogas and for preparing the organic fraction from the unsorted waste is the same in continuous flow- and batch-digestion systems, batch fermentation of the organics of municipal waste has been suggested to be a simpler and cheaper process than continuous digestion.

Batch digesters were discussed earlier, but it seems appropriate to discuss solid-state batch digestion of municipal waste here. Wise (1987) reviewed some batch digestions of farm and municipal wastes. A batch process, so far tested only on a small scale, for mechanically-sorted municipal waste has recently been described by ten Brummeler *et al.* (1988). De Baere *et al.* (1985) said that TS contents higher than 40% inhibited dry anaerobic digestion, so as in other digestions the initial TS content of the waste was set at 35%. The waste was packed into a cylindrical, unmixed, digester tank incubated at 30°C, with an inoculum of digested sewage sludge.Tests showed that the waste had to be buffered by sodium bicarbonate to prevent acidification of the material and allow good start-up. Alkali addition to a large-scale system would be costly, but it was found that a partial aerobic composting of the waste before digestion prevented acidification (Chapter 5). It seemed that about 15% of the waste solids had to be destroyed in the composting to get a good digestion, these 15% solids being easily degraded and so causing a rapid acid formation under anaerobic conditions. An effect of size of inoculum on the start-up and course of digestion was also noted. The overall time of the batch digestions was about four months and the methane yield about 801 kg organic waste.

Cecchi & Mata-Alvarez (1990) have recently summarised data on the large-scalc Europcan plants for digestion of solid municipal waste.

Plug-flow and similar digesters

A number of horizontal-cylinder tank digesters have been made with large diameter axial screw mixers which seem to overcome scum problems. One is illustrated by Demuynck *et al.* (1984), another by Zeemann *et al.* (1988). These are not plug-flow, but are horizontal stirred-tank, continuous-flow digesters. A horizontal-tank digester used some years ago (Fry, 1974) had a hand-operated rake which was passed at intervals along the surface of the digesting slurry to rake the scum to one end of the tube where it could be removed through a door. A 3·8 m^3 pilot-plant horizontal-tube digester had a hand-rotated horizontal-axle fitted with discs at intervals to break up scum (Field *et al.*, 1985). Hand-operated mechanisms cannot be used on large digesters, but some of these kinds of mixers can be adapted for intermittent motor drive. Some horizontal-cylinder digesters are made commercially in glass-fibre plastic. These have gas stirring and can be made as modular units to deal with different flows of feedstock.

These digesters, although of a horizontal tubular construction are not true 'tubular' or 'plug-flow' digesters (Chapter 2) as although the feedstock enters at one end of the tank and exits at the other, the incoming feed and the digester contents are mixed up and each portion of feed does not proceed in a 'plug' fashion along the tube. These are really horizontal stirred-tank digesters with stirring of various efficiences.

Some digesters have been constructed as plug-flow fermentors without mixers, but, again, none behave ideally. An early design was that described by Hayes *et al.* (1980), and this type of design has been used in many subsequent digesters, particularly for small-scale plant. The construction is a trough excavated in the ground and lined with some impervious plastic or, perhaps, brick or cement. The trough is covered with a plastic balloon which is anchored down into the liquid in the trough to provide a gas-tight seal. The feed slurry is pumped or run by gravity into one end of the trough and it proceeds slowly by gravity to an effluent overflow at the other end, both inlet and exit having weirs to prevent gas escape. The digester contents are not stirred, but some mixing occurs because of gas formation. Heating is by water heat-exchangers in the trough in the larger digesters run in cool climates. The nominal retention time is similar to that of stirred-tanks treating the same feedstock at the same temperature. However, there will be settlement of solids and the solids RT may be longer than the nominal, liquid, RT. Thus, since stirred tanks are not run at RT where maximum degradation of solids can take place, the longer retention of solids in the trough can give rise to greater degradation and a performance of the tubular digester apparently better than that of a stirred-tank. Some stirred-tanks have been designed for solids retention to increase the time of reaction of the bacteria and the solids without increasing the RT of the liquid, the main bulk of the feedstock. Solids retention can be increased by having some form of screen in the outlet, but such systems can clog up. Another method is to stop the mixing and allow solids to partially settle before taking off supernatant by pump or by gravity flow as feed is added. Too much build-up of solids in the tank may have to be prevented by periodic removal of sludge from a bottom valve.

Hayes *et al.* (1980) described a full-size (38 m^3) tubular digester on the lines mentioned above to treat excreta from up to 65 dairy cows at about 10% TS content. Since then many similar digesters have been built, especially on a small scale in hot countries where they are climatically heated and used for the excreta from a few cattle or pigs. Because of the material of construction of the gas holders, these are often known as 'red-mud plastic digesters'.

Some digesters made of an inclined tube with the feed end lower than the effluent end have been constructed. Chapman (1986) and Floyd and Hawkes (1986) described laboratory and small pilot-plant tests on such digesters treating farm wastes. Solids were retained, but gas production was similar to that of stirred-tank digesters. Some mechanical problems such as blockage in the overflow weir were apparent. A commercial design of plug-flow digester where the feed flowed along an inner tube and then back along a concentric outer tube was marketed some time ago. This was used for dairy- and beef-cattle wastes and plant sizes were 20 to 80 m^3. The retention time was said to be about 5 days, so, presumably, it was heated (MIRCEN, 1986). All these types of 'tubular' digesters are designed to treat slurries of about 6–14% TS, as are the stirred-tank digesters, and the same mechanical constraints on pumping of feed slurries, etc., apply to these as to the stirred-tank digesters. However, since they are built at about ground level many digesters of this type have gravity feed and overflow.

Liquified solids

In the old town-gas works coal was gasified by pyrolysis at high temperature. Pyrolysis of solid wastes for gas production using more efficient plant than that of the town gas plants is now a subject for much research. The composition of the gas produced (various mixtures of H_2, CO, CH_4, CO_2 and minor components) varies with temperature and pressure of the reaction. However, research is also going on in a number of centres on the production of gas (CO_2 and CH_4) by anaerobic digestion of lignites and peats.These materials consist mainly of lignin, and while lignin itself is not anaerobically biodegradable the aromatic monomers from which it is constructed are degradable (see Chapter 2). The lignite can be converted to aromatics by alkaline oxidation at temperatures up to about 300°C and pressures of 55 atmospheres. The solutions of aromatics can then be gassified in various of the types of digester described here. Wise (1988) has summarised some aspects of the process, and more details of present research, including biological pretreatment of the lignites are given in the report of a conference on the subject (Wise, 1987).

CONCLUSION

The possible inhibitory effects of high solids concentrations on solids degradation have been mentioned previously. However, mechanical problems rather than inhibition have usually been the limiting factors in digestion of slurries of high solids content.

Slurries of up to about 12–14% TS (usually about 4–8%) are still the most common digester feedstock and various types of digester for such feedstocks have been tested. The variations in design have been largely to provide cheaper and simpler construction and operation, or to get over operational problems of some of the earlier designs. Although many digesters seem different in design, in many cases they are essentially of the same type and produce similar results. The key operating parameters are retention time and temperature. These are interdependent and changes in one can be balanced by changes in the other, so that digesters operating over a wide range of conditions can produce similar results in pollution control and energy. For various reasons there is now much interest in digestion of 'dry' or 'solid' feedstocks and it would seem as if digestion of these at high solids-content will be preferred over digestion of the materials mixed with water or other liquids to give slurries of the more usual 5–10% TS. 'Dry digestion' may have a place on farms in arid countries where deposited animal excreta can quickly dry and there is a lack of water for making up slurries. Recycling of effluent liquid may also help in such cases, but the liquid would generally have to be obtained by gravity settlement of solids as the mechanical solids separators used with some of the digesters described here are expensive, and also need large power inputs.

The digestion of excreta, whether of animal or human origin, has now been extensively tested and the results to be expected under various conditions of retention time and temperature are known. Many full-scale digesters are running on excreta slurries and any variations in the more or less standard designs are generally, as said before, to give a better or cheaper mechanical design. In particular, the digesters for animal and human wastes advocated for peasant farmers in developing countries are in many cases expensive in terms of incomes of the farmers, and cheaper, but easy to manage, digesters are being sought.

The digestion of crop wastes and weeds has been investigated partly from the point of view of provision of alternative energy sources but probably mainly for the utilisation of wastes otherwise left to rot or burnt in the fields or in some cases to provide additional feedstock to augment animal wastes which are in short supply. The digestion of some crop wastes has been done on a large scale, but in general these feedstocks have been tested only on laboratory and pilot-plant scales. Some vegetation contains bacterial inhibitors and simple methods of pretreatment of the materials to remove the inhibitors have had to be devised (Chapter 5). Usually, vegetation needs extra nitrogen for good digestion and this is supplied by mixing the waste with excreta. But, if not inhibited, the various crop wastes and other vegetations give similar specific gas yields, the main factor controlling extent of digestion being the extent of lignification of the material. The largest problem in scaling-up digestion of such feedstocks to full-size digesters is the handling of fibrous materials previously mentioned.

Two-phase digestions, where the fermentation and acidification are separated physically from the later methanogenesis, were described in Chapter 2, and some two-phase digestions based on batch digesters were described earlier in this chapter. However, it was noted in the latter cases that after some time methanogenesis may start in the fermentation tank. With continuous-flow digestion of slurries the retention times needed to allow efficient hydrolysis of the solids are long (and the pH needs to be above 6 for fibre fermentation), so with the usual non-sterile wastes and mixed inocula methanogenic bacteria can develop in the first tank. Only with some feed solids degradable in a matter of hours rather than days has two-phase digestion of high-solids feedstocks been possible and the process has been little used. However, sewage sludges vary in composition according to the proportions of primary and aerobic sludges and other materials and some are relatively easily degradable. Some of the earlier work on two-phase systems was done with a sewage sludge. Recently the success of a pilot-plant two-phase digester for municipal sewage sludge in America has prompted the building of a full-scale plant (Miller & Ghosh, 1991), and another American full-scale sewage digester with the acidification-phase circular tank in the middle of the larger second-phase tank has also been described (Peterson, 1991).

The start-up of digestion has been mentioned in Chapter 2 and, in general, start-up of high-solids, mesophilic, continuous digestions of slurries is not difficult, provided that the loading rate is kept low for the first few weeks and a reasonable inoculum (from added material or faecal feedstock) is used. Flushing out of the digester tank with an inert gas (usually nitrogen) before the digestion starts may be desirable for safety reasons but is not usually needed by the bacteria. If the feedstock contains much easily-degradable carbohydrate, such as starch, then rather more care in keeping the initial loading rate low may be needed than if the feedstock is difficultly-degradable fibres, and, with some exceptions, where, again, easily-degradable solids are present, mesophilic batch digestions are also easy to start. Start-up of psychrophilic digestions may be more difficult, as previously described.

Although the running parameters and the gas productions from agricultural wastes have been determined in laboratory and pilot plant digesters and it has been shown that these gas productions and pollution reductions can be obtained with full-size plants, a large proportion of the digesters built on farms in various countries have been abandoned and many of those left are performing poorly. This applies to both heated, automated plants on large farms in cooler countries and to the climatically-heated plants on small farms in warm countries. Some reasons have been mentioned before; the price of 'standard' fuels has generally decreased since many of the large-farm digesters were built and the digesters no longer pay for themselves in terms of gas production, and pollution control is not really necessary. Some of the original digesters were built while design was in the experimental stage and were never updated. Others were built on a rather *ad hoc* basis. But probably a main reason is that the digesters have not been run properly. This may be due to mechanical factors not realised at the time of building, such as difficulties in providing feed slurries of constant composition and flow rate, but it can also be due to the farmer expecting that the plant will run with no attention. Although a well-designed digester should run with minimum attention, it still needs some regular monitoring and maintenance. In a sewage works or factory there are trained staff to run digesters. They are familiar with the possible mechanical and other problems that might occur and they are also available to remedy problems as they occur. On the farm the digester is usually run as a secondary job by workers employed in other activities. Monitoring and attention to maintenance and clearing of blockages, etc., tend to be put off until time is available; the digester being regarded as a secondary matter on the farm. So problems increase and become more difficult to remedy. A full-scale digester on a farm cannot be run with the temperature and feed rate and slurry composition controlled as exactly as the parameters can be controlled in the laboratory, but it should be possible to keep fluctuations small and these will not materially affect the digestion. Wellinger (1985) surveyed a number of farm digesters in Switzerland over a two-year period and found that temperature and retention time were the most important factors affecting gas production. He recommended that if farms had difficulty in keeping digester temperature reasonably constant, because of poor digester insulation or other factors, they would do better by running the digester at an attainable lower but constant temperature than by trying to keep the temperature in the higher mesophilic range. The feed slurries for these digesters were of fairly uniform composition, but the retention times varied because of poor digester mixing and other factors and the times were not always the optimal for the poultry, cattle or pig wastes recommended by research. In many cases combinations of the various factors tended to make the digestions unstable. Singh & Gupta (1990) surveyed a large number of community climatically-heated digesters in India. They found that although the specific gas productions were equal to those found in experimental digesters operating under the same conditions, the actual gas productions were low because the digesters were not being loaded to their design capacities. Also, because they were climatically heated, the temperature was impossible to control at a constant level. Cortellini *et al.* (1985) surveyed five digesters, of different designs, on Italian farms. They found that loading rates were generally not up to design values because of the difficulties in removing solids from excreta-collection systems and because of ingress of water to the feedstock. Retention times varied, as did operating temperatures. Specific gas yields then varied in and between digesters. The digester run for many years by one of thc authors and colleagues also had low gas yields caused by loading being below design rate because of dilution of excreta and difficulties in removing the waste from the pig houses (Summers *et al.* 1984). In their survey of European biogas plants Demuynck *et al.* (1984) listed a number of mechanical problems, on the lines of those just mentioned, which had led to poor performance or the abandoning of farm plants. Digesters do work, the problem is making them work under the generally low-technology conditions of the farm, wherever the farm is situated.

REFERENCES

A Chinese biogas manual (1979). Ed. A. Buren. Intermediate Technology Publications, 9 King Street, London WC2E 8HN.

Advances in biogas technology (1986). P. Chala, Indian Council for Agric. Res., New Dalhi, India.

Baader, W. (1985). *Biogas production from green materials*, Conference on Biomass conversion for energy, Freising, Fed. Rep. Germany, October.

Badger, D.M., Bogue, M.J. & Stewart, D.J. (1979). *N.Z. J. Sci.* **22**, 11.

Baeten, D. & Verstraete, W. (1988). *Biol. Wastes*, **26**, 297.

Balasubramanya, R.H., Rhandeparkar, V.G. & Sundaram, V. (1986). *Agric. Wastes*, **16**, 295.

Balasubramanya, R.H., Khandeparkar, V.G. & Sundaram, V. (1988). *Biol. Wastes*, **25**, 319.

Balzari, P. & Bozza, E. (1988). In *Agricultural waste management and environmental protection, vol. 2*, ed. E. Welter & I. Szabolcs, CIEC and FAL, Braunschweig, West Germany. p. 71.

Barry, M. & Colleran, E. (1982). *Agric. Wastes*, **4**, 231.

Biljetina, R. et al. (1987). Paper 6985, IGT Document Centre, State St., Chicago, USA.

Biljetina, R., Srivastava, V.J. & Isaacson, H.R. (1988). In *Energy from Biomass and Wastes XII*, IGT, Chicago, USA.

Biogas technology, an information package. (1985). Tata Energy Documantation and Information Centre, PB 698, Bombay House, Bombay, India.

Brade, C.E. & Noone, G.P. (1982). *J. Ist. Wat. Pollut. Cont.* no. 1, 70.

Brooks, J.D. & Meers, J.L. (1973). *J. Gen. Microbiol.* **77**, 513.

Brouard, F., Bories, A. & Sauze, F. (1983). In *Energy from biomass, 2n. EC Conference*, ed. A. Strub, P. Chartier & G. Schleser. Applied Science Publishers, London. p. 334.

Cadwallader, M.W. (1991). In *Energy from biomass and wastes XIV*. I.G.T., Chicago, USA. p. 859.

Chapman, J.M. (1986). Design and performance studies of pilot-scale inclined tube digesters operating on cattle slurry. Ph.D. thesis, Polytechnic of South Wales, Pontypridd, Wales, UK.

Cecchi, F., Marcomini, A., Parvan, P., Fazzini, G. & Mata-Alvarez, J. (1990). *Biocycle*, **31**, 42.

Cecchi, F. & Mata-Alvarez, J. (1990). *Advances in Appl. Biotechnol. Ser.* **4**, 285.

Chen, T.H., Day, D.L. & Steinberg, M.P. (1988). *Biol. Wastes*, **24**, 297.

Chynoweth, D.P., Earle, J.F.K., Bosch, G. & Legrand, R. (1990). *Biocycle*, **31**(10), 50.

Chynoweth, D.P. & Legrand, R. (1989). In *Landfill gas and anaerobic digestion of solid waste*, ed. Y.R. Alston & G.E. Richards, Harwell Laboratory, UKAEA, Oxfordshire, UK. p. 467.

Clausen, E.C. & Gaddy, J.L. (1983). In *Fuel gas systems*, ed. D.L. Wise, CRC Press, Roca Baton, USA.

Coble, C.G. & Egg, R. (1987). In *Energy from biomass and wastes X*, ed. D. Klass, Elsevier Applied Science Publishers, London. p. 1057.

Converse, J.C., Evans, G.W. & Verhoeven, C.R. (1977). *Am. Soc. Agric. Eng.* Paper no. 77-4051.

Cooney, L.C. & Wise, D.L. (1975). *Biotech. Bioeng.* **17**, 1119.

Cortellini, L., Piccinini, S. & Tilche, A. (1985). In *Energy from biomass*, 3rd. EC Conference, ed. W. Palz, J. Coombs & D.O. Hall, Elsevier Applied science Publishers, London. p. 599.

Cullimore, D.R., Maule, A. & Mansuy, N. (1985). *Agric. Wastes*, **12**, 147.

Dar, G.H. & Tandon, S.M. (1987). *Biol. Wastes*, **21**, 75.

De Baer, L., Verdonck, O. & Verstraete, W. (1985). In *Seventh symposium on biotechnology of fuels and chemicals*, Gatlinberg, USA.

Demuynck, M., Nyns, E.J. & Palz, W. (1984). *Biogas plants in Europe*, D. Reidel Company, Dordrecht, The Netherlands.

Dentel, S.K. & Gossett, J.M. (1982). *Water Res.* **16**, 707.

El Shazly, K. & Hungate, R.E. (1965). *Appl. Microbiol.* **13**, 62.

Ferrara, R., Barberis, R., Jodice, R., Vincenzino, E. & Vanni, A. (1984). *Agric. Wastes*, **11**, 79.

Field, J.A., Reneau, R.B., Kroontje, W. & Caldwell, J.S. (1985). *Agric. Wastes*, **13**, 207.

Fielden, N.E.H. (1981). *Agric. Wastes*, **3**, 75.

Fixed dome biogas plants (1987). Tata Energy Research Institute, 7 Jor Bagh, New Delhi, 110003, India.

Floyd, J.R. & Hawkes, F.R. (1986). *Agric. Wastes*, **18**, 39.

Fry, J.L. (1974). *Practical building of methane power plants.* D.A.K. Knox, Andover, Hants, UK.

Garber, W.F. (1978). *Prog. Water Technol.* **8**, 401.

Ghosh, S., Chynoweth, D.P. & Tarman, P.B. (1987). US Patent 4696746.

Gijzen, H.J., Schoenmakers, T.J.M., Caerteling, C.G.M. & Vogels, G.D. (1988*a*). *Biotechnol. Letters*, **10**, 61.

Gijzen, H.J., Zwart, K.B., van Gelder, P.T. & Vogels, G.D. (1986). *Appl. Microbiol. Biotechnol.* **25**, 155.

Gijzen, H.J., Zwart, K.B., Verhagen, F.J.M. & Vogels, G.D. (1988*b*). *Biotech. Bioeng.* **31**, 418.

Gossett, J.M., Wilson, J.C. Evans, D.S. & McCarty, P.L. (1978). *J. Water Pollut. Cont. Fed.* **50**, 533.

Goswami, K.P. (1991). *Energy Sources*, **13**, 267.

Guo Chao Sun, Yi-Zhe Wu, Shi-Jin Sha & Ke-Xin Liu. (1987). *Biol. Wastes*, **20**, 291.

Haga, K., Tanaka, H. & Higaki, S. (1979). *Agric. Wastes*, **1**, 45.

Hall, S.J., Hawkes, D.L., Hawkes, F.R. & Thomas, A. (1985). *J. Agric. Eng. Res.* **32**, 153.

Hashimoto, A.G. (1983). *Agric. Wastes*, **6**, 175.

Hashimoto, A.G. (1986). *Biotech. Bioeng.* **28**, 1857.

Hashimoto, A.G. (1989). *Biol. Wastes*, **28**, 247.

Hashimoto, A.G., Varel, V.H. & Chen, Y.R. (1981). *Agric. Wastes*, **3**, 241.

Hassan, H.M., Belyea, D.A. & Hassan, A.E. (1975). In *Managing livestock wastes*, A.S.A.E., St. Joseph, Michigan, USA. p. 244.
Hayes, T.D., Jewell, W.J., Dell'Orto, S., Fanfoni, K.J., Leuscher, A.P. & Sherman, D.F. (1980). In *Anaerobic digestion*, ed. D.A. Stafford, B.I. Wheatley & D.E. Hughes, Applied Science Publishers, London.
Hill, D.T. & Bolte, J.P. (1986). *Agric. Wastes*, **16**, 147.
Hill, D.T., Holmberg, R.D. & Bolte, J.P. (1985). *Agric. Wastes*, **14**, 37.
Hill, D.T., Young, D.T. & Norstedt, R.A. (1981). *Trans. A.S.A.E.* **24**, 731.
Hilpert, R., Winter, J. & Kandler, O. (1983). In *Energy from biomass* 2nd EC conference, ed. A. Strub, P. Chartier & and G. Schleser, Applied Science Publishers, London. p. 552.
Hobson, P.N. (1965). *J. Gen. Microbiol.* **38**, 161.
Hobson, P.N. (1975). In *Straw decay and its effects on disposal and utilisation*, ed. E. Grossbard. John Wiley and Sons, Chichester, UK. p. 217.
Hobson, P.N. (1983). *J. Chem. Tech. Biotechnol.* **33B**, 1.
Hobson, P.N. (1985). *Agric. Wastes*, **14**, 255.
Hobson, P.N., Bousfield, S., Summers, R. & Mills, P.J. (1980). In *Anaerobic digestion*, ed. D.A. Stafford, B.I. Wheatley & D.E. Hughes, Applied Science Publishers, London. p. 237.
Jerger, D.E. (1982). In *Energy from biomass and wastes VI*, I.G.T., Chicago, USA. p. 341.
Jewell, W.J. (1980). In *Anaerobic digestion*, ed. D.A. Stafford, B.I. Wheatley & D.E. Hughes, Applied Science Publishers, London. 467.
Jewell, W.J., Cummings, R.J., Richards, B.K. & Rector, D.J. (1985). In *Energy from biomass and wastes IX*, IGT, Chicago, USA. p. 669.
Kalia, A.K. & Kanwar, S.S. (1989). *Biol. Wastes*, **30**, 217.
Kaul, K.K., Kalra, M.S. & Srivastava, R. (1987). *Biol. Wastes*, **21**, 29.
Ke-Xin Liu, Guo-Chao Sun, Xian-Qiang Wu & Yi-Zhe Wu. (1987). *Biol. Wastes*, **20**, 303.
Kishore, V.V.N. (1989). *Biol. Wastes*, **30**, 199.
Llabres-Luengo, P. & Mata-Alvarez, J. (1988). *Biol. Wastes*, **23**, 25.
Lunsford, J.V. & Dunstan, G.H. (1958). *Biological Treatment of Sewage and Industrial Wastes*, **2**, 107.
Marchaim, U. (1988). In *Fifth Int. Symp. An. Digestion*, Extension of the Poster-paper Book. p. 53.
Marique, Ph., Gilles, A., Edeline, F. & Joassin, L. (1989). *Biotech. Bioeng.* **33**, 536.
Mathison, B. & Thyselius, L. (1985). In *Bioenergy 84 vol. 3*, ed. H. Egnéus & A. Ellagard, Elsevier Applied Science Publishers, London. p. 289.
McKain, N. & Hobson, P.N. (1987). *Biol. Wastes*, **22**, 147.
Miller, T.L. & Ghosh, S. (1991). In *Energy from biomass and wastes XIV,* IGT Document Centre, Chicago, USA. p. 869.
MIRCEN Anaerobic Digestion. (1986). Karolinska Institute, Stockholm, Sweden. **3**, 144.
Molnar, L. & Bartha, I. (1989). *Biol. Wastes*, **28**, 15.
Neth, (19–). *J. Agric. Sci.* **27**, 255.
Nazir, M. (1991). *Biores. Technol.* **35**, xx.
Olsen, J.E., Jorgensen, J.B. & Nansen, P. (1985). *Agric. Wastes*, **13**, 273.
Olsen, J.E. & Larsen, H.E. (1987). *Biol. Wastes*, **21**, 153.
Paus, A., Nyns, E.J. & Naveau, H. (1984). In *Anaerobic digestion and carbohydrate hydrolysis of waste*, ed. G.L. Ferrero, M.P. Ferranti & H. Naveau, Elsevier Applied Science publishers, London. p. 209.
Peterson, A.C. (1991). In *Energy from biomass and wastes XIV*, IGT Document Centre, Chicago, USA. p. 877.
Pfeffer, J.T. (1979). In *Fuel gas production from animal and agricultural residues and biomass*, 2nd Quarterly Report, Dynatec Co., 99 Erie St., Cambridge, USA.
Pfeffer, J.T. (1987). In *Energy from biomass and wastes X*, ed. D.L. Klass, Elsevier Applied Science publishers, London. p. 1149.

Powell, G.E., Hilton, M.G., Archer, D.B. & Kirsop, B.H. (1983). *J. Chem. Technol. Biotechnol.* **33B**, 209.
Ranade, D.R., Yeole, T.W., Meher, K.K., Gadre, R.V. & Godbole, S.H. (1989). *Biol. Wastes*, **28**, 157.
Reig, M., Toldra, F., Gow, J.T., Jansen, N.B. & Tsao, G.T. (1989). *Biol. Wastes*, **29**, 201.
Rijkens, B.A. (1981). In *Energy from biomass and wastes V*, I.G.T., Chicago. p. 463.
Rijkens, B.A. (1983). In *Energy from biomass* 2nd. EC conference, ed. A. Srub, P. Chartier & G. Schleser, Applied science Publishers, London. p. 572.
Robertson, A.M., Burnett, G., Bousfield, S., Hobson, P.N. & Summers, R. (1975). In *Managing livestock wastes*, A.S.A.E., St. Joseph, Michigan, USA. p. 544.
Safley, L.M. & Westerman, P.W. (1989). *Biol. Wastes*, **27**, 43.
Safley, L.M. & Westerman, P.W. (1992). *Biores. Technol.* **41**, 167.
Sayed, S., Van der Zanden, J., Wijfels, R. & Lettinga, G. (1988). *Biol. Wastes*, **23**, 117.
Sharma, S.K., Saini, J.S., Mishra, I.M. & Sharma, M.P. (1989). *Biol. Wastes*, **28**, 25.
Singh, N. & Gupta, R.K. (1990). *Biol. Wastes*, **32**, 149.
Stewart, D.J. (1980). In *Anaerobic digestion*, ed. D.A. Stafford, B.I. Wheatley, & D.E. Hughes, Applied Science Publishers, London. p. 303.
Stewart, D.J. (1983). In *Fuel gas systems*, ed. D.L. Wise, CRC Press, Roca Baton, USA. p. 85.
Summers, R. & Bousfield, S. (1980). *Agric. Wastes*, **2**, 61.
Sutter, K. & Wellinger, A. (1988). In *4th Int. CIEC Symposium, Proceedings*, Vol. 2, ed. E. Welte & I. Szabolcs. CIEC and FAL, Braunschweig, Germany. p. 61.
Svendsen, E.K. & Blackburn, T.H. (1986). *Agric. Wastes*, **16**, 47.
Swanwick, J.D., Shurben, D.G. & Jackson, S. (1969). *Water Pollut. Control*, 68, 639.
Szikriszt, G., Frostell, B., Norrmen, J. & Bergstrom, R. (1988). In *Anaerobic digestion 1988*, ed. E.R. Hall & P.N. Hobson, Pergamon Press, Oxford. p. 375.
ten Brummeler, E., Horbach, H.C.J.M. & Koster, I.W. (1991). *J. Chem. Technol. Biotechnol.* **50**, 191.
ten Brummeler, E., Koster, I.W. Zaevalkink, J.A. (1988). In *Anaerobic digestion 88*, ed. T.J. Hall & P.N. Hobson, Pergamon Press, Oxford. p. 335.
Tritt, W.P. & Kang, H. (1991). *Biores. Technol.* **36**, 161.
Van Meenen, P., Vermeulen, J. & Verstraete, W. (1988). In *anaerobic digestion 1988*, ed. E.R. Hall & P.N. Hobson, Pergamon Press, Oxford. p. 345.
Van Velsen, A.F.M. (1977). *Neth. J. Agric. Sci.* **25**, 151.
Van Velsen, A.F.M., Letinga, G. & den Ottelander, D. (1979). *Neth. J. Agric. Sci.* **27**, 255.
Varel, V.H., Isaacson, H.R. & Bryant, M.P. (1977). *Appl. Environ. Microbiol.* **29**, 374.
Verstraete, W., Baere, L. & Rozzi, A. (1981). *Trib. bebedean. Novs.* **34**, 367.
Virturtia, A.M., Mata-Alvarez, J., Cecchi, F. & Fazzini, G. (1989). *Biol. Wastes*, **29**, 189.
Walter, D.K. (1985). In *Energy from biomass and wastes VII*, IGT, Chicago, USA. p. 705.
Wellinger, A. (1985). *Process Biochem.* **20**, 131.
Wellinger, A. & Kaufman, R. (1982). *Process Biochem.* **17**, 5, 26.
Wheatley, A.D. (1978). In *Proc. Seminar anaerobic digestion of farm wastes*, ADAS, Reading, UK, p. 72.
Wise, D.L. (1987*a*). *Resources and Conservation*, **13**, 83.
Wise, D.L. (1987*b*). *Resources and Conservation*, **15**, 229.
Wise, D.L. (1988). In *Fifth Int. Symp. An. Digestion, Poster Papers*. Ed. A. Tilche & A. Rossi, Monduzzi Editore, Bologna, Italy. p. 961.
Wohlt, J.E., Frobish, A.A., Davis, C.L., Bryant, M.P. & Mackie, R.I. (1990). *Biol. Wastes*, **32**, 193.
Wong, M.H. & Cheung, Y.H. (1989). *Biol. Wastes*, **28**, 143.
Yang, P.Y. (1981). In *Energy from biomass and wastes V*, IGT, Chicago, USA. p. 307.
Yadav *et al* (1987).
Zaoui, R. (1989). In *Landfill gas and anaerobic digestion of solid waste*, ed. Y.R. Alston & G.E. Richards, Harwell Laboratory, UKAEA, Oxfordshire. p. 481.

Zauner, E. (1988). In *4th Int. CIEC Symposium, Proceedings Vol. 2*, ed. E. Welte & I. Szabolcs. CIEC and FAL, Braunschweig, Germany. p. 103.

Zeeman, G. (1991). Thesis, The Agricultural University, Wageningen, quoting a report by van Asseldonk, M.M.L. & Voermans, J.A.M. (1990) in Dutch.

Zeeman, G., Sutter, K., Vens, T., Koster, M. & Wellinger, A. (1988*a*). *Biol. Wastes*, **26**, 15.

Zeeman, G., Vens, T.J.M., Koster-Treffers, M.E. & Lettinga, G. (1986). Proc. NVA-EWPCA Water-treatment conference, Amsterdam, Sept.

Zeeman, G., Vens, T.J.M., Koster-Treffers, M.E. & Lettinga, G. (1988*b*). In *Anaerobic digestion 1988*, ed. E.R. Hall & P.N. Hobson, Pergamon Press, Oxford. p. 397.

Zeeman, G., Wiegant, W.M., Koster-Treffers, M.E. & Lettinga, G. (1985). *Agric. Wastes*, **14**, 19.

Zubr, J. (1985). In *Bioenergy 84 vol. III*, ed. H. Egne'us & A. Ellegard, Elsevier Applied Science Publishers, London. p. 295.

Zwart, K.B., Gijzen, H.J., Cox, P. & Vogels, G.D. (1988). *Biotech. Bioeng.* **32**, 719.

Chapter 7

DIGESTION OF LOW-SOLIDS FEEDSTOCKS

INTRODUCTION

As was described in the previous chapter, anaerobic digestion developed from the septic tank to treat the sludges of larger particles settled off from municipal sewage in large, stirred-tank, digesters (see Chapter 4). This pattern has generally been followed and most digesters today are part of a sewage-treatment system where the larger particles in the raw sewage water are filtered off to be anaerobically treated while the sewage waters, with only low concentrations of the smallest particles of suspended solids, flow on to be treated aerobically. Problems of disposal of the excess sludge (microbes and entrapped solids particles) from the aerobic plant have led to this sludge being anaerobically digested along with the sludge from the primary settling tanks. The sludges treated in this type of completely-mixed tank digester have a suspended solids concentration of about 4% and upwards. Since the solids are not easily degradable, the retention time of the microbes, sludge-solids and liquid, has to be more than about ten days and this allows time for the growth of the slow-growing lipid-degrading and methanogenic consortia of microorganisms. As the major proportion of the biogas is generated by degradation of the solids, the 4% or greater solids concentration allows of sufficient biogas production to provide energy to heat the digester and usually to provide energy for other systems as well. A stirred-tank digester can be run with a feedstock of lower than 4% solids suspended in a low-strength water, but the gas production per unit volume of digester is very low and barely sufficient to heat the digester and not sufficient to produce worthwhile returns in energy production for other purposes, and there may be some washout of bacteria (see, for instance, Summers *et al.*, 1984). If the stirred-tank is the only type of digester available then the digestion of low-solids wastewaters is not very feasible, particularly if the water flow is fast. In the case of the sewage, the flow of sewage water is usually great, which with a retention time of days would mean very large digesters (related to retention time and water-flow rate). But as the digesters are fed from the solids removed from the water the size of the digester is governed by the relatively slower flow-rate of the sludge.

Anaerobic treatment of the sewage water and small suspended particles would be advantageous. Anaerobic plants do not need the energy-intensive and expensive aeration of the aerobic activated-sludge plant used on most sewage works. The anaerobic treatment could, on the other hand, produce energy as biogas. Since only some 10% of the substrate is converted to microbial cell mass (the rest to energy for cell growth, etc.) in anaerobic microbial reactions as compared to about 50% for aerobic reactions, the residual anaerobic microbial sludge is less than the aerobic sludge and is more easily dewatered. This could improve sludge handling.

One of the earliest methods of controlled treatment of sewage, the septic tank patented in 1895, did treat anaerobically both the sewage water and solids. However, although septic tanks are still used for sewage from one house or a few grouped houses, it was soon realised that the effluent from large septic tanks was often too coloured and malodorous for discharge. Various experiments led to the Imhoff tank (1905) which treated only the solids separated from the sewage water and so the place of anaerobic treatment in the sewage works became that of treating solids. The trickling filter and the activated-sludge plants then became the methods of (aerobic) treatment of the sewage water. Interest in treating low-solids wastewaters by anaerobic methods did not decline, but until the development of the modern sludge-blanket (UASB), expanded-bed-, filter- and contact-digesters treatment of wastewaters did not become a really practical proposition. Some of the history of wastewater, particularly sewage water, anaerobic treatment is given by McCarty (1985).

The digesters just mentioned are all examples of the 'retained-biomass' system, the basis of anaerobic digestion of wastewaters. The problems of the stirred-tank type of digester where feedstock and microorganisms pass through the tank at the same rate, are related to the high flow-rate and generally low substrate content of the wastewater and the rate of growth of the anaerobic bacteria. If the retention times of the water and suspended bacteria are equal and shorter than the doubling times of slow-growing bacteria, then these latter may be washed out of the digester before they can divide and so a stable digestion is not attained after start-up. However, if the bacteria can be retained in the tank as a dense mass of organisms, then they can grow at a low rate while the overall fermentative activities of the mass of bacteria will be sufficient to gasify the substrates dissolved in the water passing through the organisms. Thus, the hydraulic retention time of the water can be short, and so the size of the digester small, while the rates of substrate utilisation and biogas production per unit digester volume can be high. While some retained-biomass digesters, particularly those for hot factory-wastewaters, are run at higher mesophilic temperatures, it has been found that satisfactory performance can be obtained at lower mesophilic (cool ambient) temperatures. The anaerobic treatment of sewage waters is then possible.

All the types of retained-biomass digesters, including composite systems, have been tested for digestion of sewage waters over the last eighty years, and much of the work up to 1985 is summarised in the papers in the proceedings of a workshop on sewage-water digestion held in that year (see ref. to McCarty, above). Although the digestion of sewage waters has been shown to be feasible there has so far been little large-scale adoption of the process in preference to the activated-sludge types of aerobic plants. On the other hand, digestion of factory wastewaters is extending quite rapidly. The digestion of sewage waters will then be considered along with that of factory waters in considering the types of retained-biomass digesters.

The 'feed-back' or 'contact' digester

The contact digester is similar in working to the aerobic activated-sludge plant in that a retention time of the bacteria greater than the retention time of the liquid is obtained by returning most of the bacteria in the effluent to the input flow of the tank. A diagramatic contact digester is shown in Fig. 2B. The digester itself is usually a heated stirred-tank like the single-stage tank for digestion of high-solids feedstocks. However, the effluent is passed through some form of solids separator and the separated bacteria are returned to the input side while the clarified effluent flows on; for further treatment or discharge.

An early application of the contact digester was to wastewaters from a meat-packing plant in the USA by Steffen, Fullen and co-workers. After laboratory experiments a pilot-plant and then a full-scale plant were built. The pilot-plant treated 26·5 m^3 of water per day. The full-scale plant had two digesters which each treated an average of 2880 m^3 per day, with a maximum of 3379 m^3. The retention time of the liquid was 12·23 hours based on feed volume, but a return flow of 2·4 volumes of settled sludge to 1 of raw feed decreased the RT to 3·58 hours. The digesters were tanks with mechanical stirring (Fullen, 1953; Steffen, 1953; Schroepfer *et al.*, 1955).

The difficulty with contact digesters is settling the effluent to obtain a concentrated bacterial sludge to return to the digester. The effluent is gassing and this tends to float the bacteria. In the digesters mentioned above the outflowing digester contents were drawn up into a trough above the digester by a vacuum which also degassed the sludge as it trickled from the trough over slats in a tank below. The degassed sludge ran to a gravity-settling tank from the bottom of which it was returned to the digester feed. The gas coming off the degassifier was higher in carbon dioxide than the main digester gas because of release of the CO_2 from solution. A little while later, Hemens and Shurben (1959) described a pilot-plant digester for meat-factory wastewaters in Britain. This had a different type of degassifier. A scraper arm rotated in a circular tank underneath a spiral baffle. The baffle constrained the effluent to move from the centre of the tank to the edge while the settled solids moved to the centre of the conical bottom of the tank from where they were returned to the digester. The top of the separator was open so gas was lost. It was found that separation was best when the temperature of the effluent, entering at 33°C, did not fall below 20°C in the separator. A more recent commercial digester system uses cooling of the effluent to slow metabolism and gassing while bacteria are being separated for return. The performance of a digester using this principle for treating a starch-production waste is described by Butcher (1989). Contact digesters have also been used for brewery wastes. Coulter *et al.* (1957) said that a contact digester was generally unsuitable for sewage waters. Hamdi & Garcia (1991) compared an anaerobic filter and a contact digester for treatment of olive-mill wastewaters which had been pretreated by fermentation with *Aspergillus niger*. They found that the filter gave more gas per unit COD removed (and greater removal) and the contact effluent contained more VFA than the filter effluent. The better performance of the filter was ascribed to the dense biomass being more resistant to toxic compounds in the wastewater (see Chapter 2). The filter also avoided the apparatus and energy consumption of sludge settling and return.

To ensure maximum activity of the returned bacteria theory shows that they should not be subject to mechanical or temperature shocks during the return process. The paragraphs above show that this is impossible to attain in practice, so there will always be some lag time before returned bacteria become fully active in the digester and some of the bacteria may die. It would be much better if the bacteria could be returned to the digestion without leaving the confines of the digester. This has been obtained with other retained-biomass types of digester. In fact, Pascik & Henzler (1988) working with wastewaters from a sulphite pulp bleaching plant reported that it was not possible to separate the bacteria in a contact digester at reasonable cost, so they developed a digester with the bacteria immobilised on a polyurethane support matrix which successfully treated these organo-chlorine-containing wastes.

The anaerobic filter

The anaerobic filter is similar in principle to the aerobic trickling filter used for many years in sewage works to treat dilute, low-solids sewage water. It consists of a tank filled with a porous material; stones, plastic shapes, unglazed pot in different shapes, etc. The active biomass, a mixture of hydrolytic, fermentative and methanogenic bacteria, is either trapped in the spaces between the packing material or adheres to the surfaces of the pieces of packing (see Chapter 4). The wastewater flows either upwards or downwards over the packing. In the case of the up-flow filter the wastewater does not 'trickle' over the filter matrix, but the packing is immersed in the liquid. The high concentration of bacteria allows rapid degradation of substrates in the water which can flow at rates giving hydraulic retention times of only a few hours, while, although the biomass does gradually detach from the support matrix, the average retention time of the bacteria in the filter is many days, thus allowing slow growth.

The filter can be used on its own to give a complete digestion and some of the earliest experiments on these lines were those of Young & McCarty (1968, 1969) who worked with upflow filters on a laboratory scale. The nature of the packing in a filter precludes the direct treatment of feedstocks with much particulate matter, particularly large particles, as this will block the spaces in the filter matrix. However, the filter can be used as a second stage in the treatment of particulate feedstocks. Some examples of this have been mentioned in the last chapter, where the filter can act as the methaniser part of a 'two-phase' digester, but the application to be described is rather different. An early example of this was the laboratory- and pilot-plants built by Coulter *et al.* (1957) for treatment of whole sewage waters. In this case the sewage passed upwards through a conical tank where large solids settled towards the bottom and were slowly degraded by bacteria while the liquid passed from the top to an up-flow anaerobic filter tank where dissolved materials and fine solids were degraded. The system was apparently not too successful, but a similar pilot-scale plant was developed by Pretorius (1971) in South Africa. Again, the sewage flowed upwards through a cylindrical tank with baffles to prevent overflow of floating particles. Large solids were slowly degraded while the water flowed on to an anaerobic filter. This ran at African ambient temperature (20°C) with a hydraulic retention time of about a day in each part of the digester. Effluent suspended solids were reduced to 30 mg/l and total COD to 110 mg/l.

The first part of these digesters can be used as a single digester and such a one was the tower digester of Callander & Barford (1984). Here, a 4% TS pig slurry flowed upwards through a tower. The large solids settled towards the bottom and were slowly degraded, while the liquid passed up the tower through a sludge-blanket where dissolved substrates were digested. The tower was baffled at the top to reduce overflow of bacteria and solids. The tower worked well although the 15–20 day HRT used was much the same as the HRT of a stirred-tank digester treating a similar pig waste.

There have been numerous examples of anaerobic filters being used alone to treat low-solids wastewaters on laboratory, pilot-plant and industrial scales: Bonastre & Paris (1989) produced a survey of such filters. An early example (Taylor, 1972) was of three filters treating wastewater from starch production. These filters were wooden tanks 20ft high by 30 ft in diameter filled with graded broken rock about 1 in to 3 in in size. Each treated a waste flow of 0·34 m^3 per minute and the effluent was passed directly to the town sewerage system. What is described as the world's biggest anaerobic filter has been run for some years by the Bacardi Corporation on distillery wastewaters. The tank volume is 3 500 000 gallons (13250 m^3) with one tenth of this volume as plastic packing. The filter treats some 136 tonnes of COD per day. A 4500 m^3 filter digester treating wine-distillery wastewaters and some other similar commercial plants was described by Henry (1985). A 5600 m^3 downflow filter for wine-distillery wastes is also in use (Camilleri, 1988) This uses PVC rings as support matrix. The paper also mentions similar digesters in use for treating other types of distillery wastewaters and waters from beet sugar, jam, cheese, dairy, and chemical plants. An upflow filter packed with polypropylene rings and working at about 35°C was described by Saslawsky *et al.* (1988) This has a volume of 2550 m^3 and was designed to treat 21 t/d of COD in wastewater from a sulphite pulp mill. The plant had not attained its designed performance, however, and the reasons at the time of the paper were obscure. An addition of whey to the waste to increase the easily-available carbon and to provide some trace elements was only partially successful; physical aspects, such as partial blocking of the filter might have had some effect.

Chave *et al.* (1988) reported on small-scale experiments on digestion of effluents from sulphite pulping of wood. An anaerobic filter packed with polypropylene Pall rings working in an upflow mode with recirculation was used. Diluted sulphite liquor was used and the digestion could be run with 60% liquor in water as feed. However, although the digester bacteria were adapted to the waste as far as possible, toxicity was encountered and the authors said that a detoxification step, by filtration or flocculation, would be needed to allow digestion of undiluted waste. Vogel & Winter (1988) also tested digestion of wastewaters from a sulphite pulp mill which produced wastes from various types of pulping. Upflow recirculating filters were used, different filters being packed with polyethylene/charcoal or polyethylene/clay tubes, sintered glass Raschig rings or Liapor clay beads. Biodegradability was very dependent on the origin of the wastewater and evidence for sulphate-reduction replacing methanogenesis was found. Nevertheless, COD reductions of up to 60% at HRT of 3-5 days were obtained. Winter *et al.* (1988) also used Liapor clay beads in an upflow filter treating whey. The filter performance was better than that of a parallel stirred-tank reactor with no support material.

Hamoda & Kennedy (1987) used needle-punched polyester film, sewn on to stainless steel-wire frames to make vertical channels, as a support for bacteria in a filter treating acetic acid and ammonium and other salts in solution. Biofilm concentration increased with increased organic loading-rate and decreased HRT.

Guiot & van den Berg (1984) introduced the idea of combining a filter with a sludge-blanket digester in the same tank and a number of designs of 'hybrid' digester have been made which combine two different types of digester in one tank or in close proximity. For instance, Tsonis & Grigoropoulos (1988) produced a sludge-blanket digester in a conical vessel into which fitted a stone-filled filter so that the liquid in the cone came half-way up the outside of the tubular filter. This was used in pilot-plant experiments to treat low-solids olive-oil mill wastewaters. It functioned well, with a high gas-production, although the reported experiments were not very long-term.

Coates & Colleran (1990) reported experiments on two laboratory-scale hybrid digesters fed with a solution of glucose, acetate, propionate and ethanol (plus N-sources, etc.). An intermittently-stirred digester was slower to start up than an unstirred digester, but both came to the same operating conditions. They found only 3–7% of the active biomass was on the filter, the rest in suspended granules. This proportionality, though, will probably vary with the feed substrate.

Another hybrid digester consisting of a bottom, cubic, sludge-blanket chamber with a top cube filled with polyurethane foam plates was described by Calzada *et al.* (1988). After laboratory experiments, two pilot-plants were constructed to treat coffee-processing wastewaters and vinasses from a molasses distillery. Laboratory-scale digesters also succesfully treated banana-retting wastes and diluted wastewater from a kraft paper mill. Xiushan *et al.* (1988) described a similar hybrid digester with plastic Pall rings as the filling in the top, filter, part. This was of 125 l volume and treated molasses stillage wastewaters. Guiot & van den Berg (1985) also described the treatment of sugar-processing wastewaters with a hybrid reactor. Britz *et al* (1990) tested a laboratory-scale hybrid digester of a filter above a sludge-bed for treatment of the highly-polluting liquid which escapes from landfills. In this case, the digester was a tube with the filter medium of polyethylene foam fitted as an inner layer on the top part of the tube wall. Ninety percent of the COD of 1800 mg/l in a landfill leachate was removed by the digester.

Pichon *et al.* (1988) used a hybrid digester with a filter above a sludge-bed running at mesophilic temperature for treatment of pulp-mill wastewaters containing sulphate and sulphide in high concentration. At COD/S ratios of about 10 satisfactory digestion was obtained with reductions in COD and BOD of 50 and 78% at 2–4 days HRT. At lower ratios, corresponding to 3·5 g sulphate/litre, efficiency was poor and acids built up in the digester. Huster & Mobius (1988) used a pilot-scale hybrid digester with a filter above a sludge-bed, and recirculation, to treat wastewaters from paper board making and deinking of waste paper. COD and BOD degradation decreased as the volumetric loading of the board-mill effluent increased, both becoming about 50% at a loading of 11 kgCOD/m^3/d and 7 hour HRT. A biocide used in the board production decreased digestion efficiency. Digestion of the deinking effluent was not as successful: nevertheless the authors concluded that digestion was a useful process to be followed by further treatment.

These are only some examples of hybrid digesters treating low-solids wastewaters. They differ from the sewage digesters described previously in that the sludge in the lower part is, mostly, bacteria treating substrates dissolved in the wastewaters, not a sludge of solids from the waste being degraded by attached bacteria. In these cases there was no phase separation in the plant, in both parts complete digestion processes occurred.

The liquid effluents from pyrolysis of wood (pyrolignitic acids) contain phenols and other compounds potentially toxic to anaerobic digestion and are of very high COD content. Andreoni *et al.* (1990) tested digestion of these liquors mixed with diluted pig waste (TS about 1%) in upflow filters filled with wood chips or PVC. The digesters had been running on pig waste for some years and the pig waste was used to dilute the pyrolignitic acids and to raise the pH. Up to 6.5% acids in pig waste could be digested, but 10% acids produced a drastic reduction in efficiency. However, the digester with wood chips recovered from this while the performance of the PVC-filled digester remained poor.

Filters on various scales from laboratory to full-size plant have also been used for treatment of the low-solids supernatants from settled farm-animal excreta, Kennedy & van den Berg (1982), for example, used a filter to treat settled piggery waste. This was a downflow filter packed with vertical 'rods' of U section and made of unglazed fired clay. The screened waste had a content of small suspended solids of about 18 g/l. Hydraulic retention times (at 35°C) of 1 to 8 days were tested. Performance improved with longer retention, but an intermediate retention time was suggested. Earlier, these workers had compared upflow- and downflow-filters for treatment of bean-blanching wastewaters and had compared the relative efficiencies of filters, contact digesters and stirred tanks (van den Berg & Lentz, 1979; van den Berg *et al.*, 1980). Bull *et al.* (1984) also reviewed filters, sludge-blanket, fluidised-bed, contact, and other types of digester and compared performances on the basis of published work. Weiland (1987) reviewed filters for the anaerobic treatment of various agro-industrial wastes.

Barry & Colleran (1982) used a laboratory-scale upflow filter packed with limestone chips to treat silage effluent. The digester had been treating settled pig slurry, of about 1% TS and obtained from a farm holding tank, at 3-day RT and 28°C. The silage effluent had been diluted with rainwater on the farm and had a COD about 14 500 mg/l and VFA of 4300 mg/l. Solids were 3 g/l. At 3-day RT the filter removed about 70 to 85% of the COD.

The use of a filter as the methaniser in two-phase digestion of high-solids wastes has been described (Chapter 6). Weiland & Wulfert (1986) described pilot-plant experiments on digestion of distillery effluents. These can be of low-solids depending on their origin and the degree of filtering or settling of yeasts and fermentation substrates which has taken place. The tests of Weiland & Wulfert were carried out with a stirred-tank acidification stage running at 1 day RT where about 70% of the dissolved organic matter was acidified. The effluent passed to filters filled with high-porosity plastic packings of different types, and a contact digester. Ninety two to 95% of the acid COD was removed in all filters at loading rates of 11 $kgCOD/m^3$ or higher, provided that some of the effluent water was recirculated to dilute the influent. The contact digester had a poorer COD removal and became unstable at loadings of 6 $kgCOD/m^3/d$. The authors reported that the highly-porous packings could deal with distillery wastewaters containing up to 6% suspended solids without blocking. These solids were, presumably, not degraded in the short RT of the acidifier stage.

The direction of flow of anaerobic filters (up or down) needs to be selected on the properties of the wastewaters being digested and there is no general optimum direction. Upflow generally gives a higher COD removal, while downflow reduces risks of clogging and gives better stability under shock loads (Weiland & Wulfert, 1987).

Urgulu & Forster (1992) subjected two thermophilic filters, one with a porous-glass packing, the other with non-porous ceramic Raschig rings, to various shock loadings. Thermophilic digesters have been said to be affected by sudden changes in loading rate (Duff & Kennedy, 1982) and to be affected by lower concentrations of toxic agents than mesophilic systems (Macleod & Forster, 1988). The bacteria on the filter with porous packing had greater stability to short-time (2 hours) changes in loading rate at constant HRT or to changes in loading and HRT and to a fall in temperature than the bacteria on the ceramic packing. A change in feed for 29 days from the normal simulated ice cream-wastewater to a paper-mill effluent which contained biocides resulted in poor performance of both digesters, particularly that with the ceramic packing, but both digesters gradually recovered when the easily-degraded ice-cream wastewater was again fed.

The expanded- or fluidised-bed digester

This is a development of the upflow-filter digester which provides a bigger surface area per volume of tank than the filter and a more intimate mixing of the substrates and bacteria. The bacteria are attached to millimetre-sized particles of sand, glass, carbon, etc., which are kept in suspension in the wastewater by the upflow of the water (Fig. 2b). The rate of flow of the waste itself may not be sufficient to suspend the particles properly, so the upflow rate is often increased by pump recirculation of the overflowing water from the top to the bottom of the tank. The degree of lifting of the particles determines whether it is an 'expanded' or 'fluidised' bed.

Jewell and colleagues were pioneers of these digester systems in attempts to develop an attached-biomass digester that could handle fine suspended solids, and the system was first applied to sewage (Jewell *et al.*, 1981). This was highly effective in treating sewage water from primary settling tanks at HRT of about 2 h and 20°C; 80–90% of the COD was removed and the suspended solids and BOD were reduced to about 10 mg/l. Others were not successful in repeating these results but they could have been using unsuitable bed conditions. Brown *et al.* (1985) reported laboratory and pilot-plant experiments on treatment of synthetic sewage and primary-settled sewage. They used sand, diatomaceous earth- and granular activated-carbon as bed materials. The pilot-scale digesters removed up to 75% of the COD, but this was not as good performance as required. The experiments suggested that high rates of sulphate reduction were competing with methanogenesis and reducing the performance of the digestions. Jewell and co-workers repeated their experiments later and showed that the bed particles with attached bacteria had remained active for over six years without feeding. By 1985 scaled-up plant was being built (Jewell, 1985). However, Rockey & Forster (1982) said that an expanded-bed digester at 37°C could not treat settled sewage water with an efficiency in removal of organic matter that would make digestion a competitor with existing aerobic treatments. This they attributed to the fact that they used sewage from an actual plant which varied in strength from day to day, as is normal. Other workers getting better results had used sewage of constant strength. On the other hand, the expanded-bed digester is generally reported to be very tolerant of loading and temperature shocks. Studies suggested that with feedstocks with suspended solids the solid particles were trapped between the bed particles and degraded to acids and hydrogen which were converted to methane by bacteria attached to the surfaces of the bed particles (Morris & Jewell, 1982). The retention time of the bacteria on the bed can be as long as 300 days at hydraulic retention times of about 2 hours and because of the large amount of biomass built up on the very large surface area of the small bed-particles high rates of reactions can be obtained at low temperatures (10–20°C). The expanded-bed digester has now been used for a number of feedstocks, although generally at laboratory or pilot-plant scale.

Jewell (1980) reported that an expanded-bed digester had successfully treated diluted cattle slurry of 2% solids content at an HRT of 3 hours. Ng & Chin (1988) tested sand and granular activated-carbon as support media for laboratory-scale expanded-bed digesters treating the dilute piggery wastewaters (TSS < 1%) produced on Singapore farms. Treatment was still effective at HRT down to 14 hours. Iza *et al.* (1988) described experiments with laboratory and pilot-scale fluidised-bed digesters treating sugar-beet processing and alcohol-production wastewaters. The laboratory-scale bed was of granular PVC and the pilot-scale bed natural- plus calcined-sepiolite. Both performed well, the pilot-scale digester removing 85% of the soluble COD loaded at about 40 kg/m^3/d. Like others, they found that the digester could be quickly started after being non-operational for long periods, in this case for nearly a year. They found that the bed contained two types of particles, the upper particles having greater attached bacterial growth than the lower. Sorlini *et al* (1988) found in laboratory tests that sepiolite was superior to sand as a support material in expanded beds. de Bazua *et al.* (1991) successfully used spent ion-exchange resin in a pilot-scale fluidised-bed digester treating vinasses from alcohol distillation.

Carozzi (1988) compared a fluidised-bed digester, with sand as the support, with upflow filters filled with plastic Flocor R or sintered-glass nodules, and with a contact digester, for treatment of fish-processing wastewaters. These have high COD and BOD, about 0·2% solids, and high concentrations of proteins and fats. All the digesters gave good results, but at HRT about 2 days the filters were more stable than the contact or fluidised-bed.

Suidan *et al.* (1988) described the use of expanded-bed digesters in the treatment of potentially toxic wastewaters from coal gasification and petrol refining. The bed consisted of granular activated-charcoal which not only provided a surface for growth of bacteria but adsorbed toxic compunds from the wastewaters, thus allowing the biological treatment. To prevent overloading of the carbon with toxic compounds and release of these into the digester, part of the carbon bed was replaced at intervals. Reference has been made to the ability of retained-biomass digesters to deal with potentially-toxic compounds in feedstocks because of the large mass of bacteria in the system, the long biomass retention time which allows adaptation of the bacteria, and the effects of diffusion through the microbial film: expanded-bed digesters have all these attributes.

The upflow sludge-blanket digester

The upflow anaerobic sludge-blanket (UASB) digester is a development of the digesters of Pretorius and Coulson *et al.* described in a preceding section. However, whereas the sludge in these digesters was composed of large solids from the feedstock with attached bacteria, the sludge in the UASB is predominantly bacteria and the feedstock is wastewaters of very low suspended solids content with pollutants (which can be of low- or high-concentration) mainly in solution. Besides this, the main difference in the two digesters is that the UASB has a system of baffles at the top of the tank (fed in an upflow mode) which returns floating sludge to the tank and separates sludge and liquid from biogas. Rather like the fluidised-bed, the UASB sludge is in the form of granules of millimetre size which are floated part-way up the digester by the upflow of feed water although the granules, of course, have no support particles (Fig. 2B). The formation of adequate granules during start-up of the digester can be a difficulty with UASB digesters. This is referred to later.

Early work on the laboratory development of the UASB digester was described by Lettinga *et al.* (1980) and since then many industrial-scale units have been built and the UASB has become the most common form of retained-biomass digester, particularly in the treatment of carbohydrate-containing wastewaters from fruit, vegetable and other food-processing. Paper-mill wastes (Hulshof-Pol & Lettinga, 1986) and petrochemical wastes (Nel *et al.*, 1984) are examples of other wastewaters that have been treated. The process can work at ambient or more elevated temperatures, the working temperature often being dictated by the temperature of the process wastewater being treated. Ambient temperature can also play a part here, of course, and Schellinkhout *et al.* (1985) described a 64 m^3 UASB digester treating raw domestic sewage in a town in Columbia where the ambient temperature is a fairly constant 25°C. The high temperature caused the sewage to become septic in the sewers and it was more dilute than the usual temperate-climate, large-city, sewage, but the treatment was successful in reducing BOD and COD by about 80% at a hydraulic retention time of 3–3·5 h, and it coped with the usual diurnal variations in sewage flow and composition. However, some further treatment of the digester effluent could be needed depending on local regulations for discharge. Schellinkhout *et al.* (1988) described the follow-up with a full-scale plant which was then being constructed. In a further study Grin *et al.* (1985) found that treatment efficiency of raw sewage dropped off considerably at temperatures below 12°C.

Callander *et al.* (1987) treated wood-ethanol stillage in a 10 l UASB. N and P supplements had to be added, but soluble BOD and COD removals of 86 and 93%, respectively, were obtained. Colour was reduced by 40%. The maximum COD loading was 16 kg COD/m^3/d.

Some comparisons of different types of digester have been made. For instance, de Man *et al.* (1988) described comparisons of small pilot-scale UASB and Expanded Granular Sludge Bed (EGSB) digesters for treating low-strength wastewaters (sewage, diluted vinasses) at between 8 and 30°C. The EGSB digester is similar to the UASB except that the sludge bed is expanded by increasing the upward velocity of the water flow. This can be achieved, as in the expanded particulate bed, by pump recirculation of effluent from top to bottom or by building the digester tank with a greater height to diameter ratio than that of the UASB. The latter results in too tall a tank in practice, so the former, or a combination of the two methods, is used. The EGSB digester should result in a better contact between substrates and bacteria.

This was found in practice and the EGSB digesters could give good purifications of dissolved pollutants at lower HRT than the UASB; thus the former would be better for treating wastes such as settled sewage at low ambient temperatures. The UASB digesters were better for treating wastewaters with particulate matter, such as raw sewage, because the lower fluid velocities allowed feedstock particles to remain in the digester for some time.

Edeline *et al.* (1989) compared laboratory-scale UASB and activated-carbon fluidised-bed digesters for treatment of Kraft-sulphate pulping wastewaters. Two mixtures of wastewaters were used giving COD about 4000 and 12 000 mg/l. The authors said that the fluidised-bed was more stable than the UASB and gave generally better results, with COD removals of 90%.

Much work has been done on anaerobic digestion of olive-oil mill wastewaters, and Boari *et al.* (1984) tested laboratory and pilot-scale UASB digesters for such wastes. The waste had to be diluted to COD 13–18 g/l, but with this feed COD removals of 70% at loadings up to 21·5 kg/m^3/d could be obtained. This was better than the performance of contact digesters tested by Antonacci *et al.* (1981) and Fiestas Ros de Ursinos *et al.* (1982).

Olive-mill wastewaters contain relatively large amounts of long-chain fatty acids which are broken down via hydrogenation (Chapter 2). Morelli *et al.* (1990) compared a UASB pilot-plant and a laboratory upflow-filter filled with wood chips for treatment of olive-mill wastewater diluted similarly to that in the previous paragraph. COD removal in both digesters was about 65–75%, but while the hydrogenation reaction seemed to be more efficient in the UASB, the degradation of saturated acids to volatile fatty acids was more efficient in the filter.

Athanasopoulos (1990) compared two UASB digesters of 9 and 220 litres capacities with a 9 litre fluidised-bed digester for treating currant-finishing wastewaters. The fluidised-bed was quartz sand and all the digesters ran at 35°C. The fluidised-bed treated loadings of 16 and 21 kgCOD/m^3/d with 95 and 85% removal of COD, while the 220 litre UASB removed 93% of the COD at maximum loading of 12·5 kg COD/m^3/d and the 9 litre removed 95% COD at 10·2 kg/m^3/d. These results were compared with those of previous work (Athanasopolous *et al.*, 1987, 1990) and for COD removal from these wastewaters the ranking was fluidised-bed, upflow filter, UASB, downflow filter. However, other considerations should be taken into account in deciding on a digester plant and on the basis of the careful monitoring and control required by the fluidised-bed and the risks of clogging of the upflow filter, the UASB was suggested as best choice; being easy to operate and withstanding long idle periods.

Although many UASB digesters have been, or are being, constructed for treatment of carbohydrate-containing effluents from the production of foods such as confectionery, fruit and vegetable products, etc., and of brewery wastes (e.g. Hulshoff Pol & Lettinga, 1986; Campos *et al.*, 1986), fatty wastewaters from the production of dairy products have also been treated. As described in Chapter 2, fats are hydrolysed to long-chain fatty acids in digestion and some wastewaters may also contain free fatty acids. Some of the long-chain acids inhibit methanogenic bacteria and so can inhibit digestion in UASB reactors (Koster & Cramer, 1987). However, it has been found possible to treat fatty wastewaters in UASB digesters. Goodwin *et al.* (1990) treated a synthetic ice-cream factory wastewater (based on ice-cream) in laboratory-scale UASB reactors. The experiments were mainly concerned with start-up of the digesters and were not carried on for prolonged periods. However, the waste could be treated effectively, although a long start-up time (60–70 days) was required. Addition of calcium chloride to the waste aided granulation of the bacteria and possibly helped in reduction of any fatty acid toxicity (Koster, 1987). The digester with added calcium was removing 2·67 kg total organic carbon per day (87% of added TOC) after start-up. On the other hand, Wheatley *et al.* (1984, 1988) compared UASB and upflow filter digesters for treatment of wastewaters from production of sweets and confectionery, and production of yeast. They found that both types gave similar results when the UASB was working well, but both

UASB were more difficult to start than the filters and did not cope as well with changes in feed flow, etc., and sludge from the blanket was lost. However, the authors did say that if the UASB had been constructed with baffles to separate gas, liquid and solids and had had an inoculum from a working plant, performance could have been better. They also treated sewage at ambient temperature and warm wastewaters from malt production, a brewery, a distillery, potato processing, and production of pharmaceuticals, ice cream and sweets, with laboratory- and pilot-plants. Treatment was possible, but there were some problems, which were overcome.

Baffled and similar digesters

Horizontal digesters for treating sludges are described in Chapter 6. The horizontal-tank system has been applied to wastewaters by fitting plates of various configurations inside to produce a non-linear flow of the water along the tank and to act as a support surface for bacteria or to retain suspended organisms. The Rotating Biological Contactor (RBC) is commonly used in aerobic treatment of wastewaters. In this, the aerobic microorganisms grow attached to the surfaces of a number of discs which are rotated in a trough along which the wastewater flows. As the discs slowly rotate, on an axle running along the trough, the bottom third or so of each disc, with the organisms, is plunged through the water and then through the air, where the organisms can pick up oxygen to metabolise pollutants in the water film picked up on the disc. If the discs are completely enclosed in a tank the system becomes anaerobic (AnRBC). An early example of this concept was described by Friedman *et al.* (1980). Lo & Liao (1986*a*, 1986*b*, 1989) used an AnRBC in the treatment of cattle waste, cheese whey and a mixture of the two on a laboratory scale. The feedstock for the latter digestions was a mixture of whey and dairy-cattle manure (2:1) passed through a 2 mm screen to give a liquid of up to 6·7% TS and COD about 65 000 mg/l. About 50% of the COD and TS was removed at an HRT of about 5 days. The anaerobically-treated liquid was then further treated in a series of aerobic sequencing-batch reactors.

While testing an AnRBC McCarty (1982) found that rotation of the discs was not necessary and developed the 'Baffled' digester. In this, 'walls' across the tank are built from the top and bottom so that the wastewater flowing along the tank has to go alternately under and over the baffles (the Septic and Imhof tanks were somewhat similar). The baffles tend to keep the bacteria in the tank and also help to prevent problems with floating solids. Tilche & Yang (1988) identified the formation of bacterial granules in baffled digesters. Various designs have been tested, Bachman *et al.* (1985) reported on one of the earlier, 13l laboratory-scale, models with recycle for treating a dissolved, synthetic, waste. They said that the biomass was in flocs or granules. Grobicki & Stuckey (1988) tested 10 l-scale digesters with vertical baffles on a synthetic wastewater and found the digesters very efficient and stable even under shock loading. Pande *et al.* (1988) said that the tank with vertical baffles had advantages for developing countries as the waste (either waters or slurries) could be made to flow through the tank by hydrostatic pressure rather than a pump. If heating was required, the first compartment alone needed to be heated. Even 10% TS farm slurries could flow through the tank. Yang & Chou (1985) used a tank with baffles of different sizes set at different angles to the water flow to treat dilute piggery wastewater with TVS of less than 0·2%. The microbial RT was 15–300 days at an HRT of 0·25–5 days.

Boopathy *et al.* (1988) used the original vertical baffles in a hybrid digester treating whisky distillery wastewater, while Tilche & Yang (1988) used the same design to treat molasses stillage evaporator concentrates. Boopathy & Tilche (1992) described a similar 165 litre hybrid digester. The top portions of the chambers in a vertically-baffled tank were filled with plastic Pall-ring filter media. This was used to treat acidic wastewaters from alcohol production based on molasses: 73% of the soluble COD was removed at loadings of 20 kg COD/m^3/day.

The anaerobic sequencing-batch reactor is another digester design for wastewaters based on an aerobic system. In this the wastewater flows into a vertical tank. The water then reacts with a bacterial sludge for a time. The sludge is then allowed to settle and then a large proportion of the supernatant is discharged and the tank refilled. In studies by Ng (1989) with a large laboratory system, the complete cycle time was about 6 hours, with 10 min each for filling and discharging, 120 min for settling and 220 min for reacting. The sludge was slowly stirred during the reaction period. The whole system has similarities to the fed-batch digester (Chapter 6). Ng used the digester to treat piggery wastewater of about 2–6 g VSS/l and BOD about 2–5000 mg/l.

CONCLUSION

Laboratory or pilot-plant experiments have now been done on wastewaters from most industries, from chemicals to food productions. The results show that anaerobic digestion can be a successful means of treatment of these waters and many full-size plants have been constructed on the basis of the small-scale tests. Although digesters for wastewaters have not by any means attained the numbers of digesters treating sewage sludge, digestion is increasing in popularity as a method of pollution control for wastewaters, especially high-strength wastewaters. In the foregoing paragraphs examples of a number of types of digester for low-solids waters have been given. While all types have been built on a large scale, it would seem that the UASB and filter types are most popular at the moment on the basis of stability and ease of running. The filter may be easier to run than the UASB, but the latter can cope better with solid particles in the feed which could block a filter. With this type of feed the hybrid digester, where the particles can be digested in the lower part, may have advantages over the sludge-blanket or the filter alone. On the whole, the retained-biomass rather than the returned-biomass (Contact) digester seem to be most effective.

The retained-biomass digesters can perform at low, ambient, temperatures, but many factory wastewaters are warm and digesters can be run, with increased performance, at higher than ambient temperatures with no external heat source. While warm wastes are advantageous, some of the factory wastes are unbalanced in nutrients for the digester bacteria and may need supplementation by either chemicals such as urea or carbohydrates or by mixing with other wastes. Wastes from confectionery and sweet and starch manufacture, as well as those from fruit processing and jam making, are usually low in nitrogen and may need supplementing. Some of the wastes may contain toxic, or potentially toxic, compounds. These can be a product of the factory reactions, such as sulphur and phenolic compounds in wood-pulp wastes or fatty acids in olive-oil wastes. In most cases bacterial populations adapted to these compounds can be built up as the compounds cannot be excluded from the digester. But in the case of detergents and disinfectants used in plant cleaning, biocides, etc., it may be possible and better to exclude these from the digester feedstocks. Biocides used on forest trees may reach the pulp plants and pass into the wastewaters, but these have been degraded in digesters (Salkinoja-Salonen, 1983; Benjamin, 1983).

Methods of starting-up retained-biomass digesters are described in many of the papers referred to in the previous sections. In the previous chapter the start-up of digesters for high-solids feeds was briefly described. The basic principles were an inoculum (or a feedstock containing suitable bacteria) and a low loading-rate for some time to allow a balanced population of bacteria to develop (this is shown, for instance, by Hobson & Shaw, 1974). These principles are applied to wastewater digesters: however, as in most cases wastewaters are of factory origin and contain few bacteria an inoculum of active digester bacteria is generally needed for retained-biomass digesters.

Contact digesters are started just as are stirred-tank sludge digesters with an inoculum which is added to by pumping in the feed wastewater at low rate until the digester tank is filled and it is running as a continuous-flow system. As digestion becomes stabilised the loading rate is increased to the design rate. The whole process, as with all types of digesters, can take some weeks or months.

Attached-biomass digesters can be started with an inoculum and feed waste. The inoculum and waste may be grown as a batch culture for some days to allow initial attachment of the bacteria to the support before low-rate continuous feeding is started. The nature of the support seems to have some effect on the rate of attachment of bacteria (Breitenbucher *et al.*, 1990; Pascik, 1990), and some difficulties have been noted in starting thermophilic digestions (Kawase *et al.*, 1989). One of the authors and colleagues (unpublished) started filters treating distillery wastewater with an inoculum from a piggery-waste digester. Canovas-Diaz & Howell (1987) found they could shorten the start-up time of a filter by growing an inoculum of strained digesting sewage sludge in a synthetic medium in the filter for 64 days to develop a good flora before changing the medium flow in steps over four days to the feed of deproteinised cheese whey. The start-up medium contained acetic, propionic and butyric acids with nitrogenous and other salts, and with the best formulation start-up could be cut to 35 days. Stronach *et al.* (1987) determined various factors affecting start-up of a fluidised-bed digester with a feed of synthetic meat waste. The support bed was sand and the inoculum digesting sewage sludge. The loading-rate was increased stepwise. The bed expansion seemd to affect the attachment and growth of the inoculum; a 5% expansion was detrimental to attachment, 10–30% expansion was optimum, start-up taking 56 days. Flocculating polymers increased the mass of attached bacteria but did not improve performance; perhaps the wrong bacteria became attached to the support. Keeping the system initially anaerobic by nitrogen purging helped start-up. As previously mentioned, purging digester tanks with an inert gas before and during initial filling with liquid has been advocated as a safety measure, but its aid in maintaining anaerobic conditions for the bacteria seems variable. Sludge digesters with or without an inoculum but fed with faecal sludges will rapidly use up oxygen in the digester head-space, particularly if the tank is rapidly filled to working volume. Wastewater digesters with little suspended-solids in the feed and a small inoculum may be more susceptible to oxygen from air initially in the tank headspace.

UASB digesters seem to be the most difficult to start, the problem being that not only is the correct balance of bacterial species needed but these must be induced to form granules which will then form a sludge bed that floats but does not excessively wash out from the tank. Various 'granules', from loose and fluffy flocs to compact, discrete, particles have been described in different digesters. The macroscopic stucture of the granule probably reflects the predominant types of bacteria in the granule and the mineral content. The predominant bacteria in granules appear to vary with the digester feed composition (see, for instance, Dubourguier *et al.*, 1988; Grotenhuis *et al.*, 1991). Lettinga (1990) said that an inoculum was not required for UASB digesters treating raw sewage waters: the faecal bacteria provided an inoculum which would granulate. However, as said before, most factory wastewaters are nearly sterile and an inoculum is needed. In commercial practice inocula of granulated sludge are obtained from a working digester whenever possible. When a granulated sludge is not available recourse must be had to an inoculum of non-granulated sludge, for instance, from a sewage-digester. (Sewage-digester sludge is much used for starting digesters not because of any special properties of the sludge but because it is usually the sludge most easily available in large quantities).

The rate of formation of a granular sludge can be affected by different factors, including the composition of the feed waste. An inorganic nucleus may help and the effect of calcium salts in aiding granulation (and reducing long-chain fatty acid toxicity) has already been mentioned. Dubourguier *et al.* (1988*a*) in a review of granulation found calcium to be about 2·5% of the dry weight of granules, and their own work and that of others showed the mineral contents of UASB sludge granules to be from about 10 to 55% by weight. A number of different metal salts were present, but the mineral content may be affected by the digester feed and the level of sampling of the sludge bed. The start-up of UASB digesters and the development of different types of sludge granules was reviewed by de Zeeuw in 1988 in the Proceedings of a workshop on granular sludges, and other authors discussed microbiological and practical aspects of granule composition and development. It would appear that with most, or all, feedstocks a suitable granular sludge can be developed by slowly increasing the digester loading, even if the inoculum is non-granular sludge, but Goodwin *et al.* (1992) showed in laboratory digesters of a sugar wastewater that an inoculum of granulated sludge could give a start-up period for their digesters of a few days, whereas over 60 days start-up was needed with a non-granular sewage-sludge inoculum, even though the granular sludge came from a full-scale UASB digester working on a different substrate.

The treatment of low-solids wastes and the types of digester used in the treatments have been briefly described. The general principles have been illustrated by reference to particular wastes, but there are other examples in the literature and there are other examples of hybrid digesters made by combining the basic digesters in different ways. The treatment of wastewaters is a rapidly-expanding aspect of anaerobic digestion.

REFERENCES

Andreoni, V., Bonfanti, P., Daffonchio, D., Solini, C. & Villa, M. (1990). *Biol. Wastes* **34**, 203.
Antonacci, R., Brunetti, A., Rozzi, A. & Santorri, M. (1981). *Ingegneria Sanitaria* **6**, 357.
Athanasopoulos, N. (1990). *Biol. Wastes* **32**, 161.
Athanasopoulos, N., Koutinas, A. & Papadimitriou, A. (1987). *Biol. Wastes* **22**, 129.
Athanasopoulos, N., Kouinis, J. & Papadimitriou, A. (1990). *Biol. Wastes* **32**, 91.
Bachman, A., Beard, V.L. & McCarty, P.L. (1985). *Water Res.* **19**, 99.
Barry, M. & Coleran, E. (1982). *Agric. Wastes* **4**, 231.
de Bazua, C.D., Cabrero, M.A. & Poggi, H.M. (1991). *Biores. Technol.* **35**, 87.
van den Berg, L. & Lentz, C.P. (1979). In *Proc. 34th. Purdue Ind. Waste Conf.* p. 319.
van den Berg, L., Lentz, C.P. & Armstrong, D.W. (1980). In *Proc. 35th. Purdue Ind. Waste Conf.*
Benjamin, S.L., Woods, S.L. & Ferguson (1983). *Wat. Sci. & Technology* **15** 9/10, 160–170.
Boari, G., Brunetti, A., Passino, R. & Rozzi, A. (1984). *Agric. Wastes* **10**, 161.
Bonastre, N. & Paris, J.M. (1989). *Process Biochem.* **24**, 15.
Boopathy, R., Larsen, V.F. & Senior, E. (1988). *Biomass* **16**, 133.
Boopathy, R. & Tilche, A. (1992). *Biores. Technol.* **40**, 101.
Breitenbucher, K., Siegl, M., Knupfer, A. & Radke, M. (1990). *Water Sci. Technol.* **22**, 25.
Britz, T.T., Venter, C.A. & Tracey, R.P. (1990). *Biol. Wastes* **32**, 181.
Brown, D.S., Mehrotra, S., Petrasek, A.C., Eralp, A.E. & Bishop, D.F. (1985). In *Anaerobic treatment of sewage*, ed. M. Switzenbaum, rep. No. Env. E. 88-85-5, University of Massachusetts, USA. p. 209.
Bull, M.A., Sterrit, R.M. & Lester, J.M. (1984). *Biotech. Bioeng.* **26**, 1054.
Bull, M.A., Sterrit, R.M. & Lester, J.M. (1984). *Chem. Eng. Des.* **62**, 203.
Butcher (1988). *International Biodeterioration* **25** 2, 71–77.
Callander, I.J. & Barford, J.P. (1984). *Agric. Wastes* **11**, 1.
Callander, I.J., Clark, T.A. & McFarlane, P.N. (1987). *Biotech. Bioeng.* **30**, 896.
Calzada, J.F., Roltz, C., de Arriola, M.C., Porres, C. & Valladares, J. (1988). In *Anaerobic Digestion 1988*, ed. E.R. Hall & P.N. Hobson, Pergamon Press, Oxford. p. 125.
Camilleri, C. (1988). In *Anaerobic Digestion 1988*, ed. E.R. Hall & P.N. Hobson, Pergamon Press, Oxford. p. 407.

Campos, J.R., Foresti, E. & Camacho, R.D.R. (1986). *Water Sci. Technol.* **18**(12), 87.
Canovas-Diaz, M. & Howell, J.A. (1987). *Biotech. Bioeng.* **30**, 289.
Carozzi, A. (1988). In, *Poster Papers, 5th. Int. Symp. An. Digestion*, ed. A. Tilche & A. rozzi, Monduzzi Editore, Bologna. p. 477.
Chave, E., Fuentes, J.L. & Goma, G. (1988). In *Poster Papers, 5th. Int. Conf. An. Digestion*, ed. A. Tilche & A. Rozzi, Monduzzi Editore, Bologna. p. 635.
Coates, J. & Colleran, E. (1990). *Process Biochem.* **25**, 162.
Coulter, J.B., Soneda, S. & Ettinger, M.B. (1957). *Sewage Ind. Wastes* **29**, 468.
Dubourguier, H.C., Pressier, G. & Albagnac, G. (1988). In *Granular anaerobic sludge: microbiology and biotechnology*, ed. G. Lettinga *et al.* Pudoc, Wageningen. p. 18.
Dubourguier, H.C., Archer, D.K., Albagnac, G. & Prensier, G. (1988*a*). In *Anaerobic digestion 1988*, ed. E.R. Hall & P.N. Hobson, Pergamon Press, London. p. 13.
Duff, S.J.B. & Kennedy, K.J. (1982). *Biotechnol. Lett.* **4**, 815.
Edeline, F., Lambert, G. & Fatticcioni, H. (1989). In *Energy from Biomass 4*, ed. G. Grassi, D. Pirrwitz & H. Zibetta, Elsevier Applied Science, London. p. 463.
Fiestas Ros de Ursinos, J.A., Navarro Gamero, R., Leon Cabello, R., Garcia Buendia, A.J. & Maestrojuan Saez de Jaurequi, G.M. (1982). *Grasas y Aceites* **33**, 265.
Friedman, A.A., Young, K.S., Bailey, D.G. & Tait, S.J. (1980). New observations with anaerobic fixed-film reactors. Workshop on anaerobic filters, Orlando, Florida, USA.
Fullen, W.J. (1953). *Sewage Ind. Wastes* **25**, 276.
Goodwin, J.A.S., Wase, D.A.J. & Forster, C.F. (1990). *Biol. Wastes* **32**, 125.
Goodwin, J.A.S., Wase, D.A.J. & Forster, C.F. (1992). *Bioresource Technol.* **41**, 71.
Grotenhuis, J.T.C., Smit, M., Plugge, C.M., Xu Yuansheng, van Lammeran, A.A.M., Stams, A.J.M. & Zehnder, A.J.B. (1991). Appl. Environ. Microbiol. **575**, 1942.
Grin, P., Roeersma, R. & Lettinga, G. (1985). In *Anaerobic treatment of sewage*, ed. M. Switzenbaum, Rep. No. Env. E. 88-85-5, University of Massachusetts, USA. p. 109.
Grobicki, A. & Stuckey, D.C. (1988). Poster Papers. 5th. Int. Symp. An. Digestion, ed. A. Tilche & A. Rozzi, Monduzzi Editore, Bologna. p. 267.
Guiot, S.R. & van den Berg, L. (1984). *Biotech. Lett.* **6**, 161.
Guiot, S.R. & van den Berg, L. (1985). *Biotech. Bioeng.* **27**, 800.
Gutierrez, J.L.R., Encina, P.A.G. & Fdz-Polanez, F. (1991). *Biores. Technol.* **37**, 271.
Hamdi, M. & Garcia, J.L. (1991). *Biores. Technol.* **38**, 23.
Hamoda, M.F. & Kennedy, K.J. (1987). *Biotech. Bioeng.* **30**, 272.
Hemens, J. & Shurben, D.G. (1959). *Food Trade Rev.* **29**, 2.
Henry, M.M. (1985). In, *Energy from biomass and wastes IX*. Institute of Gas Technology, Chicago, USA. p. 829.
Hobson, P.N. & Shaw, B.G. (1974). *Water Res.* **8**, 507.
Hulshoff Pol, L.W. & Lettinga, G. (1986). *Water Sci. Technol.* **18**(12), 87.
Hulshoff Pol, L.W. & Lettinga, G. (1986*a*). *Water Sci. Technol.* **18**(12), 41.
Huster, R. & Möbius,C.H. (1988). Poster Papers. 5th. Int. Symp. An. Digestion, ed. A. Tilche & A. Rozzi, Monduzzi Editore, Bologna. p. 651.
Iza, J., Garcia, P.A., Sanz, I., Hernando, S. & Fdz-Polanco, F. In *Anaerobic Digestion 1988*, ed. E.R. Hall & P.N. Hobson, Pergamon Press, Oxford. p. 155.
Jewell, W.J. (1980). In *Anaerobic digestion*, ed. D.A. Stafford, B.I. Wheatley, & D.E. Hughes, Applied Science Publishers, London. p. 467.
Jewell, W.J. (1985). In *Anaerobic treatment of sewage*, ed. M. Switzenbaum, Rep. No. Env. E. 88-85-5, University of Massachusetts, USA. p. 17.
Jewell, W.J., Switzenbaum, M.S. & Morris, J.W. (1981). *J. Wat. Poll. Cont. Fed.* **53**, 482.
Kawase, M., Nomura, T. & Majima, T. (1989). *Water Sci. Technol.* **21**, 77.
Kennedy, K.J. & van den Berg, L. (1982). *Agric. Wastes* **4**, 151.
Koster, I.W. (1987). *Biol. Wastes* **22**, 295.
Koster, I.W. & Cramer, A. (1987). *App. Environ. Microbiol.* **53**, 403.
Lettinga, G. (1990). In *Anaerobic Digestion Workshop*, ETSU, Harwell, UK, p.33.

Lettinga, G., van Velsen, L., de Zeeuw, W. & Hobma, S.W. (1980). In *Anaerobic digestion*, ed. D.A. Stafford, B.I. Wheatley & D.E. Hughes, Applied Science Publishers, London. p. 167.

Lo, K.V. & Liao, P.H. (1986*a*). *Biomass* **10**, 187.

Lo, K.V. & Liao, P.H. (1986*b*). *Biomass* **10**, 243.

Lo, K.V. & Liao, P.H. (1989). *Biol. Wastes* **28**, 91.

de Man, A.W.A., van der Lart, A.R.M. & Lettinga, G. (1988). In *Anaerobic Digestion 1988*, ed. E.R. Hall & P.N. Hobson, Pergamon Press, Oxford. p. 197.

McCarty, P.L. (1982). In *Anaerobic digestion 1981*, ed. D.E. Hughes *et al.*, Elsevier Biomedical, Amsterdam. p. 3.

McCarty, P.L. (1985). In *Anaerobic treatment of sewage*, ed. M. Switzenbaum, Univ. of Massachusetts Report Env. E. 88-85-5. p. 3.

Macleod, C. & Forster, C.F. (1988). *Microbios.* **54**, 31.

Morelli, A., Rindone, B., Andreoni, V., Villa, M., Sorlini, C. & Balice, V. (1990). *Biol. Wastes* **32**, 253.

Morris, J.W. & Jewell, W.J. (1982). In *Proc. 36th ann. Purdue ind. waste conf.* p. 621.

Nel, L.M., de Haast, J. & Britz, T.J. (1984). *Biotechnol. Lett.* **6**, 741.

Ng, W-J. (1989). *Biol. Wastes* **28**, 39.

Ng, W-J. & Chin, K.K. (1988). *Biol. Wastes* **26**, 215.

Pande, D.R., de Poli, F. & Tilche, A. (1988). Poster papers, 5th Int. Symp. An. Digestion, ed. A. Tilche & A. Rozzi. Monduzzi Editore, Bologna. p. 405.

Pascik, I. (1990). *Water Sci. Technol.* **22**, 33.

Pascik, I. & Henzler, H-J. (1988). In *Anaerobic digestion 1988*, ed. E.R. Hall & P.N. Hobson, Pergamon Press, Oxford. p. 491.

Pichon, M., Rouger, J, & Junet, E. (1988). In *Poster papers, 5th. Int. Symp. An. Digestion*, ed. A. Tilche & A. Rozzi, Monduzzi Editore, Bologna. p. 73.

Pretorius, W.A. (1971). *Water Res.* **5**, 681.

Rockey, J.S. & Forster, C.F. (1982). *Sci. Technol. Lett.* **3**, 487.

Salkinoja-Salonen, M., Hakulinen, R., Valo, R. & Apajalahti, J. (1983). *Wat. Sci. & Techn.* **15** 9/10, 149.

Schellinkhout, A., Lettinga, G., van Velsen, L., Kooijams, J.L. & Rodriquez, G. (1985). In *Anaerobic treatment of sewage*, ed. M. Switzenbaum, Report No. Env. E. 88-85-5, University of Massachusetts, USA. p. 259.

Saslawsky, J., Liziard, Y. & Chave, E. (1988). In *Anaerobic digestion 1988*, ed. E.R. Hall & P.N. Hobson, Pergamon Press, Oxford. p. 499.

Schellinkhout, A., Jakma, F.F.G.M. & Forero, G.E. (1988). Poster papers 5th. Int. Symp. An. Digestion, ed. A. Tilche & A. Rozzi, Monduzzi Editore, Bologna. p. 767.

Schroepfer, G.J., Fullen, W.J., Johnson, A.S., Ziemke, N.R. & Anderson, J.J. (1955). *Sew. Ind. Wastes* **27**, 4.

Sorlini, C., Ranalli, G., Merlo, S. & Balsari, P. (1988). In *Poster papers 5th. Int. Symp. An. Digestion*, ed. A. Tilche & A. Rozzi, Monduzzi Editore, Bologna. p. 331.

Steffen, A.J. (1953). *Biol. Treat. Sew. Ind. Wastes* **11**, 126.

Stronach, S.M., Rudd, T. & Lester, J.N. (1986). *Anaerobic Digestion Processes in Industrial Waste Treatment*, Springer-Verlag, Berlin.

Stronach, S.M., Diaz-Baez, M.C., Rudd, T. & Lester, J.N. (1987). *Biotech. Bioeng.* **30**, 611.

Suidan, M.T., Pfeffer, J.T. & Nakhla, G.F. (1988). *Anaerobic digestion 1988*, ed. E.R. Hall & P.N. Hobson, Pergamon press, Oxford. p. 249.

Summers, R., Hobson, P.N., Harries, C. & Feilden, N.E.H. (1984). *Process Biochem.* **19**, 77.

Taylor, D.W. (1972). In *Proc. 3rd Nat. Symp. Food Proc. Wastes*, New Orleans, USA.

Tsonis, S.P. & Grigoropoulos, S.G. (1988). In, *Anaerobic digestion 1988*, ed. E.R. Hall & P.N. Hobson, Pergamon Press, Oxford. p. 115.

Tilche, A. & Yang, X. (1988). In *Granular anaerobic sludge; microbiology and technology*, ed. G. Lettinga, A.J.B. Zehnder, J.T.C. Grotenhuis & L.W. Hulshoff Pol, Pudoc, Wageningen, The Netherlands. p. 170.

Urgulu, A. & Forster, C.F. (1992). *Biores. Technol.* **39**, 23.
Vogel, P & Winter, J. (1988). Poster papers, 5th Int. Symp. An. Digestion, ed. A. Tilche & A. Rozzi, Monduzzi Edotore, Bologna. p. 689.
Weiland, P. (1987). *Bioprocess Eng.* **2**, 39.
Weiland, P. & Wulfert, K. (1986). *Chem. Ind. (Düsseldorf)* **38**, 97.
Weiland, P. & Wulfert, K. (1987). In *Biotechnology in Germany*, VCH-Verlag, Weinheim.
Wheatley, A.D., Cassell, L. & Winstanley, C.I. (1984). In *Anaerobic digestion and carbohydrate hydrolysis of waste*, ed. G.L. Ferrero, M.P. Ferranti & H. Naveau, Elsevier Applied Science, London. p. 284.
Wheatley, A.D., Johnson, K.A. & Winstanley, C.I. (1988). In *Anaerobic digestion 1988*, ed. E.R. Hall & P.N. Hobson, Pergamon Press, Oxford. p. 135.
Winter, J., Zellner, G. & Vogel, P. (1988). In *Poster papers 5th Int. Symp. An. Digestion*, ed. A. Tilche & A. Rozzi, Monduzzi Editore, Bologna. p. 617.
Yang, P.Y. & Chou, C.Y. (1985). *Agric. Wastes* **14**, 221.
Young, J.C. & McCarty, P.L. (1968). Tech. Rept. No 87, Dept. of Civil Eng., Stamford University, USA.
Young, J.C. & McCarty, P.L. (1969). *Water Pollut. Cont. Fed.* **41**, 160.
Xiushan, Y., Garuti, G., Farina, R., Parisi, V. & Tilche, A. (1988). In *Poster papers 5th Int. Symp. An. Digestion*, ed. A. Tilche & A. Rozzi, Monduzzi Editore, Bologna. p. 355.
de Zeeuw, W.J. (1988). In *Granular anaerobic sludge: microbiology and technology*, ed. G. Lettiga *et al.* Pudoc, Wageningen. p. 132.

Chapter 8

USES OF DIGESTER PRODUCTS

INTRODUCTION

Whether the primary objective of digestion is pollution control or energy production the digester will produce 'products' which can be divided into liquid and solid 'effluents' and 'biogas'. If a crop is grown for energy production by digestion, then disposal of the digester feed would not have been a problem, only disposal of the effluent has to be considered and costs incurred in this have to be set against the price of the energy produced. However, for digestion for pollution control one of the reasons for digestion would have been the difficulty of disposing of the digester feedstock and in this case the costs of disposing of the digester effluent can be balanced against the costs of disposing of the untreated waste. A benefit of digestion should be a decreased cost of disposing of the waste, or, better, the possibility of disposing of a waste which untreated could not be disposed of. Since the digestion of dissolved or suspended solids does not appreciably reduce the volume of liquid digester feedstock, the effluent from most digesters will be, for all practical purposes, of the same volume as the feed. Only with solid-state digestions will the volume and weight of the effluent be appreciably different from those of the feed. To a large extent, then, the handling problems of disposing of the digester effluent will be those of disposing of the feed. However, the effluent will be reduced in pollutants and may be inproved in physical structure, and this can make disposal of the effluent easier than disposal of the feed in that the number of options for disposal may be increased. However, disposal of the effluent must be considered and the costs of disposal included in the costing of the digester plant. Set against this may be the value of the other product, the biogas, as a source of energy.

It can be argued that if the feed is a waste that had to be got rid of by dumping, for instance domestic garbage, then the cost of dumping the digested garbage can be ignored or a positive benefit might be applied as the digested garbage will be of lower weight than the feed and will make better and more stable landfills. A similar argument might be applied to sewage sludge digestion, but the benefit may lie not so much in reduction of the volume of sludge to be dumped as in the decreased polluting power of the digested sludge. Decreased polluting power of digester effluents is a main benefit of digestion, but its monetary value is not always easy to quantify. This has been one of the problems of farm-waste digestion over the last ten years. In many cases penalties for odour or other pollution caused by farm wastes have not been severe or have been non-existent. No monetary value has, then, been placed on the pollution control achieved by digestion: the digester has been costed on the value of the biogas as farm energy and this, in turn, has depended on prices and availability of other fuels. The net result has in many cases been an apparently adverse costing for digesters.

On the other hand, there is no doubt that the present increasing public awareness of pollution and increasing legislation to control pollution will increase the value of pollution control in the future cost/benefit analyses of digestion in a way that can be quantified in penalties for causing pollution. But whether or not pollution control is considered, digester effluents have to be disposed of, and if this can be done so as to make an obvious and definite monetary contribution to the digester costs so much the better. However, the positive uses for digester effluents are rather limited.

Use as fertilizer

The most common use of digested sludges has been, and still is, as fertilizer for farm crops. The benefit of the small Gobar, Chinese-type or other digesters in countries like India and China and Africa is that they provide energy and leave a residue that has the original fertilizer value in N,P,K while being free from the weed seeds and pathogens of the raw animal and human excreta. Also, the odour from farm and human wastes is much reduced by digestion (see, for instance, Pain *et al.*, 1990). In these countries, chemical fertilizers are too expensive for the small farm and the soils are often poor and can benefit both from the minerals in the digester effluents and the soil-conditioning properties of the fibre. All effluents can be used as fertilizer. The digested effluents from some industrial processes, such as palm oil production, are also used as fertilizer and irrigation water, even though such effluents may be of low solids content. (See, for instance, Lim Kim Huan, 1988; Yeoh, 1988). Here the factory is at the heart of the oil-palm plantations and the effluents can be easily piped to the palms. However, it is probably true to say that in most industrialised countries the effluents from digestion of wastewaters from food preparation and other processes are further purified and discharged to drains or water courses: the suspended solids from the treated waters may be disposed of as fertilizer. In a pilot-plant scheme for production of methane as a car fuel by digestion of energy crops in New Zealand the digester effluent was returned as fertilizer to the energy crops (Stewart, 1980). Wong & Cheung (1989) found that the effluents from digesters treating mixtures of pig-waste and cardboard or other wastes were good fertilizers when tested on rye-grass.

The digester bacteria ferment the more-easily-available carbohydrates, lipids and proteins in feedstocks and leave the more-lignified plant fibres of faecal wastes or energy crops, and similarly poorly-degradable constituents of other wastes. It is not, then, surprising that experiments have shown that digested sludges of animal, plant, or sewage origin are generally decomposed in the soil slower than are the original materials (Sweeney *et al.*, 1985; Atalay & Blancher, 1984; Moorhead *et al.*, 1987; Miller, 1974). This slow decomposition has advantages as it preserves the fibre structure as soil conditioner and leaves readily-available ammonia nitrogen to plants instead of its being used by microbes growing rapidly on sludge constituents.

While in some countries digested sludges from whatever sources are needed as fertilizers, in many industialised countries chemicals are the main fertilizers and digested sludges have to compete with these for farm use. In the case of digested sewage sludges from large cities the production has been so great and the demand so small, compared with that for chemical fertilizers, that only a proportion of sludges has been used as fertilizer, the rest has been dumped at sea or elsewhere. The increased awareness of the dangers from excessive use of chemical fertilizers and demands for 'organically grown' farm produce may, though, change this picture in the coming years, as could legislation to prevent dumping at sea.

Digested sewage sludge can be delivered by tanker to central tanks on farms and spread by pipelines from the tanks or by farm slurry tankers in the same way as farm-animal slurries. In some cases the sewage-works tankers may also spread the sludge. For household and market-garden use the sludge needs to be dried to give an easily-handled material. The conventional method of drying sewage sludges has been spreading the sludge on drying beds of clinker, sand, tiles or other porous material where it can drain and dry in the wind for many weeks. One of the authors earlier showed that digested pig-waste would dry on beds to a blackish, odourless, friable mass. Such drying is subject to the vagaries of the climate, but it is less costly and uses less energy than drying by chemical or physical coagulation and settling followed by filter or centrifuge dewatering and final drying. Dried sludges can be bagged and sold for domestic and commercial horticultural uses at a higher price than can sludge in bulk, although the market may be rather limited. However, the water from sludge drying, if it is not all evaporated, is polluting and has to be returned for aerobic or other treatment before discharge to water courses.

Dried sludges have advantages as fertilizer over the wet material in that they do not add to the ground water-content and runoff on wet fields. Drying of sludges may also allow the use of the water from the sludge on the farm or factory or as irrigation water in controlled applications. There are thus reasons for drying sludges, but the problem remains that the sludge is over 90% water and removal of water, unless done on drying beds, is difficult and costly. The sludge can be 'thickened' by allowing it to stand in tanks where the solids settle over some days and the supernatant water can be run off for further treatment or return to the digester. This is the usual practice in sewage works where the digested sludge is settled for ten or twenty days. More rapid and complete settlement can be obtained by the use of chemical coagulants. Chitin has already been mentioned, but the usual ones are tannins, ferric sulphate, ferric chloride, ferrous sulphate ('copperas'), chlorinated copperas, aluminium chloride, aluminium chlorohydrate, and synthetic polyelectrolytes. The best reagent varies with the sludge and should be selected by laboratory tests. Summers & Bousfield (1980) found that digested piggery-waste could not be flocculated by tannins, but that some inorganic flocculants could be used and that a synthetic polyelectrolyte gave rapid flocculation to give a water-clear liquid which could have been used for piggery washing water. The BOD was 387 mg/l, but this compared with 2787 for the digester effluent and represented a 98% reduction on the digester-feed. Ammonia-N was 690 mg/l, compared with 1781 for the effluent, a result of the previously-mentioned absorption of ammonia to bacteria. Although a 5-day aeration of the same effluent reduced the BOD to 210 and the ammonia to 84 mg/l, the aerated liquid was still dark-brown in colour. Camarero *et al.* (1988) also found that inorganic flocculants were not as efficient as synthetic anionic or non-ionic flocculants in coagulating pig waste. Coagulating or floculating agents add to the costs of sludge drying, but the costs of other processes such as electrolytic or air-flotation coagulation will be even greater.

The thickened sludge will still be of high water-content and will have to be further dried to provide a handlable solid. If drying is not done on beds, mechanical methods have to be used. The sludge can be brought to a damp solid by filtration in vacuum drum filters or filter presses or band filters, but, like centrifugal dewatering, all these methods are high in capital and running costs. Farm separators, where the slurry is run over various types of screens to separate solids and liquids, can be cheaper to purchase and operate than filters, but they do not remove all bacteria and fine particles. Not only does this leave a water of high pollutional power, but, as much of the fertilizer nitrogen value lies in the bacteria and fine particles, the solid is of limited fertilizer value, although the liquid may be used for irrigation and fertilization ('fertigation'). By arranging screens of smaller apertures in series a farm separator might be programmed to remove more of the solids, but this then becomes more expensive to construct and operate. Solids, mechanically-separated by any method, still have a relatively high water content (up to about 70%) and if left stacked will allow microbial growth. This can be put to use in composting the digested solids to stabilise them as a soil conditioner and horticultural peat substitute. Tarre *et al.* (1987) described a process used in Israel where the effluent from a thermophilic cattle-waste digester was passed through a coarse-mesh vibrating screen and then a finer-mesh rotating-drum separator. The liquid, containing the finest particles and dissolved nutrients, was used as fertilizer and irrigation water while the mixed solids were composted to give a peat substitute used in horticulture. Schuchardt (1988) has reviewed the separation and composting of solids from effluents of digesters treating various feedstocks. But dampness may be a problem as the original separated material cannot be bagged and sold in small quantities. Further drying to less than 15% moisture content, at which level mould growth is inhibited and the solids will be stable if bagged, will require breaking up of the separated material and hot-air or other drying. It could be possible with a large farm digester that the excess heat, from engine cooling or gas burning, could be used to dry digested solids after an initial separation. This was being done some years ago on a commercial piggery. However, the use of any of these options can only be decided on after detailed considerations of the situation at a particular site; no general recommendations can be given. The situation at a particular site may vary with time as legislation and other factors change.

There are many papers on the analyses of digested sludges and the drawbacks and values of sludges as fertilizers; some points will be illustrated from the authors' own work and a few papers, and a guide to further reading given. The value of sludges as fertilizers lies mainly in the N,P,K content and the form in which these elements exist in the sludge. A typical analysis of the solids of digested piggery slurry in terms of fertilizer value is: total N, 6·8%; NH_3-N, 2·7%; P, 2·6%; K, 1·2%; ash, 15·4%; fibre, 41·3%. This material was obtained by freeze-drying the thickened sludge (about 7–7·5% TS) obtained by lightly centrifuging digested pig slurry and would be similar to sludges obtained by gravity settling. The supernatant was a darkish liquid containing (%w/v): Total N, 0·17; NH_3-N, 0.14; P, 0·006; K, 0·08 (about 0·014 as P_2O_5 and 0·1 as K_2O). The analysis shows that the water could not be discharged to many water courses, but that it would have some fertilizer value if used as irrigation water. The analysis of the solids shows a high percentage of fibre and this could usefully affect soil structure, as previously mentioned. Much of the nitrogen of the digested slurry is in the form of ammonia which is in solution or absorbed in or on the bacterial cells. This nitrogen is available to plants and if applied to growing plants at the right season is beneficial. If it is not taken up by plants the nitrogen can be lost by volatilisation or by leaching into surface or underground waters. Although both represent a loss in value of the fertilizer, the former is hardly likely to cause pollution, but the latter can be a cause of pollution (these effects, of course, occur with undigested slurries). The nitrogen of bacterial protein will be more slowly available to plants as it is changed to ammonia by death and degradation of the microbial cells. Ammonia can also be converted to nitrate. The phosphorus and potassium are also associated with the bacterial cells and will become available to plants as the cells die. In fertilizer terms, the analysis of digested pig-slurry would class the solids as a high-nitrogen, high-phosphate fertilizer. A general-purpose fertilizer has N, P_2O_5, K_2O in 1:1:1 proportions; a high-nitrogen fertilizer has 2:1:1, and so on. The book edited by Hucker & Catroux (1981), for instance, considers the phosphorus content of sewage and farm sludges and the fate of phosphorus on the land.

Some experiments have shown digested slurries to be better than raw slurries in terms of plant growth and other properties (for instance, tests by one of the authors and colleagues (Hobson *et al.*, 1981) showed that spreading of digested slurry by tanker did less damage to grassland than did spreading of raw slurry). Digested sewage sludge acted as a soil conditioner and fertilzer on strip-mine spoil (Stuckey & Newman, 1977). However, the effects seem to vary with type of digested slurry (cattle or pig, etc.) and with crop, and general conclusions are difficult to give. Fertilizer values, runoff, and other factors in uses of raw and variously-treated slurries are discussed in reports of meetings and books such as those just mentioned and those edited by L'Hermite & Ott (1984); Dam Kofoed *et al.*,(1986); Williams *et al.*(1985); Brogan (1980) and many others.

Atalay & Blanchar (1988) found that addition to soil of digested sludge from a farm digester, with a feedstock of fescue hay and sewage sludge, increased the urease activity after plants had begun to grow in the soil. The increase in urease activity was related to a general increase in soil biological activity as indicated by carbon dioxide evolution.

The analyses of digested sludges and slurries vary with source of digester feedstock and effluent treatment. For most efficient use of digested sludges as fertilizers account must be taken of the season,plant type and stage of growth, and soil conditions and soil analysis in terms of N,P,K content, so that the right amount of sludge, of known N,P,K content, can be applied at the right time. This means that storage for digested slurry on farms must be equivalent to some two hundred days production to allow for weather and plant variations. Unfortunately, the amounts of digested sludges and slurries produced, and the limitations of storage, means that in many cases slurries have to be land-spread without regard to the factors mentioned above, and so spreading becomes more or less a way of dumping the slurry, not efficient use. The proceedings of the conference edited by Dam Kofoed *et al.* (1986) have many papers on the efficient land-spreading of digested and undigested farm and sewage sludges and losses due to volatilisation and leaching from fields.

Pathogenic organisms are unlikely to be found in digested factory wastes, but with wastes derived from animal or human excreta the question of the possible spread of pathogens always occurs. It has been shown in Chapter 2 that destruction of bacteria and viruses occurs in mesophilic, and (quicker) in thermophilic, digesters, although helminth eggs may be rather more resistant to mesophilic digestion (see also Demuynck [1984] for review). Except in the case of a very heavy contamination of the digester feed (which might possibly occur if there was a massive outbreak of disease on a farm) even if pathogens are not completely killed they should be reduced to such small numbers that the likelihood of disease transmission through spreading of digested slurry on grassland to be grazed by cattle would be small or negligible, especially after secondary storage of digester effluent which will further decrease numbers of pathogens such as Salmonella (Hobson *et al.*, 1981). Animals are usually not turned out immediately on to freshly-fertilised grass and the delay allows further death of pathogens. However, the only way to be certain that slurries, digested or otherwise, are free from pathogens is by treatment with chemicals such as lime; although, again, helminth eggs are the pathogens most resistant to chemicals (see e.g. Strauch [1984] for a short review). Gasi *et al.* (1988) reported that destruction of pathogenic bacteria in a UASB digester treating domestic sewage water (running conditions not given) was not efficient, but that chlorination of the digester effluent at 15 $mgCl_2/l$ for 20 min removed coliforms and streptococci by better than 99·99%; but *Clostridium perfringens* was reduced by only 84%. Parasites seemed to be efficiently removed by the digestion.

It has already been mentioned that heavy metals are precipitated in the solids of digested sludge and concern has been expressed over the possible contamination of food crops with heavy metals by the spreading of sewage sludges as fertilizers. The extent of heavy-metal contamination of sewage sludges depends on the source of the sewage. Severe heavy-metal contamination comes from industrial sources, and cleaning of effluents by factories before discharge to sewers will reduce this contamination. It is a common practice to add copper sulphate to pig feeds to improve animal performance. The copper is usually added at 200 ppm in the feed and the copper becomes concentrated in the digested or undigested slurry solids, mainly as copper sulphide. Analyses showed 850 ppm of copper in dry digested-solids (Hobson & Shaw, 1973). Since the copper is in an insoluble form, it was suggested some years ago that the copper would not be taken up by plants if pig slurry were used as a fertilizer and that the only danger to animals grazing on fertilized grassland would be if they were turned out immediately after the slurry had been spread and ingested pieces of slurry solids (Hobson *et al.*, [1981] based on Gracy *et al.*, 1976). Since then many millions of tonnes of pig slurries have been spread on farm land without, apparently, ill effects. It would seem that, in Britain, normal applications of pig slurry to land do not result in toxicity to crops or animals ingesting these crops (Unwin, 1980: this paper is in the proceedings of a conference devoted to copper in animal and sewage slurries and sludges; L'Hermite & Dehandschutter, 1980).

The long-term effects of spreading sludges and slurries as fertilizers have been investigated by a number of workers, but, because, apart from sewage digesters, there have been few large-scale digesters that have operated for long enough to provide sludges for field trials most work has been done on undigested sludges. Reports of meetings convened to discuss the results and predictions have been published in books. Some books, which will give a lead into the subjects, are edited by: Williams *et al.*, 1985; Kofoed *et al.*, 1986; Brogan, 1981; L'Hermite & Ott, 1984; Davis *et al.*, 1983; Welte & Szabolcs, 1988.

Digested sludges have also been used to fertilise water bodies for growth of aquatic plants which could be used either as further digester feedstocks or as animal feedstuffs. Water hyacinth, for example, reponded to digested pig waste, but digester effluents have been found to be toxic to some plants and so the effluents may need to be diluted to suit the plant (Graetz *et al.*, 1986; Moorhead, 1986).

As said before, a large proportion of the nitrogen in digested animal slurries is in the form of ammonia. This is immediately available to plants, but if not taken up can be lost by volatilisation (this is considered in some of the books quoted above). Some years ago a process for converting digested pig slurry into a slow-release nitrogen fertilizer was being developed (Watson *et al.*, 1979). In this, the sludge was combined with urea and formaldehyde at pH4 and ambient temperature for 12 hours. The granular material was then neutralised with lime and the supernatant removed. The solids were mixed with peat moss to form a stable, easily-handled, fertilizer. The release rate of the nitrogen could be varied by varying the reaction conditions. Whether this process reached a commercial scale is not known. The production of a solid manure in Chinese digesters to suit the local needs has been mentioned in Chapter 6. This was produced from a solid-state digestion and the residues or effluents from solid-state digestions will, obviously, require little treatment to produce a handleable or compostable solid fertilizer. The residue from the solid-state digestion of willow dust (Chapter 6) was utilised as a fertilizer without any dewatering (Balasubramanya *et al.*, 1986).

Another method suggested for producing a slow-release fertilizer is by precipitating N and P from effluent liquids by adjustment of the pH to 9 and adding $MgSO_4$ when the N and P are precipitated as struvite ($MgNH_4PO_46H_2O$) along with some apthitatite ($K_3Na(SO_4)_2$) and thermardite (Na_3SO_4). Struvite has been shown to be a good N:P fertilizer (Bridger *et al*, 1962; Rothbaum & Rohde, 1976). This precipitation reduces the N and P in the effluent water to very low concentrations and so reduces the possibilities of algal blooms and other pollution if the effluent is discharged to a water-body. The precipitation could be done on effluents from which solids had been separated as described previously, or on a dilute effluent from digestion of piggery wastes where large volumes of water are used for washing-down pens (Wrigley *et al.*, 1991). The economics of large-scale production have still to be worked out.

Struvite has also been produced in a pilot-scale plant for removing N and P from (non-digested but treated) municipal wastewaters by ion-exchange (Liberti *et al.*, 1986), but the plant looks too complicated for use with farm or similar digester systems. P has also been removed from the supernatant of settled (undigested) pig waste by precipitation with alum, lime or zirconium tetrachloride. Alum was best, but the sludge would have to be disposed of.

Much of the wastewater from Finnish pulp and paper mills is, or will soon be, treated by activated-sludge plants. This generates a vast amount of waste activated-sludge and Puhakka *et al.* (1992*a,b*) investigated the anaerobic digestion of this sludge and the dewatering properties of the digested sludge. The literature gives a rather confused picture of the effects of digestion on dewaterability of sludges (see Puhakka *et al.*, 1992*b* for a short review). The dewaterability, on a belt press after addition of chemical flocculants, of digested pulp-mill aerobic sludge was improved by mixing with primary sludge from the mill. The mill wastewater was low in nutrients (P and N) and these had to be added before the aerobic treatment. Recycling of the filtrate water from the digested sludge not only purified it but could reduce the cost of adding chemical sources of nutrients to primary wastewater for the aerobic plant.

A comprehensive review of the effects of digestion on the composition of wastes, the death of pathogens and weed seeds, and the use of digester effluents as fertilizers is that by Demuynck (1984) previously mentioned. It describes the effects on soils of the digester effluents and the availability of nutrients and the effects on plant growth. Recommendations for spreading and the crops most benefited by digester effluents are described.

Animal feedstuffs

The value as animal feeds of digester effluents, from whatever type of digester or feedstock, lies in the microbial protein and, for ruminants, in the ammonia nitrogen. The microbial protein will be similar in amino acid composition to rumen bacterial protein, which is similar to other bacterial proteins (Hobson, 1969). All readily-degradable carbohydrates in the feedstock will have been removed during the anaerobic digestion. The poorly-degradable residues serve only as a diluent for the nitrogenous compounds. The residues from digestion of municipal waste, for instance, will contain only the relatively small amount of bacteria grown in the digester and so will be of little use as animal feed. A further consideration is that this residue will contain small pieces of plastic and other materials, of no feed value and possibly dangerous when swallowed. Digested sewage sludges have not been much, if at all, considered as animal feeds, possibly because of their origin in human excreta but also because of their possible content of heavy metals and dangerous bits of rubbish from the drains. Some sludges from digestion of food-processing factory wastes are comprised of vegetable or fruit residues which may contain residual plant proteins as well as microbial proteins, and these should have no undesirable materials present. The principal digester effluents which might be used in animal feeds are thus those from food processing and farm animals themselves. The latter have been most tested as these are produced on the farms where they can be used.

There is much information on the use of animal excreta as components of farm feeds; a recent review is by Flachowsky & Hennig (1990). Excreta have been used raw or processed and one method of processing was introduced by Anthony (1966). This is production of 'wastelage', a type of silage made from animal excreta and grass or other vegetation. Silage making overcame two of the problems of using excreta in feeds; the fermentation produced an acid pH which would kill pathogens in the excreta and the grass or other dry material reduced the water content of the excreta slurries. Seepage from the silos also helped in this latter. Ensiling also improved the palatability of the excreta. Summers *et al.* (1980) showed that a wastelage could be successfully made from a mixture of 13% digested cattle slurry, 81% chopped barley straw and 6% molasses (to give a greater acid fermentation). After 11 weeks ensilage at ambient temperature the mixture was sweet-smelling and compared favourably with grass silage in composition and nutritive value (Table 5). Wastelage prepared from digested poultry excreta was unsuccessful: there was no rapid fall in pH and little lactic acid production, possibly due to buffering by the high ammonia content of poultry wastes. Production of wastelage from digested pig slurry was not tried as the

Table 5

The compositions of wastelage made with digested cattle waste and conventional grass silage.

	Wastelage	Silage
Dry matter	24·2	18–25
pH	4·2	<4·5
Lactic acid	4·3	2–10
Total VFA	2·3	2·4
Crude protein	11·2	10–15
Ammonia N*	22	12
GE+	19·2	19–20

Values % of material as prepared, except: *, % total N; + MJ/kgDM; pH. Values for silage are ranges from literature where available.

Table 6
Some analyses of digested farm animal-wastes.

	Pig[a]	Pig[b]	Cattle[c]
Ash	15·4	7·9–16·0	23·1
Protein	25·6+	8·8–12·5	15·1
Ammonia-N	2·7	—	
Total-N	6·8	1·4–2·0	2·4
NDF	41.3	—	—
Crude fibre	—	20·6–32·5	23·4
Lignin	11·8	—	—
Lipid	9·4	1·5–2.5	1·0

Values as %DM.
[a]Freeze-dried sludge containing bacteria and small particles. Summers *et al.* (1980).
[b]Solids from separator, small particles would be left in the liquid. Flachowsky & Hennig (1990).
[c]Average from cattle fed on different diets. + non-ammonia nitrogen times 6·25. Saxena *et al.* (1989).

copper content of pig slurries makes them unsuitable for feeding to animals other than pigs. The drawback to wastelage production is that to produce a solid sludge it needs a larger mass of hay, straw or other material than of the waste. If hundreds of tonnes of waste are being produced this makes wastelage an impossible proposition for using all the waste. There seems little point in separating the digested solids to make a wastelage, the separated solids might be used as below, although ensiling could increase palatability.

Some analyses of digested animal-wastes are shown in Table 6. These can be taken as approximate compositions, actual values will vary with conditions of digestion, feed of animal and other factors. Raw cattle-wastes are low in nitrogen and high in fibre, and this is reflected in the compositions of digested wastes. Some examples of fattening- and dairy-cattle wastes had 1·6 and 2·3% total N in the freeze-dried slurry (Summers *et al.*, 1987).

Prairie & Stickney (1988) have recently described some trials in Canada of the use of digested pig wastes in pig feeds. The digester effluent was screened to remove coarse, indigestible particles for use as soil conditioner and fertilizer. The liquid was passed through a high-speed centrifuge to recover the bacteria and small particles for the feed while the centrifugate was used for irrigation. To aid in collection of the fine particles (including the N-containig bacteria) a cationic-polymer flocculating agent was added to the liquid before the centrifugation (the best polymer and dosage were determined by experiment). The centrifuge cake was used to replace part of the soya meal in rations containing this as protein, with corn, minerals and vitamins. The feed was given to starter, growing and finishing pigs. Detailed results of the trials were given, but, overall, the results showed that digested pig-waste solids could be used in pig rations without deterioration in pig performance or feed intake. In some tests a flavouring agent was added to the feeds.

Maramba *et al.* (1983) reported that mixing of 5–10% digested pig waste with the usual feed for growing pigs resulted in slightly lower weight gains but increased feed conversion efficiency. They attributed the latter to vitamins in the digested-sludge bacteria. Digester effluents are particularly rich in vitamin B_{12}; fishmeal contains about 0·2 μg/g, digested sludge about 3 μg/g.

Ravindram *et al.* (1987) used anaerobically digested pig waste as a constituent of diets for broiler chickens. They found that although weight gain rates were depressed by above 15% waste in the feeds 20% waste gave the lowest cost per kilogram weight gain, and they concluded that digested pig waste could be used successfully in broiler feeds.

Mention has previously been made (Chapter 6) of the use of digested cattle wastes in cattle feeds in India, and these at least maintained the animals, and so could help to tide the peasant farmer over a period when grass or other feeds were in short supply, but they were not good protein feeds.

Ward (1983) reviewed the reports of a number of trials of using mesophilic and thermophilic cattle-waste-digester effluents to feed cattle. The results were variable, as have been other results. The rate of weight gain of effluent-fed animals was generally lower than that of the control animals. However, the effects seem to differ with the basal ration used and the extent of replacement of conventional protein meals. While up to about 10% of the feed has usually been replaced, Marchaim *et al.* (1982) could replace 25% of the dry matter of a feed for growing cattle by digester effluent from cattle-waste and other digestions without ill effects on cattle performance. Prior *et al.* (1986) fed digested cattle wastes to various classes of cattle and sheep as part of corn-based rations. Both the dried solids centrifuged off from the effluent and the whole effluent were fed. Centrifuged solids had less nutrient value than the whole effluent, but the incorporation of the latter in feeds was limited (effluent dry matter about 5–7% of the feed dry matter) by the need to keep the water content of the feed to about 65%. Because of the low digestibility of the fibre in the dried solids the dried effluent would probably be restricted to maintenance diets. Overall, it appeared that digester effluent could be fed to the animals without loss of performance, but the response to the effluents was not consistent in all experiments. Considering all the data in various reports, costs and benefits are difficult to quantify.

Wastelage preparation would kill, or very much reduce the numbers of, any pathogens present in the digested excreta. Although digestion itself would destroy pathogens, any feed containing raw or treated excreta could not be guaranteed as free from pathogens without constant testing, and effluents from digested excreta will contain both digester bacteria and harmless intestinal bacteria (see Chapter 2). The presence of pathogenic bacteria in small numbers might not be of consequence in feedstuffs produced on a farm for use on that farm, although recycling of pathogens should be avoided if possible. However, feed supplements made from digested excreta for sale to other farms would come under the various feed-protein bacteriological standards of different countries. These may stipulate maximum (and low) numbers of viable bacteria of any form, including faecal bacteria, as well as freedom from pathogens such as Salmonella. Such orders may effectively stop the use of feeds based on digester effluents unless these are sterilised in some way. Such sterilisation, most likely by heat, could increase the cost of the feeds sufficiently to make their preparation uneconomic. Digester residues are not high-protein supplements and can compete with vegetable and animal protein meals only on a basis of very low cost when the latter are easily-available and can be afforded by farmers. Different considerations may apply, of course, in peasant-farming systems where high-protein supplements cannot be afforded and feed vegetation may be of poor quality.

It may be possible to recycle the water after separation of solids from digester effluents. Building washing is one possibility that has been previously mentioned, but reuse as drinking water is another. In some unpublished experiments with effluent from a full-scale piggery waste digester, which had been gravity settled, the water was palatable to pigs either by itself or mixed with fresh water. Pigs taking the water showed a slight decrease in feed-conversion efficiency over short trials, but there were no other effects.

Indirect use to produce animal or human feeds

An indirect method of using digested wastes as feedstuffs is to use the digester effluent as the substrate for growth of another micro- or macro-organism. This latter can be of better feed-quality than the digested sludge and/or can be more easily purified. If the final organism is, for instance, a fish, then this may be used as human food without problems.

Earth worms grown on different wastes have been suggested as a source of protein meal for animal feeds. Neuhauser *et al.* (1980) found that freshly-centrifuged digested sewage sludge solids were lethal to worms, but that aged solids were less toxic although they did not increase the weight of worms. However, a mixture of 4 parts of the sludge solids with 1 part of horse manure was suitable for worm growth. Unpublished reports suggest that digested sludges are not, overall, very good for worm production.

Levanon *et al.* (1988) reported the use of the coarse fibre from separated digested cattle waste in mushroom compost. It gave better and quicker growth than conventional compost.

Growth of algae on digester effluents has been small-scale practice in hot countries for many years (Chung Po *et al.*, 1975). The settled effluent from small digester plants is run into shallow troughs, and about 1% of the trough volume was suggested as the daily loading rate. Troughs can be arranged in series and wind-driven rotors can be used to slowly circulate the liquid in the troughs. The harvested algae can be used as feed for pigs or poultry.

Mahadevaswamy & Venkataraman (1986) suggested an integrated scheme for anaerobic digestion of poultry droppings and use of the digester effluent in algal production. The best results were obtained when the effluent was fed to the algal pond so as to give 2% (v/v) effluent in the pond; then the growth of *Spirulina* was only slightly less than in a chemically-defined optimal medium. On the large scale the algae were harvested and dried and then fed back to the poultry. Overall, when the value of the biogas for cooking and lighting and the value of the algae as poultry feed were taken into consideration the system was economically viable for small-farm use. Effluent-grown algae could also be used as a supplementary feedstock for the digester thus increasing the biogas production as well as purifying the digester effluent.

Calzada & Rolz (1983) found that a blue-green alga or water hyacinth would grow on the effluent from digestion of coffee pulp or pulp juice. The algal or plant growth was here suggested as a secondary treatment to reduce the COD of the effluents if they could not be used as fertilizers, but the alga could be used as an animal feed and the water hyacinth as a feedstock for digestion. de la Nöue & Basséres (1989) tested three algae, a *Chlorella* sp., *Scenedesmus obliquus* and *Phormidium bohneri*, for growth on digested pig waste. All three grew on the waste, but this was considerably diluted for the trials (see earlier sections). The biomass produced and the effects on pollutional parameters of the waste varied with algal species and no one could be considered completely 'best'. The primary purpose was further purification of the digester effluent before final discharge, but the recovery of the algae for animal feed was also kept in mind. The best alga for N and P removal was a *Chlorella* sp., but *Phormidium bohneri* would be easier to remove from the algal ponds.

Phang & Ong (1988) reported good yields of a natural mixture of algae cultivated in ponds fed with digested palm oil-mill effluent. Light transmission through the darkish effluent appeared to be the factor limiting algal growth.

Since algae depend on sunlight for growth one of the problems of using digester effluents is that the brown colour of the wastes tends to exclude sunlight unless the algal ponds are comparatively shallow. This is also one of the reasons for diluted digester effluents having been used. However, even if light penetration limits algal growth from the point of view of the algae being the only crop from the waters, fertilising with digester effluents increases the growth of many plants and micro- and macro-organisms in ponds. This effect can be used in fish farms where animal wastes have been traditionally used for increasing production of fish such as carp. Natarajan & Varghese (1980) found poultry waste to be the best fertilizer for ponds stocked with catla and common carp, followed by digested sewage sludge solids and cow-manure. Kaur *et al.* (1987) tested digested cattle slurry as a fertilizer for carp ponds. The growth rate of the fish on digested slurry was about three times that of fish fed on raw cattle slurry. Small-scale experiments had previously determined the optimum addition of digester slurry to be 52·6 litres/hectare/day. Additional feeding with rice bran and groundnut cake further increased carp weight gains. Another of the integrated schemes of Mahadevaswamy & Venkataraman (1988) involved digestion of rabbit droppings and use of the digester effluent as fertilizer for ponds producing major carp and common carp. Weight gains of the carp fed with digester slurry were only slightly less than those of carp fed with a concentrate made of groundnut cake, fish meal, rice bran and rice powder, with vitamins and minerals. Economic analysis showed that the system could be profitable for rabbit farmers when use of the biogas was taken into consideration. A report in the Nepal Biogas

Newsletter (1987) tells of a farmer who had a Gobar digester but found that he had to replace the gas dome after 5 and 10 years because of corrosion. He had then built a modified fixed-dome, Chinese-type digester. From about seven cattle plus calves he produced gas for family cooking, thus saving on firewood as well as having a better fuel than the wood. Ten percent of the digester effluent was used to fertilise a fishpond, from which he got fish to sell as well as for his family, while the remaining effluent was channelled to the fields to fertilise various crops in rotation.

The possibility of growing aquatic plants such as water hyacinth in ponds fed with digester effluent has been mentioned. The plants might not be much use a feedstuffs, but they could be used as an extra feedstock for the digester and their growth would also reduce pollutants in the effluent.

These indirect uses of effluent provide an upgraded protein for animal or human feeds. It is possible to produce a practically pure microbial protein from effluents. The sludge from a digester treating very low suspended-solids wastewater would be very largely bacterial protein which could be recovered comparatively simply. More complex recovery procedures, involving differential separations and washing of the coarse solids, would be needed to separate bacteria from digested cattle wastes or similar digester feedstocks. In each case the bacteria would have to be dried if not being directly incorporated in feeds, and the material might have to be sterilised. Whether the economics of such processes would be all right depends on many factors.

Biogas

Another product of digestion is biogas, and this is generally regarded as the most important product. For energy-from-biomass systems it is, of course, but in pollution-control the importance of the biogas can be overemphasised, as previously noted.

However, biogas is produced, and it is usually used in some way. The gas 'as-produced' composition can vary from about 50% methane–50% carbon dioxide to about 80% methane. Gas of about 50% methane is produced, for instance, from cattle wastes, and 80% methane from fatty meat wastes. The most usual gas is 65–70% methane. In addition, there can sometimes be small amounts of nitrogen from air (Chapter 2), with smaller amounts (usually less than 1%) of hydrogen sulphide and hydrogen from microbial metabolism, and possibly traces of other hydrocarbons. The gas is also wet and can contain traces of ammonia and volatile fatty acids coming partly from volatilisation from the gassing digester contents and partly from fine water droplets carried along with the gas. Water droplets and vapour are (in temperate climates) usually condensed out of the gas in condensate pots as the gas cools in the pipelines leading from the digester. The condensing water can dissolve VFA and ammonia and other constituents of the biogas; similar dissolution in gas-holder water has already been mentioned. Gas from landfills can contain traces of chlorinated hydrocarbons and other chemicals not usually found in digester gas.

Gas with 50% or more methane will burn in gas burners or engines without further purification, although the presence of H_2S in the gas would preclude its being used (without H_2S removal) in a house in some countries, purely on safety grounds; hydrogen sulphide is extremely toxic. Because of toxicity and the explosive nature of methane (about 5 to 15% in air is potentially explosive) engine- and boiler-houses on digester sites should be well-ventilated.

The heating value of the biogas is proportional to its methane content and dry biogas of 65% methane has a lower heating value of about 600 Btu/ft^3 (6·2 kWh/m^3) at STP. One thousand cubic feet of biogas is thus equivalent to about 600 cubic feetof natural gas, 3·9 UK gal petrol, 5·5 gal propane, and so on. Such gas can be burnt in natural-gas gas rings, boiler burners, etc., with the jets slightly increased in size (about 1·4 times, depending on the methane content of the gas). Use of the gas in burners for direct or indirect (boiler) space heating, cooking and lighting takes place all over the world with digesters of all sizes.

The other use of biogas is in engines for generation of electricity or motive power. The engines are usually run as Combined Heat and Power (CHP) units, the engine block, oil and exhaust being water-cooled and the water used for space-heating, as well as heating the digester if this is required.

Gas turbines have been used in very large installations, but dual-fuel (biogas plus a few percent diesel oil for ignition) compression-ignition engines are usually used in large sewage works, etc. Spark-ignition engines, using biogas alone, are usually used in small installations (about 15–100 kW).

Hydrogen sulphide can cause some problems of wear in engines, and valves are usually stellited and oil changed frequently to reduce this. There are various economic arguments for removal of hydrogen sulphide to get longer engine life compared with more frequent overhaul or short-engine renewal. Most farms and similar relatively small digesters have used unpurified gas, the H_2S content usually being only a fraction of a percent and not really affecting engines or boiler flues. Preventing condensation in boiler flues, where possible, can help the life of boilers. Various modifications are made to engines for biogas running, but there is a number of firms producing engines and complete CHP generating sets as package deals ready to install with the digester.

Many methods used for purifying natural gas, while theoretically possible for digesters, are in practice feasible and economical only on a large scale far above the production rate and pressure of the usual digester gas.

Water-scrubbing is the simplest method of removing acid and ammonia vapours, sulphide, and at least part of the carbon dioxide, from biogas, but even this needs a pressurised water spray in a tank up which gas can pass, and a means of disposal of the dirty water generated.

Hydrogen sulphide can be removed (as it was from the old town-gas) by passing biogas through iron oxide-sponge beds. Such beds usually need to be used on a large scale, but smaller-scale systems have been suggested. Ravishanker & Hills (1984), for instance, describe detailed laboratory experiments on mathematical relationships in design of iron-oxide beds. Membranes (e.g. Kayhanian & Hills, 1988) and solvents (e.g. Hobson & Feilden, 1982) have been suggested or used for removing carbon dioxide and other gases from biogas. Tests have also been made on the biological removal of hydrogen sulphide from biogas. As examples, photosynthetic bacteria such as *Chlorobium thiosulfatophilum* (Cork *et al.*, 1983) have been used, but these require light and hence a large trasparent culture vessel or other system. Chemoautotrophic bacteria do not require light but they do require an oxidising agent. If this was bubbled air or oxygen there could be problems of oxygen getting into the biogas with possible disastrous results. Nitrate obviates this problem but is more costly.

Gadre (1989) described a laboratory-scale apparatus in which biogas from digesting distillery waste ascended a packed column against a downflow of mineral medium which had picked up oxygen in an open reservoir at the bottom of the column before being pumped to the top of the column. The apparatus was shielded from light. Solution of oxygen in the medium obviated problems of oxygen getting into the biogas. The inoculum was mixed autotrophic bacteria developed from soil taken from a lagoon of digested distillery waste. The culture reduced hydrogen sulphide in the biogas from an average of 2% to 0.6%, but it was thought possible to improve on these figures. Biogas from sources such as this contains more sulphide than does biogas from farm and other wastes; whether biological treatments would reduce a few tenths of a percent of sulphide, in an apparatus that could be used (and afforded) with a farm or other small digester is problematical.

A few of the very biggest farm or factory anaerobic digestion plants have been designed to produce gas of pipeline (natural gas) quality by use of chemical and physical methods used in the purification of large volumes of natural gas, and in these cases the objective was to sell gas to the public suppliers. Most digester installations use the gas as it is produced. The lower volumetric efficiency of the gas as fuel and the corrosion caused by impurities are smaller problems mechanically and economically than the purification of the gas.

Only brief mention has been made here of the properties and uses of biogas. Detailed consideration of points noted here and many others concerned with the safety, mechanics and chemistry of biogas use (and storage) is given in the Biogas Utilisation Handbook (1988). Use of biogas from the small, climatically-heated, farm digesters is considered in some of the books on these digesters mentioned in Chapters 4 and 6. A possible small use of biogas in developing countries is as a fumigant to control insects in stored grains and similar foodstuffs (Mohan & Gopalan, 1992).

CONCLUSION

Biogas is a digester output, but digesters also produce liquid and solid effluents and use of these can contribute to pollution control and to the economics of digester use.

The uses of digester effluents are limited by the composition of the material and the siting of the digester. Fertilization is the use to which the largest volumes of effluent may be put, but the cheapest and most efficient use in this way is limited to digesters sited on farms (or on plantations, etc.) which have adequate land and crops for balancing the volume of effluent and the needs of the crops. In such cases, where effluent can be stored if necessary, not only is the best return in crop production likely to be obtained but the problems of pollution from land run-off will be minimised. However, some pig and poultry units have little land, and if land is not available then recourse must be made to transporting the effluent to other areas where fertilizer is needed, or some other use for the effluent must be found. A centralised digester taking slurries from surrounding farms and returning digested slurry to the farms as it is needed has been suggested. The economics of transporting slurry limit the radius of collection and distribution, and the bacteriological problems of distributing slurries have been mentioned previously. Use as animal feed or for algal or fish growth are possibilities, but the amounts of effluent which can be used in this way are limited. It was shown above that perhaps 10% of an animal's feed could be substituted by effluents, so it is obvious that not all the waste from a colony of animals or birds could be recycled. Whether the excess material could be sold to other farms depends on the many considerations already mentioned. The area needed for ponds limits the volume of effluent that can be used in secondary productions as the amounts fed to algal or fish ponds are relatively small. Another factor limiting algal growth as a means of disposal is climate: algal ponds for treatment of sewage or other effluents have generally been recommended for warm and sunny climates where year-round and good growth of the algae can be expected. Disposal in fish ponds is limited by human food preferences: carp and talapia are not much eaten in countries like Britain. In such areas it might be suggested that fish be grown to produce fish meal, which is a good protein feed for animals and birds. Production of the meal would need processing plant and, again, the question of cost arises: can the meal be produced at a price competitive with ordinary commercial fish meal? But the question involves not just the production price of the fish meal. The production of fish (or algae or plants) will act as a further purification step for the digester effluent and this might allow the water to be disposed of to a river. Such purification might be a requirement of the waste treatment and so the production of fish meal, for example, could be regarded as helping to pay for the water treatment and so would be a positive return even if on overall cost the meal production was not commercially viable.

There are, thus, many factors to be taken into consideration when thinking of disposal and uses of digester effluents. Some uses may be ruled out by climate or the scale of the operation or the situation of the digester. In many cases it may be necessary to combine two or more methods either to deal with all the effluent orto get an economic return on the process, and monetary quantification may be difficult. It may be possible to make some general recommendations for a particular area and kind of factory or farm: say that fish ponds be combined with fertilizer use of effluent on small farms in a certain area of a country, but, overall, general recommendations cannot be made. Each situation must be separately considered and conclusions specific for that site drawn up.

REFERENCES

Anthony, W.B. (1966). Proc. Nat. Symp. Anim. Waste Manag. ASAE Pub. no. AP-0366, p. 109.

Atalay, A. & Blancher, R.W. (1984). *J. Environ. Qual.* **13**, 341.

Atalay, A. & Blancher, R.W. (1988). *Biol. Wastes*, **26**, 49.

Balasubramanya, R.H., Khandeparkar, V.G. & Sundaram, V. (1986). *Agric. Wastes*, **16**, 295.

Biogas Utilisation Handbook. (1988). US Dept. of Energy, Southeastern Region Biomass Utilisation Program,Tennessee Valley Authority, Muscle Shoals, Alabama 35660, USA. Pub. Georgia Tech Research Institute, Atlanta, Georgia, USA.

Bridger, G.L., Salutsky, M.L. & Starosta, R.W. (1962). *J. Agric. Food Chem.* **10**, 181.

Brogan, J.C. (1980). Ed., *Nitrogen losses and surface runoff from landspreading of manures.* Martinus Nyhoff, The Hague.

Calzada, J.F. & Rolz, C. (1983). In *Proc. Third Int. Symp. An. Digestion*, Cambridge, Mass, USA. p. 315

Camerero, L., Urbano, C., Merino, J.M., Elias, A., Romero, F. & Diaz, J.M. (1988), In, *Poster Papers* (*5th. Int. Symp. An. Dig.*) ed. A.Tilche & A Rozzi. Monduzzi Editore SpA, Bologna. p. 371.

Chung Po, Wang, W.H., Chen, S.K., Hung, C.M. & Chang, C.I. (1975). In *Managing Livestock Wastes, Proc. 3rd Int. Symp. Piggery Wastes; Am. Soc. Agric. Eng.* p. 238.

Cork, D.J., Garunas, R & Sajjad, A. (1983). *Appl. Environ. Microbiol.* **45**, 913.

Coutare, P., Cluis, D. & Bastien, C. (1988). *Water Air & Soil Pollut.* **42**, 353.

Dam Kofoed, A., Williams, J.H. & L'Hermite, P. (1986). *Efficient land use of sludge and manure.* Elsevier Applied Science Publishers, London.

Davis, R.D., Hucker, G. & L'Hermite, P. (1983). Ed. *Environmental Effects of organic and inorganic contaminants in sewage sludge.* D. Reidel Publishing Co., Dordrecht.

De la Noüe, J. & Bassères, A. (1989). *Biol. Wastes*, **29**, 17.

Demuynck, M. (1984). *Utilisation in agriculture of anaerobically digested effluents.* Commission of the European Communities, Brussels.

Flachowsky, G. & Hennig, A. (1990). *Biol. Wastes*, **31**, 17.

Gadre, R.V. (1989). *Biotechnol. Bioeng.* **34**, 410.

Gasi, T.M.T., Pacheco, C.E.M., Amaral, L.A. & Vieira, S.M.M. (1988). In *Poster Papers, Fifth Int. Symp. An. Digestion*, ed. A. Tilche & A. Rozzi, Monduzzi Editore SpA, Bologna, p. 875.

Gracey, H.I., Stewart, T.A. Woodside, J.D. & Thompson, R.H. (1976). *J. Agric. Sci. Camb.* 87, 617.

Graetz, D.A., Krottje, P.A. & Norstadt, R.A. (1986). In *Conference on research applications of aquatic plants for water treatment and resource recovery. Abstract* APC-65, p. 126.

Hobson, P.N. (1969). *Process Biochem.* **4**, 53.

Hobson, P. N., Bousfield, S. & Summers, R. (1981). *Methane production from agricultural and domestic wastes.* Applied Science Publishers, London.

Hobson, P.N. & Feilden, N.E.H. (1982). *Prog. Energy Combust. Sci.* **8**, 135.

Hobson, P.N. & Shaw, B.G. (1973). *Water Res.*, **7**, 437.

Hucker, T.W.G. & Catroux, G. (1981). Ed. *Phosphorus in sewage sludge and animal waste slurries.* D. Reidel Publishing Co., Dordrecht.

Kaur, K., Sehgal, G.K. & Sehgal, S.H. (1987). *Biol. Wastes*, **22**, 139.

Kayhanian, M. & Hills, D.J. (1988). *Biol. Wastes*, **23**, 1.

Levanon, D., Danai, O. & Masaphy, S. (1988). *Biol. Wastes*, **26**, 341.

L'Hermite, P. & Dehandschutter, J. (1981). Ed. *Copper in animal wastes and sewage sludge.* D. Reidel Publishing Co., Dordrecht.

L'Hermite, P. & Ott, H. (1984). Ed. *Processing and use of sewage sludge.* D. Reidel Publishing Co., Dordrecht.

Liberti, L., Limoni, N., Lopez, A., Passino, R. & Bari, G. (1986). *Water Res.* **20**, 735.
Lim Kim Huan (1988). In *Anaerobic digestion 1988*, ed. E.R. Hall & P.N. Hobson, Pergamon Press, Oxford, p. 303.
Mahadevaswamy, M. & Venkataraman, L.V. (1986). *Ag. Wastes*, **18**, 93.
Mahadevaswamy, M. & Venkataraman, L.V. (1988). *Biol. Wastes*, **25**, 249.
Maramba, F.D., Judan, A. & Obias, E. (1983). In *Third Int. Symp. on Anaerobic Digestion, Proceedings*, Third Int. Symp. on Anaerobic Digestion, 99 Erie Street, Cambridge, USA.
Marchaim, U. *et al.* (1982). In *Anaerobic Digestion* 1981, Elsevier, Amsterdam, p. 203.
Miller, R.H. (1974). *J. Environ. Qual.* **3**, 376.
Mohan, S. & Gopalan, M. (1992). *Biores. Technol.* **39**, 229.
Moorhead, K.K. (1986). Ph.D. Thesis, University of Florida, Gainsville, Florida, USA.
Moorhead, K.K., Graetz, D.A. & Reddy, K.R. (1987). *J. Environ. Qual.* **16**(1), 25.
Natarajan, M. & Varghese, T.J. (1980). *Ag. Wastes*, **2**, 261.
Neuhauser, E.P., Kaplan, D.L., Malecki, N.R. & Hartenstein, R. (1980). *Ag. Wastes*, **2**, 43.
Pain, B.F., Misselbrook, T.H., Clarkson, C.R. & Rees, Y.J. (1990). *Biol. Wastes* **34**, 259.
Phong, S.-M. & Ong, K.-C. (1988). *Biol. Wastes* **25**, 177.
Prairie, D.M. & Stickney, A.R. (1988). In *Anaerobic Digestion 1988*, ed. E.R. Hall & P.N. Hobson, Pergamon Press, Oxford.
Prior, R.L., Hashimoto, A.G., Crouse, J.D. & Dickeman, M.E. (1986). *Agric. Wastes*, **17**, 23.
Puhakka, J.A., Alavakeri, M. & Shieh, W.K. (1992*a*). *Biores. Technol.* **39**, 61.
Puhakka, J.A., Alavakeri, M & Shieh, W.K. (1992*b*). *Biores. Technol.* **39**, 69.
Ravindram, V., Rajaguru, A.S.B., Rajapakse, A.R.K. & Dassanayake, R.M.P.H. (1987). *Biol. Wastes*, **21**, 257.
Ravishanker, P. & Hills, D. (1984). *Agric. Wastes*, **11**, 167.
Rothbaum, H. & Rohde, A. (1976). *N. Z. J. Expt. Agric.* **2**, 405.
Saxena, K.K., Nath, K. & Srvastava, S.K. (1989). *Biol. Wastes*, **28**, 73.
Schuchardt, F. In *Anaerobic Digestion 1988*, ed. E.R. Hall & P.N. Hobson, Pergamon Press, Oxford, p. 449.
Stewart, D.J. (1980). In *Anaerobic Digestion*, ed. D.A. Stafford, B. Wheatley & D.E. Hughes, Applied Science Publishers, London, p. 303.
Strauch, D. (1984). In, *Processing and use of sewage sludge*, ed. P. L'Hermite & H. Ott, D. Redeil Publishing Co., Dordrecht, p. 220.
Stuckey, D.J. & Newman, T.S. (1977). *J. Environ. Qual.* **6**, 271.
Summers, R. & Bousfield, S. (1980). *Ag. Wastes*, **2**, 61.
Summers, R., Bousfield, S. & Hobson, P.N. (1980). In *Anaerobic Digestion*, ed., B.A. Satfford, B. Wheatley & D.E. Hughes, Applied Science Publishers, London. p. 409.
Summers, R, Hobson, P.N., Harries, C.R. & Richardson, A.J. (1987). *Biol. Wastes*, **20**, 43.
Sweeney, D.W., Graetz, D.A. & Sastain, J.B. (1985). *Agronomy Abstracts*, **37**.
Tarre, S., Raviv, M. & Shelef, G. (1987). *Biol. Wastes* **19**, 299.
Unwin, R.J. (1981). In, *Copper in animal wastes and sewage sludge*, ed. P. L'Hermite & J. Dehandschutter, D. Reidel Publishing Co., Dordrecht, p. 102.
Ward, R.F. (1983). In, *Fuel gas developments*. Ed. D.L. Wise, CRC Press, Roca Baton, Florida, p. 161.
Watson, J., O'Shea, J., Spillane, T.A. & Conolly, J.F. (1979). *Conserv. and Recycling*, **2**, 269.
Welte, E. & Szabolcs, I. (1988). Ed. *Conference on agricultural waste management and environmental protection*, Proceedings, volumes 1 and 2. C.I.E.C., Vienna.
Williams, J.H., Guidi, G. & L'Hermite, P. (1985). Ed. *Long-term effects of sewage sludge and farm slurry applications*, Elsevier Applied Science Publishers, London.
Wong, M.H. & Cheung, Y.B. (1989). *Biol. Wastes*, **28**, 143.
Wrigley, T.J., Webb, K.M. & Venkitachalm, H. (1991). *Bioresource Technol.* **41**, 117.
Yeoh, B.G. (1988). In *Anaerobic digestion 1988*, ed. E.R. Hall & P.N. Hobson, Pergamon Press, Oxford, p. 315.

Chapter 9

RUNNING AND CONTROL OF DIGESTERS

INTRODUCTION

It has been mentioned in previous chapters that digestion, particularly sludge or slurry digestion, is essentially a self-controlling process once the digestion has been started and brought to a steady-state. In addition, again particularly with sludges, the nature of the feedstock and digester contents precludes use of much of the continuous-monitoring equipment of industrial fermentations. For instance electrodes for pH or other measurements rapidly foul-up in the presence of sulphides, proteins, etc. Many digesters, those on farms for instance, are on sites where manpower is not available for continuous supervision of digesters and analytical facilities are virtually non-existent. Thus, the majority of digesters in use in the world today are run with the minimum of control and supervision. However, there are digesters on sites where analytical sevices and manpower are available or where they can be provided by some centralised system and control can be more extensive. There are, thus, one or two controls or measurements which must be made in running a digester; there are very many more which are possible. Whether these latter are done depends on the use to which digester liquid and gaseous products are put, the manpower and facilities, and the money originally available to build the digester.

In the present chapter some simple, or more complex, methods of control and supervision of digesters are discussed which can be applied to digesters in different situations.

Safety

The main constituents of biogas, methane and carbon dioxide, are not poisonous but are suffocating if they replace much of the air in a confined space. However, some minor constituents of the gas, particularly hydrogen sulphide, are extremely poisonous at concentrations of a few ppm. Engine and pump houses should, therefore, be well ventilated. If a digester, or even a sump in which are pipes or valves containing gas or digester effluents which might leak, has to be entered for repairs then the digester must be emptied and the empty tank (or sump) well opened and preferably blown out with compressed air for up to some days until not only has the gas in the air-space of the vessel been removed but any sludges left in the digester tank have ceased to produce gas. If available, residual gases can be tested for with hand-held colour-change gas detectors or more sophisticated apparatus, but the instruction manuals for the Chinese and Indian digesters mentioned in other chapters contain suggestions for using a rat or a bird as a gas detector before entering a well-ventilated digester for repair purposes. These precautions also apply to tanks and pipes containing some digester feeds: there have been many instances of workers on intensive farms entering or emptying tanks and being overcome by gases released from stored animal effluents which have begun to digest. Even with these precautions, it is best if anyone entering an empty digester or feed tank is provided with an air respirator: it is obvious that a gas mask can protect against low concentrations of toxic gases in air but cannot protect against suffocation in the absence of oxygen. Instructions for a Chinese digester given in one of the books mentioned before suggest that anyone entering an empty and ventilated digester should breath through a hose led to the atmosphere.

An additional hazard is explosion and because of the previously mentioned low explosive concentration of methane (about 5–15% in air) it is possible for explosive concentrations to build up from very small leaks in pipes or tanks. The ventilation previously mentioned, with gas pipe runs in the open as far as possible, together with 'No smoking or naked lights' near digesters and in buildings, can help to reduce risks. However, it is obvious that boiler- or engine-houses must contain sources of ignition of gases. The effects of these can be reduced by fitting flame traps in gas lines to boilers and engines and having gas cut-outs if pilot-lights go out or the apparatus ceases to function in other ways. Spark-proof switch gear can be used. Such safety equipment is generally built into commercial boilers and engine set-ups. Gas can leak from condensate traps or the simple liquid pressure-relief systems on digesters or the liquid seals on gas-holders mentioned in previous chapters. Having these in the open reduces risks. Blow-off through seals may be the only way of getting rid of excess gas, but the provision of automatic flares on big installations has been previously noted. There may be local regulations about the distance from buildings of digesters and ancillary equipment, but these distances should be as great as possible commensurate with efficient movement of feedstocks and effluents and delivery of power as hot-water or electricity or gas from digester to buildings.

Physical safety aspects should be as on any factory site with guard-rails, and back rails for permanent ladders and walkways on tanks. On farms and factories fork-lifts, tractors and other potentially dangerous vehicles should be precluded from the site as far as possible. Pipe runs should be suitably painted in standard colours for gas, hot water, etc. Safety features and the site in general should be checked at intervals.

Physical control systems

Most digesters are designed to treat a particular flow of waste or other feedstock and are sized so that the daily volume of feedstock can be dealt with continuously. The extent of digestion is assymptotic with Retention Time (Fig. 6) and the design RT is usually such that a slight deviation from this RT, by inadvertant feeding at more or less than the design rate, should make little difference to the composition of the effluent. Nevertheless, it will affect gas production and variations in feed volumes should be avoided if possible. For the peasant-farm digester feed rate may be left to the run-off from the animal shed or may be set at so many buckets or other volumes a day, but the bigger digesters will be pump fed; the pump being controlled by a timer. The timer may, as previously described, be over-ruled by the effluent pump or some device sensing blockages or overfilling of the digester. A check on the feed volumes can be made by installing a magnetic flow-meter on the feed pipe, or the level in a calibrated feed tank may be noted at intervals. It may be necessary to link a feed-tank mixer to the feed pump in such a way that the mixer goes on some minutes (a time determined by experiment) before the feed pump to ensure that a uniform feed is delivered. Although computers may give the most accurate control, because digesters are often wet and messy places simple mechanical clock timers, etc., may be more robust and give sufficient accuracy. The feed-tank itself may be filled by manual observation or it may be filled automatically from some other tank (say a piggery waste-channels) by a pump under the control of level switches in the feed tank. In this case the rates of production of feedstock are small compared with the capacity of the various tanks and channels, but if the feed flow is large compared with the volume of feed- and other-tanks then some form of warning and/or the ability to stop production of the feedstock may have to be built into the system to prevent overflows. The recycle flows on contact or some other retained-biomass digesters may be intermittent and under timer control.

While the peasant-farm stirred-tank digesters are hand mixed, if mixed at all, bigger digesters will be mechanically- or gas-mixed and timer controls will be needed for this. If the digester is water-heated then a temperature-sensing device or devices will have to be installed at a suitable place in the tank to control the flow of hot water to the heat-exchangers in the digester. The hot water will come from a boiler or engine and the digester flow may be diverted from the main flow of water to building heating. If hot water is provided only for digester heating it may be necessary to arrange for the digester thermostat to switch on the boiler. Digester thermostats should, obviously, be designed to reduce hunting. A simple, or in some cases a recording, thermometer should be installed to make continuous or spot checks on the digester temperature. Temperature-sensing is generally used to control the temperature of the digester contents within a degree or so of the design temperature and the temperature control is not linked to any other system. Alatiqi *et al.* (1990) have investigated a multi-variable control system that controls the digester temperature (in mesophilic and thermophilic ranges) and the organic carbon content of the digesting sludge for optimising sewage-sludge digestion in terms of steady biogas production and sludge degradation.

A main, or in many cases the only, parameter used to check digester function is gas production, so a gas meter is needed. On the Chinese, fixed-volume digesters a simple water manometer indicates whether or not sufficient gas is present to run lights, etc. Bigger digesters will have cumulative meters which can be read by an attendant or which may give some form of printout. Ordinary household dry meters or wet meters can be used, but more elaborate calorimetry and vortex-eddy meters are used on many sewage digesters. Total volume over a long period rather than immediate rate of gas production is required, as the latter can vary considerably as mixers go on and off or feed is introduced at intervals.

The operation of ancillary equipment such as a separator for the effluent from a farm digester will be made as automatic as possible and this may mean linking, in this instance, the operation of the separator to that of the digester effluent pump.

Boilers and engine-generators will probably be supplied by the manufacturers as packages which include cut-outs for low gas supply, overheating, generator overloading, etc. Where the engine or the boiler supplies heat to the digester and buildings the boiler may be cut out while the engine is running and automatically cut in when the engine is out of use. At a digester plant run on a piggery by one of the authors and colleagues gas production was not sufficient to run the engine continuously, so gas was allowed to accumulate in the holder for some hours, say overnight, and the engine then switched on manually and the generator switched to take over the piggery electricity supply. The engine then ran until it was stopped automatically at a low gas volume and the boiler then took over heat supply. The boiler in this case was linked to an oil-fired boiler by simple difference in temperature settings so that if the digester boiler or engine did not supply hot water to the system the oil boiler took over. The engine carburettor controls compensated for varying electricity load on the generator. In more sophisticated systems the generator may be running at constant load with the system either feeding electricity into the mains or taking from the mains depending on the power demands. Farms are one site where power demands can vary considerably throughout the day as washers, feed mixers, ventilating fans and so on cut in and out. One of the preliminaries to digester installation can be replanning the use of electricity so that power demands are levelled out as much as possible. The engine will be most efficient running at near maximum and constant load.

This is a brief summary of the type of physical controls needed on digesters, some other design features have been mentioned in previous chapters. The basic safety factors are common to all digesters, what is needed beyond this depends on the type and location of the digester, the manpower and other facilities available to run it and the money to be spent on the system. Of course, if research into design and running of the digester is one, or the main, objective of the system then control and recording of running parameters will be much more extensive than in other cases. The design of the digester may also take into account the necessity of obtaining samples of feed, effluent and digester contents at many points or of multiple measurements of temperature and so on.

Analytical control systems

Analytical controls may be used just to ensure that the digester feed is not accidentally varied in composition thus keeping biogas production uniform: or they may be to ensure that feedstock streams are being properly mixed to provide a uniform feed or perhaps to dilute out some toxic compound; or to ensure that a proper pH adjustment of, or nutrient addition to, the feed is being made. Analytical controls may also be used to monitor correct function of the digestion and to predict the onset of digestion failure. If a digester effluent is being discharged to sewers or a river, then analyses of the discharge may be needed to ensure its correct quality or as a basis for charging for further sewage-works treatment. If secondary treatment of the digester effluent is done on-site, then again, analysis may be needed to ensure that the effluent remains in composition within the capacity of the secondary system. As mentioned in other chapters, chemical and bacteriological controls may be needed if digester effluents are to be used off-site or used for animal or fish feeds, etc., or to be used optimally as fertilizer.

The facilities for digestion analyses vary considerably from site to site and probably for many sites are non-existent. On the other hand, where a digester is being used for research purposes or is on a factory or large-farm site then full laboratory services may be available. Local authority river boards or sewage works may provide analytical services either at a charge or as part of their surveys of waste discharges. The contract for installing a digester may include some short- or long-time provision for digester monitoring by the company making the digester. In this section some analyses which may be made on digester feeds and effluents will be discussed.

The desirability of a constant feed volume has been mentioned and the digester will have been designed to take the average volume of feed provided by the waste-producing farm or factory plus or minus some slight variation. Where larger variations in feed volume may occur these can be smoothed by balancing-tanks as, or in addition to, the digester feed-tank. Control of feed volume should be relatively easy; control of feed consistency may be more difficult. For the small, simple, farm digester consistency of feed may be based only on the design instructions that a certain volume of, say, cattle excreta should be mixed with a certain volume of water or another waste to make the digester input. It has been previously mentioned that provision of feedstock of uniform solids content has been a constant problem with the larger farm digesters. It may be that nothing can be done about the problem and the digester has to be run as if the feed conformed to design criteria. But where there is a possibility of correcting the feed solids content, or even as a record of feed variation for alteration of the farm-waste system, then some analysis of feedstocks may be made. Correcting the solids content might be brought about, on a pig farm for instance, by better mixing of low-solids sow-house effluent with fattening-pig excreta of high solids-content, or by better mixing of some slurry tanks or channels. Proper determination of TS requires heating and weighing crucibles, but one of the authors devised a simple procedure involving a hand centrifuge and graduated tubes that could give a reasonably accurate, on-farm, determination of piggery-waste TS in a few minutes. Concentrations of dissolved feed constituents may also vary, particularly with some factory wastes. Such variation in feed composition will affect gas production and may affect the composition of the effluent: if a wastewater digester is running near the maximum loading rate with a short HRT, an increase in feed COD could be more likely to lead to digester failure than in the case of a sludge digester. However, if factory wastes are of fairly uniform composition then adherence to design criteria may suffice for digester running; but more analytical facilities may be available than on the farm and it may be easier to adjust feed composition by mixing waste streams. However, in the case of some large intensive farms analytical facilities for animal feed preparation may be available and these can be used for proper determination of feedstock and effluent constituents. All the parameters mentioned in the Appendix (TS, VS, etc.) are determined by standard methods given in texts used by the river boards and water authorities in different countries. Commonly used methods are those of the American Public Health Association (Standard Methods, 1976 and updates). The dissolved or suspended solids of the feed and effluent may be analysed for constituents in more detail and the content of sugars, starch, cellulose, protein, etc., determined. Such analyses are very often done by methods used in animal-feed analysis; suggestions for methods are given in the paper by Colin *et al.* (1983), for instance. Determinations of metals, antibiotics, detergents, etc. will all require special laboratory methods. The gas production may be just a record of volume per day or the gas may be analysed by simple or more accurate absorption methods and the colour-change methods mentioned previously or by the more sophisticated electronic hand-held meters or by gas-chromatography: this latter, and some of the other methods, needs a laboratory. The methane content of the biogas should remain fairly constant if digestion is proceeding properly. The low content of hydrogen in digester gas can be measured only with special apparatus. Hydrogen analysis has been suggested as an check on digester running, increasing hydrogen concentration predicting possible digestion failure (Mosey & Fernandes, 1984). However, difficulties have been found in on-site measurements of hydrogen and research into the application of the method is continuing (Switzenbaum, Giraldo-Gomez & Hickey, 1990). The measurement of the trace amounts of hydrogen or, possibly, carbon monoxide (Hickey & Switzenbaum, 1990) might eventually be used as the basis for a feedback control system for some digesters. Hydrogen measurements have been used to control laboratory digesters (Whitmore *et al.*, 1987; Collins & Paskins, 1987) but the monitors were subject to interference and more development is needed before they can be used routinely. An aid to reliable control by hydrogen concentration may be the miniature fuel-cell for rapid measurement of dissolved hydrogen recently developed by Pauss *et al.* (1990). Dochain *et al.* (1991) have begun the testing of a control scheme based on the hydrogen concentration, but this will need long-term testing on full-scale digesters before its application can be fully assessed.

The pH of digester contents can be measured on samples taken from the digester and equilibrated with carbon dioxide in the digester gas. Measurement of pH, and other analyses, should be done at the same time(s) in relation to digester feeding if this latter is at long intervals, say once a day. The pH should then remain fairly constant with time and any deviation could be an indication of digestion failure or coming failure. However, pH is not a sensitive control parameter, in many cases buffer actions make pH changes very slow and not allied to changes in acid concentrations. Ammonia is measured with laboratory apparatus, for instance by distillation. Ammonia electrodes have been investigated (Le Duy & Samson, 1982) but are not in general use. Various methods of keeping analytical electrodes clean have been tried, including air or water jets, ultrasonics and brushing. These are most likely to be usable with wastewater feedstocks and brushing was found to be most effective in polluted waters (Colville, 1986).

The concentration of volatile fatty acids should remain at a constant value depending on the retention time (see Chapter 3) and the measurements are, as above, made at the same time each day. The principal acid should be acetic. An increase in total VFA could precede digester failure, and the possible connections of branched-chain acids and increased propionic acid as indicators of future failure have been mentioned earlier. Total VFA can be measured by distillation and titration methods, but analysis of the individual acids needs gas chromatography: both these analyses need laboratory equipment. Hickey & Switzenbaum (1990) found that CO was produced in conversion of acetate to methane and suggested that CO in the digester gas could be used for monitoring acid in the digester.

'Alkalinity' is a measure of the buffer capacity of the digesting liquid against increased acidity and so sufficient alkalinity is related to digestion stability. The principal ion involved is bicarbonate as ammonium or metal salts, and alkalinity is generally measured in calcium carbonate equivalents. Carbon dioxide is a product of feed degradation and so alkalinity is generated by the microbial reactions. The concentration of bicarbonate in solution is related to the carbon dioxide content of the gas above the liquid. Changes in the bicarbonate will then be reflected in changes in the gas composition (Rozzi, Di Pinto & Brunetti, 1985). Automatic titrators have been used to continuously measure alkalinity (Wheatley *et al.*, 1987; Powell & Archer, 1989) but these depend on pH electrodes which rapidly foul-up. However, alkalinity is generally measured by titration of samples of the filtered digester contents with standard acid to pH 4. This gives the Total Alkalinity and values of 1000 mg/l as $CaCO_3$ or higher are needed to keep digester pH about 7 or rather higher. Substances other than bicarbonate, such as ammonia and the organic acids, can act as buffers. Titration to pH 5·75 gives the alkalinity caused by bicarbonate (called 'Partial Alkalinity' or 'PA'), while further titration to pH 4·3 gives the VFA alkalinity, called 'Intermediate alkalinity' or 'IA' (Ripley *et al.*, 1985). The ratio of IA to PA should give an indication of the acid content and buffering capacity of the digestion and a value of IA/PA between 0·1 and 0·3 has been suggested as indicative of stable digestion (Environmental Protection Agency, 1979). Again, laboratory equipment is needed to make these measurements.

Problems of measuring the bacteria (Biomass) concentration in a digester have been noted in Chapter 2, and any attempt at measuring bacterial numbers by culture or specific analyses (e.g. DNA, protein, lipids, ATP, microcalorimetry;

Hickey *et al.*, 1991) requires a specialised laboratory. A simpler method, but which still requires a laboratory, is to measure the activity of the digester contents in production of biogas under standard conditions. This can give a value which can be equated with the relative activities of the bacteria at different moments in time. A standard amount of the digester contents is incubated in a flask or serum bottle (see Chapter 4) with and without excess acetate and the gas production is measured over a standard time. A suitable standard methodology, although in this case meant for the assessment of the biodegradability of compounds in digesting sludge, is described in 'Methods for the examination of waters and associated materials' (1988). Other methods are given in papers quoted in Chapter 4 and elsewhere.

CONCLUSION

It will be evident from the above paragraphs that there is a large number of analyses that could be done on digesters, but that in practice the analyses are limited. One factor that should be noted in any consideration of analyses of digester contents has been mentioned in Chapter 3, and that is that digesters generally do not attain the stability of the laboratory or factory continuous-culture. The values of analyses will fluctuate from day to day. Experience with a particular feedstock and digester type will suggest suitable limits of parameters for stable running, but, particularly with sludge digesters of long RT, anything but a catastrophic change in a value might be ignored unless it is part of a trend over some days to higher or lower than normal values. This may then indicate coming failure or a change to unoptimised running. The cause of, say, the decreased gas production may then be sought in a change in feedstock or some mechanical fault leading to changed loading and so on. Consideration of the circumstances may suggest causes of the malfunction, it could be, for instance, that some new additive has been made to the feed of farm animals, or that there has been an accidental discharge of disinfectant to the digester feed. It is also possible that a leak has developed in some pipe or tank, an instance of this is described in a previous chapter. The immediate action to a definite change in gas production would be to stop loading. If some toxic material has got into the feed this will prevent further build-up in the digester and renewing the feed in the feed-tank and restarting the digester at a low feed rate will allow the toxic substance to be washed out of the digester and the bacterial population to build up again as the loading is gradually increased. Stopping loading will also cut down gas production until a mechanical fault can be rectified.

With a farm digester it is likely that gas production is the only thing that can be measured on site. Gas composition can be judged only by the functioning of boilers or engines: the leak previously mentioned caused a gas low in methane and the failure of a boiler to ignite was a first symptom of the problem. A simple test of absorption of the gas in a graduated container of alkali to determine the residual gas (assumed to be methane) might be aranged for on-site use. If no obvious solution to the problem can be found then it will be necessary to call in outside advice and possibly further analyses. This may be provided by the makers of the digester.

If gas production is all right the source of malfunctions of engines or boilers should be determined by the functioning of the standard engine or boiler controls and warning lights for overheating, failure of ignition systems, etc.

In many years of running of farm-waste digesters there were no problems attributable to failure of the bacteria; problems were all mechanical and associated with pump and pipe blockages or breakdowns, and boiler and engine faults. These problems cannot be predicted by analyses and cannot be rectified by computer or other control and data-logging. The problems and their solutions are purely observational and manual. However, although the repair of a broken pipe is manual, the immediate effects of the breakdowns may be mitigated by control systems, in that, for instance, failure of one pump can cut out another (as previously indicated) and so prevent catastrophe. If economics allow the building in of spares then relatively simple control systems may also switch in another pump, boiler or engine in place of a failed one, or change to mains electricity, and so on.

Many, or most, modern industrial and laboratory fermentors are computer-controlled. This type of control can measure continuously various parameters of the fermentation and can adjust pH by acid or alkali additions or dissolved oxygen by changing stirring and air-sparging rates from minute to minute. Although, perhaps, a feed pretreatment such as adjustment of pH of a factory wastewater might be computer controlled, simple pH meter-control would be just as good. Digesters themselves cannot be controlled in such ways and, as said before, in any case the sites of many digesters are unsuitable for delicate instruments such as computers. Computer-logging of data aspects such as feed rates and gas production has been used on some experimental digesters, but this is generally to provide a source of figures which can later be used to calculate average gas productions per day or per unit of feed, and so on, not for digester control. The limitations to models of digestion were noted in Chapter 3, but developments in control theory of complex biological processes are taking place and it seems possible that future developments such as self-learning neural networks and expert systems will eventually provide a robust control model. A detailed theoretical approach to these models with a description of practical applications to anaerobic digestion has recently been given by Bastin & Dochain (1990). However, it seems likely that from the point of view of expense and the other problems associated with sludge digestions that such control systems will be applicable only to wastewater digesters on well-equipped factory sites.

A time when detailed analyses of digester function may be necessary is at start-up, although here again startup of large farm digesters has been done without measurements other than gas production and flammability and the peasant-farm digesters have to be started with only such observations. The general method of starting and provision of suitable inocula has been described previously. The start-up regime of gradually increased loading may be based on experience with similar digestions or may be controlled by observations of gas production alone or with VFA and other analyses and suitable settings of feed pumps and feed composition (if this latter is possible). While laboratory digestions have shown a good effect of pH control in making start-up of digestions quicker and more reliable (e.g. Pandey *et al.*, 1990), in general, additions of alkali to control digester pH are not possible with full-scale plants and reversal of a trend to acidity is controlled by decreasing feed rate until a natural correction takes place. As start-up may be part of the contract of the makers of the digester, it should be possible to do analyses that may not be feasible once the digester is commissioned. Help for the farmer in building and starting up the small digesters in China and India is part of the plans for increasing digester usage.

Commissioning of sludge digesters usually presents few problems; not so many large-scale solid-state digestions have been started and problems with them may still have to appear. Starting of retained-biomass digesters is probably the most difficult. Mention has been made of the desirability of an inoculum of granulated sludge for UASB digesters. The granules are able to maintain their structure and adapt to new wastes (Hickey *et al.*, 1991). Microscopic, in addition to chemical, analyses of the sludge may be needed as well as determinations of sludge concentration and activity. If granulated sludge is not available then start-up may take longer and require stricter control of the digestion. Addition of inoculum in small quantities rather than in one dose, and recycling of effluent, may help attachment of bacteria and start-up of filter digesters where the biomass is partly attached and partly trapped in the filter media. The biomass has to be completely attached to the media granules in fluidised- or expanded-bed digesters and a number of additions of inoculum, combined with control of bed expansion, for instance by recycle rate, helps start-up. Microscopic as well as chemical analyses will be needed.

The start-up of two-phase digesters follows similar principles, but here the first phase may have to be run alone until a suitable hydrolytic and fermentative biomass is built up and the effluent contains sufficient acids to feed the methanogenic phase. As previously mentioned, with the solid-state first phase the effluent may have to be recirculated for some time before being passed to the methaniser: in other cases the first tank may be run as a batch system until the hydrolytic bacteria have developed. Depending on the substrates used an inoculum may or may not be required for one or both phases; a mixture of straw and cattle excreta in the first stage might provide a sufficient inoculum for both phases, for instance.

It will be evident from this brief discussion that the term 'anaerobic digester' covers a vast range of process types and substrates, and plant situations. Much of the technique of start-up and running of digesters depends on experience and in the case of sewage digesters experience in plant running and facilities available for plant control are extensive. For other types of digester, experience, and facilities, may be much less. However, as more large-scale digesters are built the records will grow and the more successful digestion will become.

REFERENCES

Alatiqi, I.M., Dadkhah, A.A., Jabr, N.M. (1990). *Chem. Eng. J.* **43**, B81.

Bastin, G. & Dochain, D. (1990). *On-line Estimation and Adaptive Control of Bioreactors*, Elsevier, Amsterdam.

Colin, F., Ferrero, G.L., Gerletti, M., Hobson, P.N., L'Hermite, P., Naveau, H.P. & Nyns, E.-J. (1983). *Agric. Wastes*, **7**, 183.

Collins, L.J. & Paskins, A.R. (1987). *Wat. Research* **21**, 1567.

Colville, R.F. (1985/86). *Measurement and Control* **18**, 396.

Dochain, D., Perrier, M. & Pauss, A. (1991). *Ind. Eng. Chem. Res.* **30**, 129.

Environmental Protection Agency (1979). *Process design manual for sludge treatment and disposal*, EPA Centre for Environmental Research Information Technology Transfer. EPA 625/1-79-011, Washington, USA.

Hickey, R.F. & Switzenbaum, M.S. (1990). *Environ. Sci. Technol.* **24**, 1642.

Hickey, R.F., Wu, W.M., Veiga, M.C. & Jones, (1991). *Wat. Sci. Technol.* **24**, 207.

Le Duy, A. & Samson, R. (1982). *Biotechnol. Letters* **4**, 303.

Methods for the Examination of Waters and Associated Materials. (1988). *The Assessment of Biodegradability in Anaerobic Digesting Sludge 1988.* H.M.S.O. Books, London.

Mosey, F.E. & Fernandes, X.A. (1984). In *Microbiological methods for environmental biotechnology*. Ed. J.M. Grainger & J.M. Lynch, Academic Press, London. p.159.

Pandey, A., Radhika, L.G. & Ramakrishna, S.V. (1990). *Biol. Wastes*, **33**, 143.

Pauss, A., Samson, R., Guiot, S. & Beauchemin, C. (1990). *Biotechnol. Bioeng.* **35**, 492.

Powell, G.E. & Archer, D.B. (1989). *Biotech. Bioeng.* **33**, 570.

Ripley, L.E., Boyle, W.C. & Converse, J.C. (1985). Proc. 40th. *Industrial Waste Conference*, Purdue, Indiana. Butterworth, Boston, USA. p. 141.

Rozzi, A., di Pinto, A.C. & Brunetti, A. (1985). *Env. Technol. Letters* **6**, 594.

Standard methods for the examination of Water and Wastewater. (1976 *et seq.*). American Public Health Association (APHA), Washington, USA.

Switzenbaum, M.S., Giraldo-Gomez, E. & Hickey, R.F. (1990). *Enzyme Microbiol. Technol.* **12**, 722.

Wheatley, A.W., Winstanley, C.I. & Johnson, K.A. (1987). In *Anaerobic Digestion: Results of Research and Demonstration Projects*, ed. M.P. Ferranti, G.L. Ferrero & P. L'Hermite, Elsevier Applied Science, London. p. 250.

Whitmore, T.N., Lloyd, D., Jones, G. & Williams, T.N. (1987). *Appl. Microbiol. Biotechnol.* **26**, 383.

Appendix

GLOSSARY OF TERMS

Types of digester

CSTR Continuous-flow stirred-tank reactor: i.e. digester.
UASB Upflow anaerobic sludge-blanket (digester).
EGSB Expanded granular sludge-bed (digester).
AnRBC Anaerobic rotating biological contactor (digester).

Terms used in digester running

RT, *Retention Time.* The mean time for which any portion of the feed will remain in a continuous-flow digester (digester volume divided by feed flow-rate). For a stirred-tank digester with perfect mixing the RT of the liquid, bacterial and feed suspended-solids portions of the feed and digester contents will be identical. Usual units hours or days. (Θ– is sometimes used to denote Retention Time).

D, *Dilution Rate.* The reciprocal of Retention Time, generally used for laboratory continuous-cultures or in modelling. Units reciprocal hours or days.

SRT, *Solids Retention Time.* The mean time for which any portion of the feed solids and/or the digester bacteria remain in a continuous-flow digester. This can be equal to the RT (above), but in a retained-biomass digester for dissolved substrates solid (i.e. bacterial) RT will be longer than liquid RT. For some retained-biomass digesters, or for some other digesters described for digestion of high-solids feedstocks, RT of the feed solids can also be longer than that of the liquid.

HRT, *Hydraulic Retention Time.* The RT of the feed liquid in the digester. For the perfect stirred-tank, RT, SRT and HRT are equal. For some solid-state digesters there is effectively no HRT as the little water is not separable from the solids. Recirculation of effluent liquid can alter HRT based solely on feed and digester volumes.

Loading Rate. Rate at which potential substrate is added to a digester. Units, amount of substrate (kg) per unit volume of digester (m^3) per unit time (usually day). Occasionally the rate is expressed as amount of substrate per unit weight of bacterial mass (MLSS, see below) per unit time. The amount of substrate may be expressed in a number of ways; e.g. COD, BOD, TS, VS. 'Potential substrate' is used above as not all the COD, etc. added may be degradable.

Gas Production Rate. This can be expressed in units of litres or cubic metres per kg TS, VS, COD, etc. added to the digester or digested, or units per unit digester volume per day.

Terms used in descriptions of wastes and their treatment

BOD, *Biochemical Oxygen Demand.* The oxygen uptake by aerobic bacteria growing in a liquid waste determined under standard conditions. Usually a five-day incubation (BOD or BOD_5) but sometimes a three- (BOD_3) or seven- (BOD_7) day incubation. A measure of the material, mostly in solution, liable to cause immediate pollution and microbial growth in a water-course or on land. Usual units mg/litre (ppm [Parts Per Million] may be used as mg/l). For convenience, in loading rates particularly, large amounts of BOD or COD may be expressed as kg.

COD, *Chemical Oxygen Demand.* The oxygen uptake in a chemical oxidation of the waste (solids and solutes). A measure of the material liable to cause long-term, as well as immediate, pollution. Usual units mg/litre.

TS, *Total Solids.* The material left after evaporation of water from a sample. Usual units %(wt/vol) or mg/litre.

VS, *Volatile Solids.* The material burnt off on incineration of the TS, i.e. the organic material in the sample. Usual units % of TS.

SS or VSS, *Suspended Solids* or *Volatile Suspended Solids.* The total (SS) or organic (VSS) material in suspension, not solution, in the waste and removed by centrifugation or filtration (SS), followed by incineration (VSS). Usual units % (wt/vol) or mg/litre or VSS % of SS.

MLSS, *Mixed Liquor Suspended Solids*, sometimes MLVSS, *Mixed Liquor Volatile Suspended Solids.* The suspended solids used as a measure of the bacterial mass in a digester treating wastewaters with low or no feed solids. Usual units as SS or VSS.

Ash, The inorganic material left after incineration of the TS or SS. Usual units % of TS or SS, or mg/l.

VFA, *Volatile Fatty Acids.* Short-chain organic acids usually measured as acetic acid in mg/litre for total VFA, but the total may contain propionic, butyric and sinilar acids up to about C6 in chain length. Acids may also be measured individually in the same units.

NH_3N, *Ammonia Nitrogen.* Total ammonia in the sample (in solution or absorbed on to bacteria or particles). Measured as unionised plus ionised ammonia but usually reported as ammonia nitrogen, in mg/litre. Some analyses distinguish between unionised ammonia and ammonium and give concentrations as either, or both, entities.

Alkalinity. A measure of the buffering capacity (i.e. the resistance to change in pH and the production of acid conditions by formation of VFA in anaerobic digestion). Measured as the concentration of calcium carbonate which would have the equivalent capacity to neutralise acids.

Ions (metallic and non-metallic). Usual units mg/litre.

Media. 'Media' is the plural of 'medium', the liquid or solid in which laboratory cultures of bacteria are grown. It has now become convention to use 'media' as a singular term for the solid matrix on which grow the bacteria in an anaerobic filter or other retained-biomass digester.

These parameters were developed for description of sewage sludges and waters, hence most of the units refer to concentrations per volume of wastewater or wet sludge. With the present treatment of solid wastes of low water content some of the parameters are described in weight for weight units. Ammonia, VFA and ions may also be described in 'molar' or 'equivalent' units in watery wastes. The names of the parameters are those originally used and these are still used by most workers, but in some cases other names (e.g. 'fixed solids' for 'ash') may be used.

Some symbols for specific gas productions and other kinetic aspects of anaerobic digestions are recommended in the paper by Colin et al. (1983), but these have not been generally adopted.

REFERENCE

Colin, F., Ferrero, G.L., Gerletti, M., Hobson, P.N., L'Hermite, P., Naveau, H.P. & Nyns, E.J. (1983). *Agric. Wastes*, **7**, 183.

INDEX